NON-BINARY ERROR CONTROL CODING FOR WIRELESS COMMUNICATION AND DATA STORAGE

NON-BINARY ERROR CONTROL CODING FOR WIRELESS COMMUNICATION AND DATA STORAGE

Rolando Antonio Carrasco

Newcastle University, UK

Martin Johnston

Newcastle University, UK

John Wiley & Sons, Ltd

This edition first published 2008.
© 2008 John Wiley & Sons, Ltd.

Registered office
John Wiley & Sons Ltd, The Atrium, Southern Gate, Chichester, West Sussex, PO19 8SQ, United Kingdom

For details of our global editorial offices, for customer services and for information about how to apply for permission to reuse the copyright material in this book please see our web site at www.wiley.com.

Library of Congress Cataloging-in-Publication Data

Carrasco, Rolando Antonio.
  Non-binary error control coding for wireless communication and data storage / Rolando Antonio Carrasco, Martin Johnston.
      p. cm.
  Includes bibliographical references and index.
  ISBN 978-0-470-51819-9 (cloth)
  1. Error-correcting codes (Information theory)   2. Wireless communicaton systems.   3. Data transmission systems.   4. Computer storage devices.   5. Analog electronic systems.   I. Johnston, Martin, 1977–   II. Title.
  TK5102.96.C37 2008
  005.7'2–dc22

                                                                                                                2008029104

A catalogue record for this book is available from the British Library.

ISBN 978-0-470-51819-9 (HB)

Typeset in 11/13pt Times by Aptara Inc., New Delhi, India.
Printed in Singapore by Markono Print Media Pte Ltd.

My wife Gladys, my daughters Gladys and Carolina, my grand children
Charles, Serena and Sophie, family and friends

Rolando Carrasco

My parents, grand parents, family and friends

Martin Johnston

Contents

Biographies

Rolando A. Carrasco was born in 1945 in Santiago, Chile. He obtained his Bachelors BEng(Hons) degree in Electrical Engineering from Santiago University, Chile in 1969 and his PhD in Signal Processing from the University of Newcastle Upon Tyne in 1980. He was awarded the IEE Heaviside Premium in 1982 for his work in multiprocessor systems. Between 1982 and 1984 he was employed by Alfred Peters Limited, Sheffield (now Meditech) and carried out research and development in signal processing associated with cochlear stimulation and response. He was with Staffordshire University from 1984, and joined Newcastle University in 2004 as Professor of Mobile Communications. His principle research interests are the digital signal processing algorithm for data communication systems, mobile and network communication systems, speech recognition and processing. Professor Carrasco has over 200 scientific publications, five chapters in telecommunications reference texts and a patent to his name. He has supervised 40 successful PhD students. He has acted as principle investigator of several EPSRC projects, a BT research project and Teaching Company schemes. He has been the local chairman on several international conference organizing committees. He is a member of several organizing committees, a member of the EPSRC College and a member of the EPSRC assessment panel. He also has international collaborations with Chilean, Spanish and Chinese universities. Professor Carrasco is a fellow of the Institution of Engineers and Technology (IET).

Martin Johnston was born in Wordsley, England in 1977. He received his BSc(Hons) degree in Physics with Electronics from the University of Birmingham in 1999, his MSc degree in Electronic Engineering from Staffordshire University in 2001 and was awarded his PhD in the Design and Construction of Algebraic–Geometric Codes from Newcastle University in 2006. His research interests include the design of non-binary error-correcting codes for wireless and data storage applications and low-complexity decoding algorithms. He is currently a Research Associate at Newcastle University investigating the design of new error-correction coding schemes for high-density magnetic storage channels. Dr Johnston is a member of the Institute of Engineering and Technology (IET).

Preface

With the increasing importance of digital communications and data storage, there is a need for research in the area of coding theory and channel modelling to design codes for channels that are power limited and/or bandwidth limited. Typical examples of such channels are found in cellular communications (GSM, UMTS), fixed and mobile broadband wireless access (WiMax) and magnetic storage media.

Our motivation to write this book is twofold. Firstly, it is intended to provide PhD students and researchers in engineering with a sound background in binary and non-binary error-correcting codes, covering different classes of block and convolutional codes, band-width efficient coded modulation techniques and spatial-temporal diversity. Secondly, it is also intended to be suitable as a reference text for postgraduate students enrolled on Master's-level degree courses and projects in channel coding.

Chapter 1 of this book introduces the fundamentals of information theory and concepts, followed by an explanation of the properties of fading channels and descriptions of different channel models for fixed wireless access, mobile communications systems and magnetic storage.

In Chapter 2, basic mathematical concepts are presented in order to explain non-binary error-correction coding techniques, such as Groups, Rings and Fields and their properties, in order to understand the construction of ring trellis coded modulation (ring-TCM), ring block coded modulation (ring-BCM), Reed–Solomon codes and algebraic–geometric codes.

Binary and non-binary block codes and their application to wireless communications and data storage are discussed in Chapter 3, covering the construction of binary and non-binary Bose–Chaudhuri–Hocquengem (BCH) codes, Reed–Solomon codes and binary and non-binary BCM codes.

Chapter 4 introduces the construction methods of algebraic–geometric (AG) codes, which require an understanding of algebraic geometry. The coding parameters of AG codes are compared with Reed–Solomon codes and their performance and complexity are evaluated. Simulation results showing the performance of AG and Reed–Solomon codes are presented on fading channels and on magnetic storage channels.

Chapter 5 presents an alternative decoding algorithm known as list decoding, which is applied to Reed–Solomon codes and AG codes. Hard-decision list decoding for these codes is introduced first, using the Guruswami–Sudan algorithm. This is then followed by soft-decision list decoding, explaining the Kötter-Vardy algorithm for

Reed–Solomon codes and modifying it for AG codes. The performance and complexity of the list decoding algorithms for both Reed–Solomon codes and AG codes are evaluated. Simulation results for hard- and soft-decision list decoding of these codes on AWGN and fading channels are presented and it is shown that coding gains over conventional hard-decision decoding can be achieved.

A more recent coding scheme known as the low-density parity check (LDPC) code is introduced in Chapter 6. This is an important class of block code capable of near-Shannon limit performance, constructed from a sparse parity check matrix. We begin by explaining the construction and decoding of binary LDPC codes and extend these principles to non-binary LDPC codes. Finally, the reduction of the decoding complexity of non-binary LDPC codes using fast Fourier transforms (FFTs) is explained in detail, with examples.

Chapter 7 begins with an explanation of convolutional codes and shows how they can be combined with digital modulation to create a class of bandwidth-efficient codes called TCM codes. The construction of TCM codes defined over rings of integers, known as ring-TCM codes, is explained, and the design of good ring-TCM codes using a Genetic algorithm is presented. It is then shown how ring-TCM codes can be combined with spatial-temporal diversity, resulting in space-time ring-TCM (ST-RTCM) codes, and design criteria are given to construct good ST-RTCM codes for multiple-input–multiple-output (MIMO) fading channels. Simulation results of ST-RTCM codes are presented for MIMO fading channels, including urban environments such as indoor, pedestrian and vehicular scenarios.

Finally, Chapter 8 presents another recent coding scheme known as the turbo code. This important class of code was the first to achieve near-Shannon limit performance by using a novel iterative decoding algorithm. Binary turbo encoding and decoding are explained in this chapter, with a detailed description of the maximum *a posteriori* (MAP) algorithm and simpler decoding algorithms derived from the MAP algorithm, such as the max-log MAP and log MAP algorithms. The chapter finishes by introducing non-binary turbo codes, which are a new area of research that have received very little attention in the literature. Non-binary turbo encoding is explained and the extension of the binary turbo decoder structure in order to decode non-binary turbo codes is shown. For further information visit http://www.wiley.com/go/carrasco_non-binary

Acknowledgements

We are deeply indebted to the many reviewers who have given their time to read through this book, in part or in full. We are also very grateful for all the discussions, support and encouragement we have had from Prof. Paddy Farrell during the writing of this book. Our thanks go to our colleagues, to the many research assistants and students over the years, to our families and friends, to Bahram Honary, Mario Blaum, Mike Darnell, Garik Markarian, Ian Wassell and Fary Ghassemlooy for the useful discussions at the conferences we participated in. We are grateful for the contributions of Javier Lopez, Ismael Soto, Marcella Petronio, Alvero Pereira, Li Chen and Vajira Ganepola to some of the chapters in the book. Finally, we would like to thank Sarah Hinton and Sarah Tilley at John Wiley and Sons for their guidance in the initial preparation and the submission of the final version of the book.

1

Information, Channel Capacity and Channel Modelling

1.1 Introduction

In this chapter, an introduction to the fundamental aspects of information theory is given, with particular attention given to the derivation of the capacity of different channel models. This is followed by an explanation of the physical properties of fading channels and descriptions of different channel models for fixed wireless access, universal mobile telecommunication systems (UMTS) for single-input–single-output (SISO) and multiple-input–multiple-output (MIMO), and finally magnetic recording channels. Therefore, this chapter presents the prerequisites for evaluating many binary and non-binary coding schemes on various channel models.

The purpose of a communication system is, in the broadest sense, the transmission of information from one point in space and time to another. We shall briefly explore the basic ideas of what information is and how it can be measured, and how these ideas relate to bandwidth, capacity, signal-to-noise ratio, bit error rate and so on.

First, we address three basic questions that arise from the analysis and design of communication systems:

- Given an information source, how do we determine the 'rate' at which the source is transmitting information?
- For a noisy communication channel, how do we determine the maximum 'rate' at which 'reliable' information transmission can take place over the channel?
- How will we develop statistical models that adequately represent the basic properties of communication channels?

For modelling purposes we will divide communication channels into two categories: analogue channels and discrete channels. We wish to construct a function that

Non-Binary Error Control Coding for Wireless Communication and Data Storage   Rolando Antonio Carrasco and Martin Johnston
© 2008 John Wiley & Sons, Ltd

measures the amount of information present in an event that occurs with probability p. A lower probability means that we obtain more information about the event, whereas if the event is entirely predictable ($p = 1$) then we obtain no information about the event since we already have knowledge of the event before we receive it.

1.1.1 Information Theory [1]

- How can the symbols of communication be transmitted? Technical or Syntactic level?
- How precisely do the transmitted symbols carry the desired meaning? Semantic problem.
- How effectively does the received meaning affect conduct in the desired manner?

1.1.2 Definition of Information [1, 2]

Information: Knowledge not precisely known by the recipient, or a measure of unexpectedness.

Given an information source, we evaluate the *rate* at which the source is emitting information as:

$$\text{Rate} = \frac{\text{symbols}}{\text{second}} \quad \text{OR} \quad \text{Rate} = \frac{\text{bits}}{\text{second}}.$$

However, given a noisy communication channel, how do we determine the maximum rate at which *reliable* information transmission can take place over the channel?

1.2 Measure of Information [1–3]

We should briefly explore basic ideas about what information is, how it can be measured, and how these ideas relate to bandwidth and signal-to-noise ratio. The amount of information about an event is closely related to its probability of occurrence. Messages containing knowledge of a high probability of occurrence (i.e. those indicating very little uncertainty in the outcome) convey relatively little information. In contrast, those messages containing knowledge of a low probability of occurrence convey relatively large amounts of information. Thus, a measure of the information received from the knowledge of occurrence of an event is inversely related to the probability of its occurrence.

Assume an information source transmits one of nine possible messages $M_1, M_2, \ldots, M_9$ with probability of occurrence $P_1, P_2, \ldots, P_9$, where $P_1 + P_2 + \cdots + P_9 = 1$, as shown in Figure 1.1.

According to our intuition, the information content or the amount of information in the ith message, denoted by $I(M_i)$, must be inversely related to P_i. Also, to satisfy

Figure 1.1 Messages and their associated probabilities.

our intuitive concept of information, $I(M_i)$ must satisfy:

$$I(M_i) > I(M_j) \text{ if } P_i < P_j$$
$$I(M_i) \rightarrow 0, P_i \rightarrow 1$$
$$I(M_i) \geq 0 \text{ when } 0 \leq P_i \leq 1.$$

We can explain the concept of independent messages transmitting from the same source. For example, the received message 'It will be cold today and hot tomorrow' is the same as the sum of information received in the two messages 'It will be cold today' and 'It will be hot tomorrow' (assuming that the weather today does not affect the weather tomorrow).

Mathematically, we can write this as:

$$I(M_i \text{ and } M_j) \triangleq I(M_i, M_j) = I(M_i) + I(M_j),$$

where M_i and M_j are the two independent messages. We can define a measure of information as the logarithmic function:

$$I(M_i) = \log_x \left(\frac{1}{P_i} \right), \quad \text{where } x \text{ is the base } 2, e, 10, \ldots \quad (1.1)$$

The base x for the logarithm in (1.1) determines the unit assigned to the information content:

$$x = e \quad \text{nats}$$
$$x = 2 \quad \text{bits}$$
$$x = 10 \quad \text{Hartley,}$$

$P_1 + P_2 + P_3 + \cdots + P_q = 1$ where P_q is the probability of the message occurring and q is the index value, that is $1, 2 \ldots$

The information content or the amount of information $I(M_k)$ in the kth message with the set kth probability (P_k) boundary values is:

1. $I(M_k) \rightarrow 0 \quad \text{as} \quad P_k \rightarrow 1$.
   Obviously, if we are absolutely certain of the outcome of an event, even before it occurred.

2. $I(M_k) \geq 0$ when $0 \leq P_k \leq 1$.

   That is to say, the occurrence of an event $M = M_k$ either provides some or no information content.

3. $I(M_k) > I(M_i)$ if $P_k < P_i$.

   The less probable an event is, the more information we gain when it occurs.

4. $I(M_k M_i) = I(M_k) + I(M_i)$.

   If $I(M_k)$ and $I(M_i)$ are statistically independent messages.

5. Mathematically we can prove that:

6. $I(M_k \text{ and } M_j) = I(M_k) + I(M_j)$, where M_k and M_j are two statistically independent messages. The logarithmic measure is convenient.

Example 1.1: A source puts out five possible messages. The probabilities of these messages are:

$$P_1 = \frac{1}{2} \quad P_2 = \frac{1}{4} \quad P_3 = \frac{1}{8} \quad P_4 = \frac{1}{16} \quad P_5 = \frac{1}{16}.$$

Determine the information contained in each of these messages.

Solution:

$$I(M_1) = \log_2 \left(\frac{1}{(^1/_2)} \right) = 1 \text{ bit}$$

$$I(M_2) = \log_2 \left(\frac{1}{(^1/_4)} \right) = 2 \text{ bits}$$

$$I(M_3) = \log_2 \left(\frac{1}{(^1/_8)} \right) = 3 \text{ bits.}$$

$$I(M_4) = \log_2 \left(\frac{1}{(^1/_{16})} \right) = 4 \text{ bits}$$

$$I(M_5) = \log_2 \left(\frac{1}{(^1/_{16})} \right) = 4 \text{ bits}$$

Information total = 14 bits.

1.2.1 Average Information

Text messages produced by an information source consist of sequences of symbols but the receiver of a message may interpret the entire message as a single unit. When we attempt to define the information content of symbols, we are required to define the average information content of symbols in a long message, and the statistical dependence of symbols in a message sequence will change the average information content of symbols.

Let a transmitter unit consist of U possible symbols, $s_1, s_2, \ldots, s_u$, in a statistically independent sequence. The possibility of occurrence of a particular symbol during a symbol time does not depend on the symbol transmitted by the source previous in time. Let $P_1, P_2, \ldots, P_u$ be the probability of occurrence of the U symbols, in a long message having N symbols. The symbol s_1 will occur on average $P_1 N$ times. The symbol s_2 will occur $P_2 N$ times, and the symbol s_i will occur $P_i N$ times. Assuming an individual symbol s to be a message of length 1, we can define the information content of the ith symbol as $\log_2(\frac{1}{P_i})$ bits. The $P_i N$ occurrence of s_i contributes an information content of $P_i N \log_2(\frac{1}{P_i})$ bits.

The total information content of message is then the sum of each of the U symbols of the source:

$$I_{\text{total}} = \sum_{i=1}^{U} N P_i \log_2 \left(\frac{1}{P_i} \right) \text{ bits.} \tag{1.2}$$

The average information per symbol is measured by dividing the total information by the number of symbols N in the message:

$$H = \frac{I_{\text{total}}}{N} = \sum_{i=1}^{U} P_i \log_2 \left(\frac{1}{P_i} \right) \text{ bits/symbol.} \tag{1.3}$$

The average information H is called the source entropy (bits/symbol).

Example 1.2: Determine the entropy of a source that emits one of three symbols, A, B, C, in a statistically independent sequence, with a probability of $\frac{1}{2}, \frac{1}{4}$ and $\frac{1}{4}$, respectively.

Solution: The information contents of the symbols are:

  one bit for A

  two bit for B

  two bit for C

$$H = \frac{1}{2} \log_2 \left(\frac{1}{\left(\frac{1}{2}\right)} \right) + \frac{1}{4} \log_2 \left(\frac{1}{\left(\frac{1}{4}\right)} \right) + \frac{1}{4} \log_2 \left(\frac{1}{\left(\frac{1}{4}\right)} \right).$$

So the average information content per symbol on the source entropy is:

$$H = \frac{1}{2}\log_2(2) + \frac{1}{4}\log_2(4) + \frac{1}{4}\log_2(4).$$
$$H = 1.5 \text{ bits/symbols}$$

If we have a fixed time, say r_s symbols/s, then by definition the average source of information rate R in bits per second is the *product* of the average information content per symbol and the symbol rate r_s, as shown below:

$$R = r_s \cdot H \text{ bits/s.}$$

1.2.2 The Entropy of a Binary Source

For 'unbiased' results, the probability of logic 0 or logic 1 is:

$$P(1) = P(0) = 0.5$$
$$\text{Entropy} = H = -\sum_{i=1}^{M} P_i \log_2 P_i$$
$$H = -P_1 \log_2 (P_1) - P_2 \log_2 (P_2)$$
$$H = -P_1 \log_2 \left(\frac{1}{2}\right) - P_2 \log_2 \left(\frac{1}{2}\right)$$
$$H = 0.5 + 0.5 = 1.$$

This agrees with expectation. Now consider a system $P_1 \neq P_2$. We note that, for a two-state system, since $P_1 + P_2 = 1$ we have $P_2 = 1 - P_1$. We can then write:

$$H = P_1 \log_2 \left(\frac{1}{P_1}\right) + (1 - P_1) \log_2 \left(\frac{1}{1 - P_1}\right) \text{ bits.} \tag{1.4}$$

Table 1.1 shows how the entropy H varies for different values of P_1 and P_2.

Table 1.1   The variation of entropy for different values of P_1 and P_2.

P_1	$P_2 = 1 - P_1$	$P_1 \log_2 \left(\frac{1}{P_1}\right)$	$(1 - P_1)\log_2 \left(\frac{1}{1-P_1}\right)$	H
0	1	0	0	0
0.2	0.8	0.46	0.25	0.72
0.4	0.6	0.528	0.44	0.97
0.5	0.5	0.5	0.5	1
0.6	0.4	0.44	0.528	0.97
0.8	0.2	0.25	0.46	0.72
1	0	0	0	0

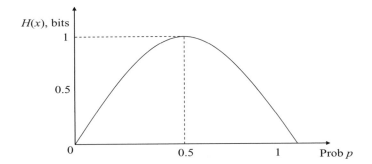

Figure 1.2   Entropy function of a binary source.

From Table 1.1 the following observations can be made:

The entropy $H(x)$ attains its maximum value, $H_{max} = 1$ bit, when $P_1 = P_2 = 1/2$; that is, symbols 1 and 0 are equally probable.

When $P_1 = 0$ or 1, the entropy value of $H(x) = 0$, resulting in no information.

The entropy is plotted in Figure 1.2.

The next two examples show how to determine the entropy functions for an event with a uniform and Gaussian probability distribution.

Example 1.3:   Determine the entropy of an event x with a uniform probability distribution defined by:

$$P(x) = \frac{1}{2.X_0}, -X_0 < x < X_0$$
$$P(x) = 0, \quad \text{elsewhere}$$

The uniform distribution is illustrated in Figure 1.3.

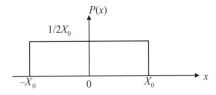

Figure 1.3   A uniform distribution.

Determine the entropy sources [1, 3]:

$$H(x) = -\int_x P(x) \log P(x) \, dx$$

$$H(x) = \int_{-X_0}^{X_0} \frac{1}{2X_0} \log\left(2X_0\right)\, dx$$

$$H(x) = \frac{\log\left(2X_0\right)}{2X_0} \int_{-X_0}^{X_0} dx$$

$$H(x) = \frac{\log\left(2X_0\right)}{2X_0} \left(X_0 - (-X_0)\right)$$

$$H(x) = \frac{\log\left(2X_0\right)}{2X_0} \left(2X_0\right) = \log\left(2X_0\right)$$

Example 1.4: Determine the entropy of an event x with a Gaussian probability distribution defined by:

$$P(x) = \frac{1}{\sqrt{2\pi}\,\sigma} e^{-\frac{x^2}{2\sigma^2}} \quad \text{where } \sigma \text{ is the standard deviation.}$$

The distribution is normalized so that the area under the pdf is unity

$$\int_{-\infty}^{\infty} P(x)\, dx = 1.$$

The variance σ^2 of the distribution is given as:

$$\int_{-\infty}^{\infty} x^2 P(x)\, dx = \sigma^2.$$

Therefore, the entropy function is:

$$H(x) = \int_{-\infty}^{\infty} \frac{1}{\sigma\sqrt{2\pi}} e^{-\frac{x^2}{2\sigma^2}} \left(\frac{x^2}{2\sigma^2} + \ln \sigma\sqrt{2\pi} \right) dx$$

$$H(x) = \frac{1}{2} + \ln\left(\sigma\sqrt{2\pi}\right)$$

$$H(x) = \ln e^{\frac{1}{2}} + \ln\left(\sigma\sqrt{2\pi}\right)$$

$$H(x) = \ln\left(\sigma\sqrt{2\pi e}\right) \text{ nats or } H(x) = \log_2\left(\sigma\sqrt{2\pi e}\right) \text{ bits}$$

1.2.3 Mutual Information

Given that we think of the channel output y as a noisy version of the channel input x value, and the entropy H is a measure of the prior uncertainty about x, how can we measure the uncertainty about x after observing the y value?

The conditional entropy of x is defined as [4]:

$$H(X|Y = y_k) = \sum_i P(x_i|y_k) \log_2 \left[\frac{1}{P(x_i|y_k)}\right], \quad k = 0, 1, 2, \ldots, \; i = 0, 1, 2, \ldots$$

$$(1.5)$$

This quantity is itself a random variable that takes on values:

$$H(X|y = y_0), \ldots, H(Y|y = y_{k-1})$$

with probabilities of:

$$P(y_0), \ldots, P(y_{k-1}) \text{ respectively.}$$

So the mean entropy is:

$$\overline{H} = \sum_k H(X|y = y_k) P(y_k)$$

$$\overline{H} = \sum_k \sum_i P(x_i|y_k) P(y_k) \log_2 \left[\frac{1}{P(x_i|y_k)}\right].$$

$$\text{Or} \quad \overline{H} = \sum_k \sum_i P(x_i, y_k) \log_2 \left[\frac{1}{P(x_i|y_k)}\right] \qquad (1.6)$$

This quantity $\overline{H}$ or $H(X|Y)$ is called conditional entropy. It represents the amount of uncertainty remaining about the input after the channel output has been observed. Since the entropy $H(X)$ represents our uncertainty about the channel input before observing the channel output, it follows that the difference is:

$$H(X) - \overline{H(X|Y)}.$$

This must represent our uncertainty about the channel input, which is resolved by observing the channel output. This important quantity is called the mutual information of the channel. Denoting the mutual information by $I(X, Y)$, we may thus write:

$$I(X, Y) = H(X) - \overline{H(X|Y)}. \qquad (1.7)$$

Similarly:

$$I(Y, X) = H(Y) - \overline{H(Y|X)}, \qquad (1.8)$$

where $H(Y)$ is the entropy of the channel output and $H(Y|X)$ is the conditional entropy of the channel output given the channel input.

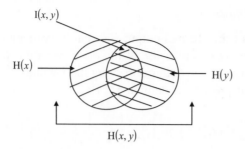

Figure 1.4   A Venn diagram illustrating the relationship between mutual information and entropy.

Property 1   The mutual information of a channel is symmetric, that is:

$$I(x, y) = I(y, x).$$

Property 2   The mutual information is always nonnegative, that is:

$$I(x, y) \geq 0.$$

Property 3   The mutual information of a channel is related to the joint entropy of the channel input and output by:

$$I(x, y) = H(x) + H(y) - H(x, y)$$

and is shown in Figure 1.4.

Example 1.5:   Determine the mutual information of the binary symmetric channel in Figure 1.5 where the associated probabilities are $P(x = 1) = 0.7$, $P(x = 0) = 0.3$, $P(y = 1|x = 1) = 0.8$, $P(y = 1|x = 0) = 0.2$, $P(y = 0|x = 1) = 0.2$ and $P(y = 0|x = 0) = 0.8$.

Solution:

$$H(x) = P(x = 0)\log_2 \frac{1}{P(x = 0)} + P(x = 1)\log_2 \frac{1}{P(x = 1)}$$

$$H(x) = 0.3 \times \log_2 \frac{1}{0.3} + 0.7 \times \log_2 \frac{1}{0.7}$$

$$H(x) = \log_2 10 - \left(0.3 \times \log_2 3 + 0.7 \times \log_2 7\right)$$

$$H(x) = 0.881 \text{ bits}$$

So, the conditional entropy of x:

$$H(x|y = 0) = \frac{24}{38}\log_2 \frac{38}{24} + \frac{14}{38}\log_2 \frac{38}{24}$$

$$H(x|y = 1) = \frac{3}{31} \log_2 \frac{31}{2} + \frac{28}{31} \log_2 \frac{31}{28}$$

$$H(x|y) = 0.38 \ H(x|y = 0) + 0.62 \ H(x|y = 1)$$
$$= P(y = 0)H \ (x|y = 0) + P(y = 1)H(x|y = 1)$$
$$= 0.645.$$

Therefore the mutual information of the binary symmetric channel shown in Figure 1.5 is:

$$I(x, y) = H(x) - \overline{H(X|y)} = 0.881 - 0.645 = 0.236 \ \text{bits/symbol.}$$

1.3 Channel Capacity

Successful electrical/optical communication systems depend on how accurately the receiver can determine the transmitted signal. Perfect identification could be possible in the absence of *noise*, but noise is always present in communication systems. The presence of noise superimposed on signal limits the receiver ability to correctly identify the intended signal and thereby limits the rate of information transmission. The term 'noise' is used in electrical communication systems to refer to unwanted electrical signals that accompany the message signals. These unwanted signals arise from a variety of sources and can be classified as man-made or naturally occurring. Man-made noise includes such things as electromagnetic pickup of other radiating signals. Natural noise-producing phenomena include atmospheric disturbances, extraterrestrial radiation and internal circuit noise.

Definition: The capacity C of the channel is the maximum mutual information, taken over all input distribution of x. In symbols [3, 4]:

$$C = \max_{P(x_i)} I(x, y) \quad \text{where} \quad P(0) = P(1) = \frac{1}{2}. \tag{1.9}$$

The units of the capacity C are bits per channel input symbols (bits/s).

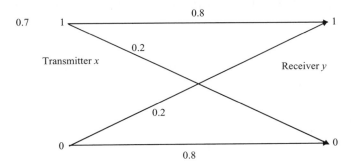

Figure 1.5 Binary symmetric channel of Example 1.5.

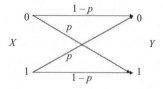

Figure 1.6   A binary symmetric channel.

1.3.1 Binary Symmetric Input Channel

Figure 1.6 shows the binary symmetric channel (BSC).

If we assume a uniform input distribution $P(x=0)=P(x=1)=1/2$ then $H(x)=1$. Furthermore, we have the set of transition probabilities [5]:

$$P(y=0\,|\,x=0) = P(y=1\,|\,x=1) = 1-p$$
$$P(y=0\,|\,x=1) = P(y=1\,|\,x=0) = p.$$

The conditional entropy is given by:

$$H(x\,|\,y) = -\left[\sum_k \sum_i P(y_i\,|\,x_k)P(x_k)\log_2 P(x_k\,|\,y_i)\right]$$

$$H(x\,|\,y) = -\left[\frac{1}{2}(1-p)\log_2 P(x=0\,|\,y=0) + \frac{1}{2}p\log_2 P(x=0\,|\,y=1)\right.$$
$$\left. + \frac{1}{2}p\log_2 P(x=1\,|\,y=0) + \frac{1}{2}(1-p)\log_2 P(x=1\,|\,y=1)\right].$$

The distribution of y is determined as follows:

$$P(y=0) = P(y=0\,|\,x=0)P(x=0) + P(y=1\,|\,x=1)P(x=1)$$
$$\therefore\ P(y=0) = \frac{1}{2}(1-p) + \frac{1}{2}p = \frac{1}{2}\qquad,$$

and $P(y=1)$ is:

$$P(y=1) = P(y=1\,|\,x=0)P(x=0) + P(y=1\,|\,x=1)P(x=1)$$
$$= \frac{1}{2}\ .$$

The result is not surprising since the channel is symmetric. The joint distribution is determined as follows:

$$P(x=0\,|\,y=0) = \frac{P(x=0,\,y=0)}{P(y=0)} = \frac{P(y=0\,|\,x=0)p(x=0)}{P(y=0)}$$
$$P(x=0\,|\,y=0) = P(y=0\,|\,x=0) = 1-p$$
$$P(x=0\,|\,y=1) = P(x=1\,|\,y=0) = p$$

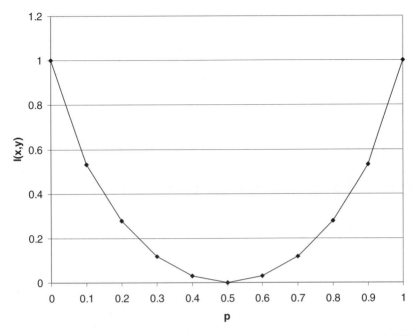

Figure 1.7   The mutual information of the BSC as a function of symbol error probability.

Thus:

$$H(x \mid y) = -\left[\frac{1}{2}(1 - p)\log_2(1 - p) + \frac{1}{2}p\log_2 p + \frac{1}{2}(1 - p)\log_2(1 - p) \right]$$
$$= -p\log_2 p - (1 - p)\log_2(1 - p)$$
$$I(x, y) = 1 + p\log_2 p + (1 - p)\log_2(1 - p)$$

The mutual information as a function of the symbol error probability p is shown in Figure 1.7.

1.3.2 Binary Erasure Channel (BEC)

The BEC output alphabets are 0 or 1, plus an additional element, denoted as e, called the erasure. This channel corresponds to data loss. Each input bit is either transmitted correctly with probability $1 - p$ or is erased with probability p. The BEC is shown in Figure 1.8.

The channel probabilities are given by [1, 5]:

$$P(y = 0 \mid x = 0) = P(y = 1 \mid x = 1) = 1 - p$$
$$P(y = e \mid x = 0) = P(y = e \mid x = 1) = p.$$

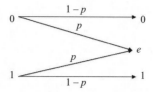

Figure 1.8 The binary erasure channel.

The conditional entropy is given by:

$$H(x \mid y) = -\left[\sum_k \sum_i P(x_k \mid y_i)P(x_k)\log_2 P(x_k \mid y_i)\right]$$

$$H(x \mid y) = -\left[\frac{1}{2}(1-p)\log_2 P(x=0 \mid y=0) + \frac{1}{2}p\log_2 P(x=0 \mid y=e)\right.$$

$$\left. +\frac{1}{2}p\log_2 P(x=1 \mid y=1) + \frac{1}{2}(1-p)\log_2 P(x=1 \mid y=e)\right],$$

$$H(x \mid y) = -\frac{1}{2}(1-p)\log_2 P(x=0 \mid y=0) - \frac{1}{2}p\log_2 P(x=0 \mid y=e)$$

where we used the fact that X is equiprobably distributed. The distribution of y is determined as follows:

$$P(y=0) = P(y=0/x=0)P(x=0) = \frac{1}{2}(1-p)$$

and similarly for $P(y=e)$, we have

$$P(y=0) = \frac{1}{2}, \quad P(y=e) = p$$

and:

$$P(x=0 \mid y=0) = \frac{P(x=0, y=0)}{P(y=0)} = \frac{P(y=0 \mid x=0)p(x=0)}{P(y=0)}$$

$$= \frac{\frac{1}{2}(1-p)}{\frac{1}{2}(1-p)}$$

and similarly:

$$P(x=0 \mid y=0) = 1$$

$$P(x=0 \mid y=e) = \frac{1}{2}.$$

Thus, $H(x \mid y) = -(1-p)\log_2(1) - p\log_2\left(\frac{1}{2}\right) = p$ and $I(x, y) = H(x) - H(x \mid y) = 1 - p$.

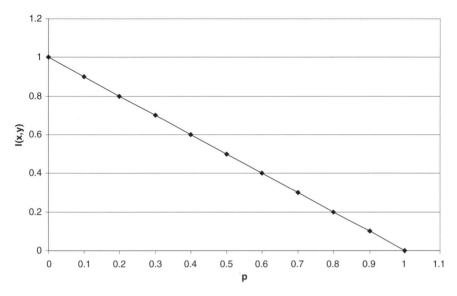

Figure 1.9   Mutual information of the binary erasure channel.

The mutual information of the binary erasure channel as a function of symbol error probability is plotted in Figure 1.9.

1.3.3 The 4-ary Symmetric Channel [1, 5]

The 4-ary Symmetric Channel (SC) is shown in Figure 1.10.

Let us consider the case of the 4-SC, where both the input X and the output Y have four possible values from the alphabet $A = [\alpha_0 \ldots \alpha_3]$. Since we are sending 4-ary symbols over the channel, we take the logarithm in equations $[H(x), I(x, y)]$ to the

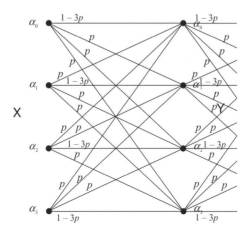

Figure 1.10   The 4-ary symmetric channel.

base 4. The same holds for the capacity defined previously. Assume a uniform input distribution where $P(x = a) = \frac{1}{4}$ for all $a \in A$. Hence, $H(x) = 1$, and we can set the transition probabilities:

$$P(y = a \mid x = a) = 1 - p, \; P(y = a \mid x = b) = \frac{p}{3} \text{ for } a \neq b, \text{ where } a, b \in A.$$

The conditional entropy is given by:

$$H(x \mid y) = -\left[\sum_k \sum_i P(y_i \mid x_k) \, P(x_k) \log_4 P(x_k \mid y_i) \right]$$

$$= -4 \left[\frac{1}{4} p \log_4 P\,(x = a \mid y = b) + \frac{1}{4}(1 - p) \log_4 P\,(x = a \mid y = a) \right].$$

For $a \neq b$, where $a, b \in A$, the distribution of y is determined as follows:

$$P(y = a) = 3P(y = a \mid x = b)P(x = b) + P(y = a \mid x = a)P(x = a)$$

$$= 3 \times \frac{1}{4}\frac{p}{3} + \frac{1}{4}(1 - p) = \frac{1}{4}.$$

The joint distribution is determined as follows:

$$P(x = a \mid y = a) = \frac{P(x = a, y = a)}{P(y = a)} = \frac{P(y = a \mid x = a)P(x = a)}{P(y = a)} = 1 - p.$$

Similarly:

$$P(x = a \mid y = b) = \frac{P(x = a, y = b)}{P(y = b)} = \frac{P(y = b \mid x = a)P(x = a)}{P(y = b)} = \frac{p}{3}.$$

For all $a, b \in A$, where $a \neq b$, we have:

$$H(x \mid y) = -4 \left[\frac{1}{4} p \log_4 \frac{p}{3} + \frac{1}{4}(1 - p) \log_4(1 - p) \right]$$

$$= -p \log_4 \frac{p}{3} - (1 - p) \log_4(1 - p),$$

and:

$$I(x, y) = 1 + p \log_4 \frac{p}{3} + (1 - p) \log_4(1 - p).$$

The mutual information as a function of the symbol error probability p is shown in Figure 1.11.

Notice that when the capacity is 1, the 4-ary symbol per channel use for $p = 0$. The mutual information is zero for $p = \frac{3}{4}$.

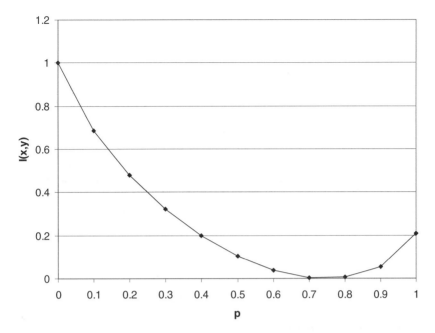

Figure 1.11 The mutual information of the 4-ary SC as function of symbol error probability.

1.3.4 Binary Input Capacity of a Continuous Channel (Channel Capacity)

Definition: The capacity C of a discrete memoryless channel is defined as the maximum mutual information $I(x, y)$ that can be transmitted through the channel:

$$I(X, Y) = H(Y) - H(Y \mid x) = H(x) - H(x \mid y)$$
$$= H(x) + H(y) - H(x, y).$$

The capacity is:

$$C = \max \ I(x, y)$$
$$C = \max \left[\sum \sum P(x, y) \log_2 \frac{P(x, y)}{P(x)P(y)} \right].$$

If $P(x) = \frac{1}{\sqrt{2\pi S}} e^{-\frac{x^2}{2S}}$ is a Gaussian distribution for the signal S then $P(y) = \frac{1}{\sqrt{2\pi (S+N)}} e^{\frac{y^2}{2(S+N)}}$ is a Gaussian distribution for the signal S and the noise N.

We will make use of the expression for information $I(x, y) = H(y) - H(y \mid x)$ and hence we require to find $H(y \mid x)$. $P(y \mid x)$ will be Gaussian distributed about a particular value of x, as seen in Figure 1.12.

$$P(y \mid x) = \frac{1}{\sqrt{2\pi N}} e^{\frac{(y-x)^2}{2N}}.$$

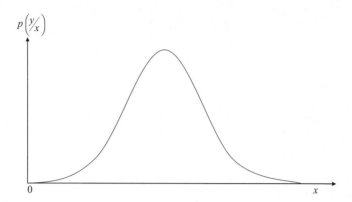

Figure 1.12   The Gaussian distribution of the conditional probability $P(y \mid x)$.

It has been shown that the entropy is independent of the DC value and thus:

$$H(y) = \log_2 \sqrt{2\pi e N}.$$

Since this is independent of x:

$$H(y) = \log_2 \sqrt{2\pi e(S + N)}$$

$$I(x, y) = \log_2 \sqrt{2\pi e\,(S + N)} - \log_2 \sqrt{2\pi e N}$$

$$= \log_2 \sqrt{\frac{2\pi e\,(S + N)}{2\pi e N}}.$$

$$= \log_2 \sqrt{\frac{S + N}{N}} = \log_2 \sqrt{1 + \frac{S}{N}}$$

This is the information per sample. Thus in time T information is:

$$\text{Total information} = 2BTI$$

$$= 2BT \log_2 \sqrt{1 + \frac{S}{N}} = BT \log_2 \left(1 + \frac{S}{N}\right).$$

Maximum information rate is:

$$C = B \, \log_2 \left(1 + \frac{S}{N}\right) \text{ bits/s,}$$

where B is the bandwidth of the channel. The practical and non-practical channel capacities of the AWGN channel are shown in Figure 1.13.

1. When the channel is noise free, permitting us to set $p = 0$, the channel capacity C attains its maximum value of one bit per channel use.

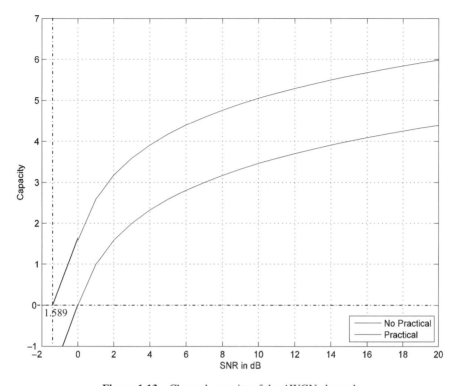

Figure 1.13 Channel capacity of the AWGN channel.

2. When the conditional probability of error $p = 1/2$ due to noise, the channel capacity C attains its minimum value of zero.

$$S = \text{Average Power (Symbols)}$$
$$N = \text{Average Noise Power.}$$

Example 1.6: Determine the capacity of a low-pass channel with usual bandwidth of 3000 Hz and $S/N = 10$ dB (signal/noise) at the channel output. Assume the channel noise to be Gaussian and white.

Solution:

$$C = B \log_2 \left(1 + \frac{S}{N} \right)$$
$$= 3000 \ \log_2 (1 + 10) \cong 10 \ 378 \ \text{bits/s.}$$

Signalling at rates close to capacity is achieved in practice by error correction coding. An error correction code maps data sequences of k bits to code words of n

symbols. Because $n > k$, the code word contains structured redundancy. The code rate, $r = k/n$, is a measure of the spectral efficiency of the code. In order to achieve reliable communications, the code rate cannot exceed the channel capacity ($r \leq c$). The minimum theoretical signal-to-noise ratio (S/N) required to achieve arbitrarily reliable communications can be found by rearranging the equation for the capacity of the AWGN channel [2].

$$S/N \geq \frac{1}{2r} \left(2^{2r} - 1 \right).$$

This is the minimum S/N required for any arbitrary distinction for the input signal. Shannon's proof of the channel coding theorem [1, 2] used a random coding argument. Shannon showed that if one selects a rate $r < c$ codes at random, the bit error probability approaches zero as the block length n of the code approaches infinity. However, random codes are not practically feasible. In order to be able to encode and decode with reasonable complexity, codes must possess some sort of a structure.

1.3.5 Channel Capacity in Fading Environments

In the case of a single input and single output (SISO) fading channel, the received signal at the kth symbol instant is $y(k) = h(k)x(k) + n(k)$, where $h(k)$ is the impulse response of the channel, $x(k)$ is the input signal to the channel and $n(k)$ is additive white Gaussian noise. To ensure a compatible measure of power, set $E(|h(k)|^2) = 1$ and $E\{|x(k)|^2\} \leq E_s$. The capacity in the case of the fixed fading channel with random but unchanging channel gain is given below [6]:

$$C = B \log_2 \left(1 + |h|^2 \rho \right), \quad \text{where } \rho = \frac{E_s}{\sigma_n^2}, \tag{1.10}$$

where E_s is the symbol energy, σ_n^2 is the n-dimensional variance and h is the gain provided by the channel. An interesting point to note is that in a random but fixed fading channel the theoretical capacity can be zero, when the channel gain is close enough to zero to make data rate impossible. In this case, the possible scenario is determining what the chances are that a required capacity is available. This is defined by Outage probability P_{out} as the probability that the channel is above a threshold capacity C_{thres} given by following equation [6]:

$$P_{out} = P(C > C_{thres}) = P \left(|h|^2 > \frac{2C_{thres} - 1}{\rho} \right) \tag{1.11}$$

$$P_{out} = 1 - e^{-\frac{2C_{thres} - 1}{\rho}}, \tag{1.12}$$

where (1.13) is valid for Rayleigh fading.

In the case of a time-varying channel, the channel is independent from one symbol to the next and the average capacity of K data symbols is:

$$C_K = \frac{1}{K} \sum_{k=1}^{K} \left\{ \log_2 \left(1 + |h|^2 \rho \right) \right\}. \tag{1.13}$$

Based on the law of large numbers, as $K \to \infty$ the term on the right converges to the average or expected value, therefore:

$$C = E_h \left\{ \log_2 \left(1 + |h|^2 \rho \right) \right\}, \tag{1.14}$$

where the expectation is taken over the channel values h. (1.14) is nonzero; therefore with a fluctuating channel it is possible to guarantee the existence of an error-free data rate.

Increasing the transmitted and received antenna, the capacity of the channel increases by $N_T N_R$-fold, where N_T is number of transmit antenna and N_R is number of receive antenna. The capacity of the AWGN channel in case of multiple-input–multiple-output is approximately given by (1.15):

$$C \approx B \log_2 \left(1 + N_T \cdot N_R \cdot \rho \right) \quad \text{where } \rho = E_s / \sigma_n^2. \tag{1.15}$$

Figure 1.14 shows the channel capacity with the increase in the number of transmitted and received antennas plotted using (1.15).

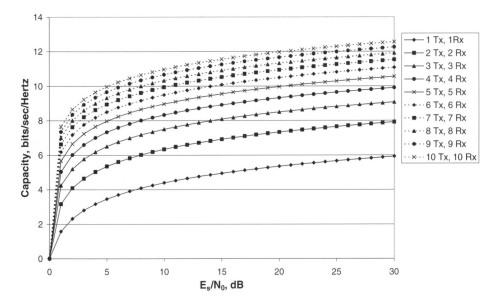

Figure 1.14   Channel capacity in bits/s/Hz with increasing number of transmit and receive antennas.

1.4 Channel Modelling

This section begins by describing the fundamentals and characteristics of the channel propagation models used in WiMAX or IEEE 802.16d (Fixed Broadband) systems and UMTS cellular communications. Propagation models are the fundamental tools for designing any wireless communication system. A propagation model basically predicts what will happen to the transmitted signal while in transit to the receiver. In general, the signal is weakened and distorted in particular ways and the receiver must be able to accommodate for these changes if the transmitted information is to be successfully received. The design of the transmitting and receiving equipment and the type of communication service that is being provided will be affected by these signal impairments and distortions. The role of propagation modelling is to predict the performance of the system with these distortions and to determine whether it will successfully meet its performance goals and service objectives.

The section ends with a description of magnetic storage channel modelling for longitudinal and perpendicular storage devices. In this situation, information is retrieved not in a different *location* from where it was originally sent, as in radio communications, but at a different *time* from when it was originally stored.

1.4.1 Classification of Models

In wireless channels, a narrowband implies that the channel under consideration is sufficiently narrow that the fading across it is flat (i.e. constant). It is usually used as an idealizing assumption; no channel has perfectly flat fading, but the analysis of many aspects of wireless systems is greatly simplified if flat fading can be assumed. Early communication systems were narrowband systems in which median signal level prediction and some description of the signal level variability (fading) statistics were the only model parameters needed to adequately predict the system performance. However, modern communications systems use wideband, hence achieving higher data rates.

In communications, wideband is a relative term used to describe a wide range of frequencies in a spectrum. A system is typically described as wideband if its bandwidth is much greater than its centre frequency or carrier frequency. For such systems, narrowband prediction of signal levels and fading alone does not provide enough information to predict system performance. In fact the concept of propagation models is now enlarged to include the entire transfer function of the channel. These new propagation models, known as channel models, now include parameters such as signal time dispersion information and Doppler effect distortion arising from motion of the mobile device. Time dispersion causes the signal fading to vary as a function of frequency, so wideband channels are often known as frequency-selective fading channels.

These propagation channel models are broadly classified into three main categories: Theoretical, Empirical and Physical.

1.4.1.1 Theoretical Models

The channel models in this group are based on theoretical assumptions about the propagation environment. The channel parameters based on these model assumptions do not define any particular environment or model any specific channel conditions. The theoretical models cannot be used for planning and developing any communication systems as they do not reflect the exact propagation medium the signal would be experiencing. The theoretical models can be nontime dispersive or time dispersive. Nontime-dispersive channel models are those in which the duration of the transmitted signal is the same on arriving at the receiving end. However, in time dispersion the signal extends in time so that the duration of the received signal is greater than that of the transmitted signal. The theoretical modelling of the time-dispersive channel has been presented in [6–8]. The theoretical time-dispersive channel can also be modelled by the tapped delay line structure, in which densely-spaced taps, multiplying constants and tap-to-tap correlation coefficients are determined on the basis of measurements or some theoretical interpretation of how the propagation environment affects the signal [7, 8].

1.4.1.2 Empirical Models

Empirical models are those based on observations and measurements alone. These models are mainly used to predict the path loss, but models that predict rain fade and multipath have also been proposed [7]. The problem can occur when trying to use empirical models in locations that are broadly different from the environment in which the data is measured. For example, the Hata Model [9] is based on the work of Okumura, in which the propagation path loss is defined for the urban, suburban and open environments. But models like Hata and ITU-R are widely used because they are simple and allow rapid computer calculations.

Empirical models can be subclassified in two categories, namely nontime dispersive and time dispersive, as described. The time dispersive provides information relating to the time-dispersive characteristics of the channel, that is the multipath delay spread of the channel. Examples of this type are channel models developed by Stanford University Interim (SUI) for use in setting up the fixed broadband systems. These types of channel model are extensively used for WiMAX or IEEE 802.16 system development [10]. This chapter is mainly concerned with the time-dispersive empirical channel model (WiMAX specification), which will be explained in detail later on.

1.4.1.3 Physical Models

A channel can be physically modelled by attempting to calculate the physical processes which modify the transmitted signal. These models rely on basic principles of physics rather than statistical outcomes from the experiments. Physical channel models are also known as deterministic models, which often require a complete 3D map of the propagation environment. These models are not only divided into nontime dispersive and time dispersive, but are also modelled with respect to the site specification.

The works published in [6, 8] are examples of site-specific, time-dispersive channel models. These models are principally identified as ray-tracing, a high-frequency approximation approach that traces the route of electromagnetic waves leaving the transmitter as they interrelate with the objects in the propagation environment. A deterministic ray-tracing propagation model was used to predict the time delay and fading characteristics for the channel in a hypothetical urban area. Using this propagation model, the channel response throughout the urban area was described in terms of the signal level, root-mean-square (RMS) delay spread, and the fading statistics at each point in the service area. These time-dispersive models provide not only multipath delay spread of the channel but also information related to the angle of arrival (AoA) of the signal.

However, those physical models which do not have signal time delay information are known as nontime-dispersive channel models. These propagation models are specifically applicable to propagation prediction in the fixed broadband wireless systems.

1.5 Definition of a Communications Channel and its Parameters

The main factor affecting the design of a fixed wireless access system is the nature of the channel available, which affects the behaviour of electromagnetic waves propagating through it. The previous subsection presented the different types of channel modelling procedures and their subcategories. Before dealing with the specification and technical parameters of this model, a general background of the channel is presented along with the definitions of parameters like delay spread, path loss, fading, Doppler effect and so on.

The main processes of a communication system consist of a source, a transmitter, a channel, a receiver and a destination, as shown in Figure 1.15, where b_k are the message bits, x_n are the modulated symbols, r_n are received symbols, $\hat{x}_n$ are the demodulated symbols and $\hat{b}_k$ are the detected message bits. The transmitter takes information from the source and converts it into a form suitable for transmission.

The wireless communication channel consists of the medium through which the RF signal passes when travelling from the transmitting antenna to the receiving antenna. The medium causes the transmitted signal to be distorted in several ways, as previously mentioned.

In the absence of a line-of-sight between the transmitting antenna and the receiving antenna, some of the transmitted signal finds a path to the receiving antenna by

Figure 1.15 Simple communication system.

reflecting or refracting from whatever is blocking the direct line-of-sight between the two antennas. This action is known as a *multipath* signal scenario as the many transmitted signals undergo different degrees of dispersion as they traverse multiple paths to reach the receiving antenna.

Eventually some of the signal paths recombine vectorially at the receiving antenna, producing a signal the amplitude of which is dependent upon the phases of the individual component waveforms. This is the concept of signal fading, which is a purely spatial phenomenon. If a mobile receiving antenna is moving relative to the environment and/or the transmitting antenna, the incoming phases of the signals will vary, producing a signal whose amplitude varies with spatial movement of the mobile relative to the environment. Although fading is a spatial phenomenon, it is often perceived as a temporal phenomenon as the mobile device moves through the multipath signal field over time.

To determine a channel model, mathematical descriptions of the transmitter, the receiver and the effect of the environment (walls) on the signal must be known. A linear channel can be totally described by its impulse response, that is by what the received signal would look like if the transmitted signal were an impulse. The impulse response is the response of the channel at all frequencies, that is, once the impulse response of the channel is known, one can predict the channel response at all frequencies.

Let $x(t)$ be the signal transmitted from an antenna through the channel $h(t)$, and $y(t)$ be the signal received at the receiving side. Assuming no delay, multipath signals and no other noise present in the system, the channel can be considered as a linear system with $x(t)$ as an input and $y(t)$ as an output. This relationship between the input and output in the time domain is represented in (1.16) and shown in Figure 1.16.

$$y(t) = h(t) \otimes x(t). \tag{1.16}$$

The channel impulse response $h(t)$ is obtained by applying the impulse function to the channel and can be represented as:

$$h(t) = \sum_{i=0}^{\infty} A_i e^{j\phi_i} \delta(t - \tau_i), \tag{1.17}$$

where A_i is the magnitude of the impulse response at delay τ_i with phase ϕ_i, and $\delta(t)$ is the Dirac delta function [5]. The system in Figure 1.16 is modelled as a linear

Figure 1.16   Channel input-output in time domain.

time variant filter with impulse response $h(t, \tau)$ and frequency response $H(f, t)$. The frequency response can be obtained by taking the Fourier transform of $h(t)$:

$$H(f, t) = \int_{-\infty}^{\infty} h(t)e^{-j2\pi ft} dt. \tag{1.18}$$

The channel output $y(t)$ at a particular time $t = \tau$ is the convolution of the impulse response $h(t)$ with the input signal $x(t)$, that is:

$$y(t) = x(t) \otimes h(t) = \sum_{\tau=0}^{\infty} h(\tau)x(t - \tau), \tag{1.19}$$

where $\otimes$ represents the convolution operation. The final output $y(t)$ can be obtained by adding independent noise $n(t)$ to (1.19):

$$y(t) = x(t) \otimes h(t) + n(t) = \sum_{\tau=0}^{\infty} h(\tau)x(t - \tau) + n(t). \tag{1.20}$$

However, in a wireless environment the transmitted signals are affected in different ways, for example by the reflection that occurs when the signal hits a surface. At the receiving side the signal arrives from different paths, which add delay at the receiver, and it can be affected by moving objects in terms of a Doppler shift.

1.5.1 Doppler Shift

Doppler frequency effects occur whenever there is relative movement between the transmitting and receiving antennas. It manifests itself as a change in the frequency of the received signal. For a receiver tuned to a particular frequency, this phenomenon has the effect of reducing the received signal energy as the receiver's front end is no longer tuned to the centre frequency of the signal and is therefore not operating efficiently.

The Doppler frequency shift in each propagation path is caused by the rate of change of signal phase (due to motion). Referring to Figure 1.17 below, if a mobile device, moving from point A to point B at a velocity v, is receiving a signal from the signal source, the distance travelled, d, can be found by $v\Delta t$, where Δt is the change in time from point A to point B.

From this it can be shown geometrically that the extra distance that the wave has to travel to get from the signal source to point B(Δl) with respect to point A is: $d \cos(\alpha)$.

The change in phase of the received signal at point B relative to point A is given by [8]:

$$\Delta\phi = -\frac{2\pi}{\lambda}\Delta l = -\frac{2\pi v \Delta t}{\lambda} \cos \alpha. \tag{1.21}$$

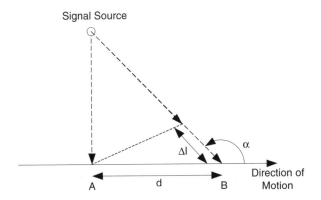

Figure 1.17   Example of the effects of Doppler shift [8].

With the change in frequency of the received signal at point B relative to point A given by:

$$f = -\frac{1}{2\pi}\frac{\Delta\phi}{\Delta t} = \frac{v}{\lambda}\cos\alpha. \tag{1.22}$$

it can be seen that the change in path length is governed by the angle between the direction of motion and the received wave. It should also be noted that when the mobile antenna is moving closer to the signal source, a positive change in frequency (Doppler) is caused, and conversely, if the mobile antenna is moving further away from the signal source it causes a negative change in frequency.

If the mobile antenna is moving in the same plane as the signal source (either to or from it) then the frequency shift is given by:

$$f = \pm v/\lambda. \tag{1.23}$$

This information is required for modelling the channel using the ray-tracing method, which calculates all the angles of the received signals in order to calculate the received signal strength over time. This method becomes impractical for large numbers of received signals due to the computational burden. The number of signal reflections also limits this method, and in practice only signals with two reflections are considered.

The movement of the receiving antenna relative to the transmitting antenna, coupled with the large number of reflections and received signal paths, causes the resultant RF signal envelope at the receiving antenna to appear random in nature. Therefore, statistical methods must be employed in order to produce a mathematically tractable model, which produces results in accordance with the observed channel properties.

1.5.2 Scattering

At the receiving antenna, the incoming signal consists of a superposition of individual waves. Each individual wave has the following characteristics: amplitude, frequency, phase, polarization angle, vertical angle of arrival and horizontal angle of arrival, all relative to the signal at the transmitter.

If we assume that both the transmitting and receiving antennas are similarly polarized and operating on the same frequency, only the amplitude (A_n), phase (ϕ_n), vertical angle of arrival (β_n) and horizontal angle of arrival (α_n) need to be considered. Figure 1.18 shows an individual wave (n) relative to the point of reception.

The values for A_n, ϕ_n, β_n and α_n are all random and statistically independent. The mean square value for the amplitude (A_n) is given by:

$$E\{A_n^2\} = \frac{E_0}{N}, \qquad (1.24)$$

where E_0 is a positive constant and N is the number of received waves at the point of reception.

Clarke [7] makes a generalization which assumes that the height of the transmitting antenna is approximately the same as that of the receiving antenna, so (assuming Clarke's model) the vertical angles of arrival (β_n) can be set to zero. In practice this is found to be a good approximation.

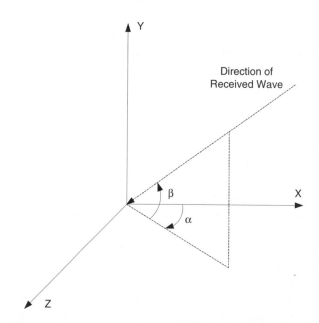

Figure 1.18   An individual wave relative to the point of reception.

The phase angles (ϕ_n) are uniformly distributed within the range of 0 to 2π, but the probability density functions for the angles of arrival α_n and β_n are generally not specified.

At the point of reception, the field resulting from the superposition of the incoming waves (n) is given by:

$$E(t) = \sum_{n=0}^{N} E_n(t). \tag{1.25}$$

If an unmodulated carrier is transmitted, the received signal ($E_n(t)$) can be expressed at the point of reception (x_0, y_0, z_0) as follows [8]:

$$E(t) = A_n \cos\left(\omega_0 t - \frac{2\pi}{\lambda}(x_0 \cos(\alpha_n)\cos(\beta_n) + y_0 \sin(\alpha_n)\cos(\beta_n)\right.$$
$$\left. + z_0 \sin(\beta_n)) + \phi_n\right). \tag{1.26}$$

If the receiving antenna moves in the xy plane at an angle γ (relative to the x-axis) with a velocity v, then after a unit time, the coordinates of the received signal can be expressed as:

$$v\cos\gamma, v\sin\gamma, z_0. \tag{1.27}$$

Which means that the received signal $E(t)$ can be expressed as:

$$E(t) = I(t)\cos(\omega_c t) - Q(t)\sin(\omega_c t), \tag{1.28}$$

where $I(t)$ and $Q(t)$ represent the in-phase and quadrature components of the signal respectively, and can be expressed as:

$$I(t) = \sum_{n=1}^{N} A_n \cos(\omega_n t + \theta_n) \tag{1.29}$$

and:

$$Q(t) = \sum_{n=1}^{N} A_n \sin(\omega_n t + \theta_n), \tag{1.30}$$

where ω_n equals $2\pi f_n$, and f_n represents the Doppler shift in frequency experienced by the individual wave n. The terms ω_n and θ_n can be expressed as:

$$\omega_n = \frac{2\pi v}{\lambda}\cos(\gamma - \alpha_n)\cos(\beta_n) \tag{1.31}$$

and:

$$\theta_n = \frac{2\pi z_0}{\lambda}\sin(\beta_n) + \phi_n. \tag{1.32}$$

The previous equations reduce to Clarke's two-dimensional model if β is taken to be zero.

If the number of received waves (N) is very large (in practice 6, but theoretically much larger), then according to the central limit theorem the components $I(t)$ and $Q(t)$ are independent Gaussian processes and are completely characterized by their mean and autocorrelation functions. Because $I(t)$ and $Q(t)$ both have zero mean values, it follows that $E\{E(t)\}$ is also zero. Also, $I(t)$ and $Q(t)$ have variance values ( ) that are the same as the mean square value (average power), thus the probability density function of $I(t)$ and $Q(t)$ can be expressed as:

$$p_x(x) = \frac{1}{\sigma\sqrt{2\pi}} \exp\left(-\frac{x^2}{2\sigma^2}\right),$$ (1.33)

where $x = I(t)$ or $Q(t)$, and $\sigma^2 = E\{A_n^2\} = \dfrac{E_0}{N}$.

1.5.3 Angle of Arrival and Spectra

For a system in motion (that is, receiver antenna movement relative to the transmitting antenna), the individual components of the received signal will experience a change in frequency due to the Doppler effect, depending upon their individual angles of arrival. This change in frequency is given by [8]:

$$f_n = \frac{\omega_n}{2\pi} = \frac{v}{\lambda}\cos\left(\gamma - \alpha_n\right)\cos\left(\beta_n\right).$$ (1.34)

All frequency components within the received signal will experience this Doppler shift in frequency and, as long as the signal bandwidth is relatively small (compared to the receive signal path bandwidth), it can be assumed that all the individual component waves will be affected in the same way.

The RF spectrum of the received signal can be obtained by using the Fourier transform of the temporal autocorrelation function in terms of the time delay (τ):

$$E\{E(t)E(t+\tau)\} = E\{I(t)I(t+\tau)\}\cos(\omega_c\tau) - E\{I(t)Q(t+\tau)\}\sin(\omega_c\tau)$$
$$= a(\tau)\cos(\omega_c\tau) - c(\tau)\sin(\omega_c\tau).$$ (1.35)

Aulin [11] showed that the correlation properties can therefore be expressed by $a(\tau)$ and $c(\tau)$:

$$a(\tau) = \frac{E_0}{2}E\{\cos(\omega\tau)\}$$

$$c(\tau) = \frac{E_0}{2}E\{\sin(\omega\tau)\}.$$ (1.36)

In order to simplify further we make the assumption that all the signal waves arrive in the horizontal plane (α) with equal probability, so that:

$$p_\alpha(\alpha) = \frac{1}{2\pi}. \tag{1.37}$$

Which means that by Fourier transforming the following, the power spectrum can be obtained:

$$a(\tau) = \frac{E_0}{2} \int_{-\pi}^{+\pi} J_0(2\pi f_m \tau \cos \beta) p_\beta(\beta) \, d\beta. \tag{1.38}$$

1.5.4 Multipath Channel and Tapped Delay Line Model

The fixed or mobile radio channel is characterized by the multipath propagation. The signal offered to the receiver contains not only a direct line-of-sight radio wave, but also a large number of reflected radio waves. This can be even worse in urban areas, where obstacles often block the line-of-sight and a mobile antenna receives a collection of various delayed waves. These reflected waves interfere with the direct wave, which causes significant degradation of the performance of the link. If the mobile user moves, the channel varies with the location and time, because the relative phase and amplitude of the reflective wave change. Multipath propagation seriously degrades the performance of the communication system. The adverse effects produced by the medium can be reduced by properly characterizing the medium in order to design the transmitter and receiver to fit the channel. Figure 1.19 shows the signal arriving at the receiver from different paths, which include the direct line-of-sight (LOS) path and nonline-of-sight (NLOS) paths.

As the different variants of the same signal arrive at different times, some are delayed relative to one another. This time dispersion is known as multipath delay spread. This delay spread is an important parameter in channel modelling and is commonly measured as root mean square (RMS) delay spread. For reliable communication over these channels without any interference reduction techniques, the transmitted data rate should be much smaller than the coherence bandwidth. This type of channel, when

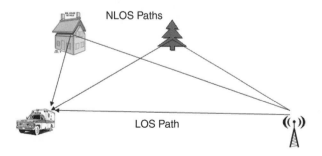

Figure 1.19 Signal arriving at mobile station from different paths.

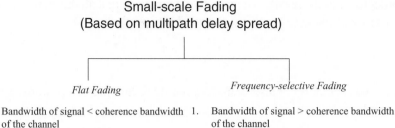

Figure 1.20   The two classes of fading channels.

the transmitted bandwidth of the signal is much smaller than the coherent bandwidth, is known as a flat channel or narrowband channel. When the transmitted bandwidth of the signal is nearly equal to or larger than the coherent bandwidth, the channel is known as a frequency-selective channel or broadband channel. The relationship between flat fading and frequency-selective fading is shown in Figure 1.20.

The multipath delay profile is characterized by τ_{rms}, which is defined as [6]:

$$\tau_{\mathrm{rms}}^2 = \sum_j P_j \tau_j^2 - (\tau_{\mathrm{avg}})^2, \tag{1.39}$$

where $\tau_{\mathrm{avg}} = \sum_j P_j \tau_j$, τ_j is the delay of the jth delay component of the profile and $P_j = $ (power in the jth delay component)/(total power in all components).

The channel output $y(t)$ of (1.22) can be realized as a tapped delay line, as shown in Figure 1.21, where $A_i(t)$ is the fading amplitude of the ith tap and T_i are the delays of the ith tap. Each tap represents a scattered ray multiplied by a time-varying fading profile coefficient $A_i(t)$. The relative tap delays are dependent on the type of channel

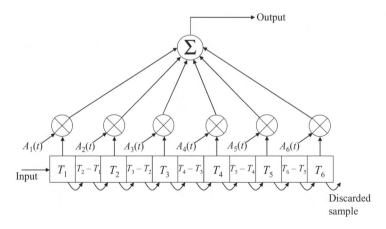

Figure 1.21   An example of a tapped delay line.

model. These values are given later for the fixed broadband wireless-access channel model and the urban UMTS mobile radio channel models.

1.5.5 Delay Spread

Because of the multipath reflections, the channel impulse response of a wireless channel looks like a series of pulses. In practice the number of pulses that can be distinguished is very large and depends on the time resolution of the communication or measurement system.

The system evaluations process typically prefers to address a class of channels with properties that are likely to be encountered, rather than one specific impulse response. Hence it defines the (local-mean) average power, which is received with an excess delay that falls within the interval $(T, T + dt)$. Such characterization for all T gives the 'delay profile' of the channel. The delay profile determines the frequency dispersion; that is, the extent to which the channel fading at two different frequencies f_1 and f_2 is correlated (Figure 1.22).

The maximum delay time spread is the total time interval during which reflections with significant energy arrive. The RMS delay spread T_{RMS} is the standard deviation (or root-mean-square) value of the delay of reflections, weighted proportionally to the energy in the reflected waves. For a digital signal with high bit rate, this dispersion is experienced as frequency-selective fading and intersymbol interference (ISI). No serious ISI is likely to occur if the symbol duration is longer than, say, ten times the RMS delay spread. The RMS delay spread model published in [14] follows a lognormal distribution and the median of this distribution grows as a power of distance. This model was developed for rural, urban and mountainous environments (Figure 1.23).

$$\tau_{rms} = T_1 d^e y, \tag{1.40}$$

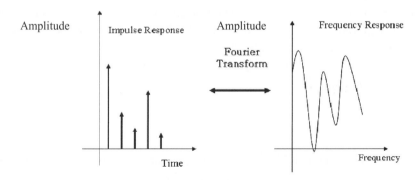

Figure 1.22 Example of the impulse response and frequency transfer function of a multipath channel.

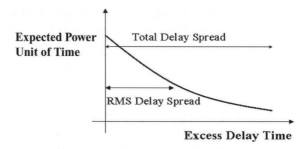

Figure 1.23   Expected power per unit of time.

where τ_{rms} is the RMS delay spread, d is the distance in km, T_1 is the median value of the $\tau_{rms\ at}\ d = 1$ km, e is an exponent that lies between 0.5 and 1.0, and y is the lognormal variable.

The narrowband received signal fading can be characterized by a Ricean distribution. The K factor is the ratio between the constant component powers and the scattered power. The model that represents the K factor by extensive experimental data taken at 1.9 GHz is given below [12]:

$$K = F_s F_h F_b K_o d^\gamma, \tag{1.41}$$

where F_s is a seasonal factor: $F_s = 1.0$ in summer (leaves) and 2.5 in winter (no leaves); F_h is the receive antenna height factor, $F_h = (h/3)^{0.46}$ (h is the receive antenna height); F_b is the beamwidth factor, $F_b = (b/17)^{-0.62}$ (b in degrees); K_o and γ are regression coefficients, $K_o = 10$; $\gamma = -0.5$; u is a lognormal variable with 0 dB mean and a standard deviation of 8.0 dB.

1.5.6 The Fixed Broadband Wireless Access Channel

The Doppler spectrum of the fixed wireless system is different from that in the mobile wireless access systems. In fixed wireless systems the Doppler spectrum of the scattered component is mainly scattered around $f = 0$ Hz. The shape of the spectrum is different from the mobile channels. The spectrum for mobile and fixed wireless channels is given below [10].

The above fixed access channel power spectrum density (PSD) is calculated by using the equation given below:

$$S(f) = \begin{cases} 1 - 1.72 f_o^2 + 0.785 f_o^4 & |f_o| \leq 1 \\ 0 & |f_o| > 1 \end{cases} \quad \text{where,} \quad f_o = \frac{f}{f_m}. \tag{1.42}$$

Figure 1.24 is based on (1.42) and is the rough approximation of the Doppler PSD, which has the advantage that it is readily available in most existing RF simulators. This Doppler spectrum is representative of fixed mobile (point-to-point) wireless

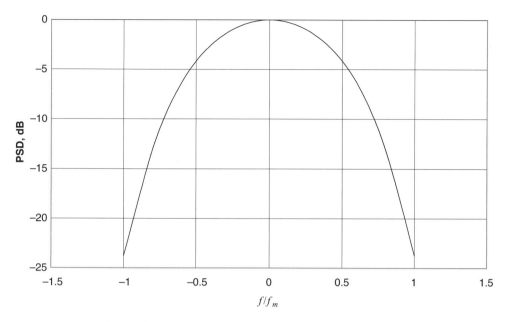

Figure 1.24 Doppler spectrum of a fixed wireless access channel.

channels and so does not represent the Doppler characteristics of a nonstationary mobile wireless channel.

1.5.7 UMTS Mobile Wireless Channel Model

Three different propagation environments for the UMTS mobile wireless channel model are considered: Indoor, Pedestrian and Vehicular. These environments were chosen because they are the main scenarios that cellular mobile radios experience in normal operation.

The Indoor environment model is characterized by having no Doppler frequency shift, as the velocities concerned within the indoor environment are well below walking pace ($\ll$4 miles per hour (MPH)), which produces either zero or a negligible shift in frequency. Within this environment there is no direct line-of-sight path from transmitter to receiver, so the signal propagates via many different paths as the signals are reflected, refracted or diffracted. Only six paths are simulated as the power contained within the strongest six rays are deemed strong enough to be included; the rest are very low power and thus will not affect the results appreciably. If they were included they would just serve to slow the simulation down.

The Pedestrian environment model is characterized by having a small Doppler frequency shift, as the velocities concerned within the pedestrian environment are around walking pace ($\approx$4 MPH), which produces a small Doppler shift in frequency. Within this environment there is limited/no direct line-of-sight path from transmitter to receiver, so the signal propagates via many different paths as the signals are reflected, refracted or diffracted. Only four paths are simulated as only the power contained within the strongest four rays are deemed strong enough to be included; the rest are very low power and thus will not affect the results appreciably.

The Vehicular environment model is characterized by having a larger Doppler frequency shift as the velocities concerned within the vehicular environment are reasonably large (≈ 70 MPH), producing a larger Doppler shift in frequency. Within this environment there is no direct line-of-sight path from transmitter to receiver, so the signal propagates via many different paths as the signals are reflected, refracted or diffracted. Only six paths are simulated as only the power contained within the strongest six rays deemed strong enough to be included; the rest are very low power and thus will not affect the results appreciably.

A Doppler spectrum is given in (1.43), which is the classic spectrum for mobile radio channels, and is used by Jakes [8] and Clarke [7]:

$$S(f) = \begin{cases} \dfrac{E_0}{4\pi f_m} \dfrac{1}{\sqrt{1 - \left(\dfrac{f}{f_m}\right)^2}} & |f| \le f_m \\ 0 & \text{elsewhere} \end{cases} , \qquad (1.43)$$

where E_0 = energy constant. Although this channel is representative of the mobile wireless channels, there are problems with this representation as the power spectral density becomes infinite at $f_C \pm f_m$. In order to find a more realistic (and useable) representation, Aulin [11] described the following:

$$S(f) = \begin{cases} 0 & \forall \; |f| > f_m \\ \dfrac{E_0}{4\sin\beta_m} \dfrac{1}{f_m}, & \forall \; f_m\cos\beta_m \le |f| \le f_m \\ \dfrac{1}{f_m}\left[\dfrac{\pi}{2} - \arcsin\dfrac{2\cos^2\beta_m - 1 - (f/f_m)^2}{1 - (f/f_m)^2}\right], & \forall \; |f| < f_m\cos\beta_m \end{cases} .$$

$$(1.44)$$

This claimed to be realistic for small values of β_m (angle of signal arrival) and is particularly useful in providing analytic solutions. The problem with this model is that there are sharp discontinuities at $\pm\beta_m$, which causes an unrealistic response. Therefore a model is required which produces the classic Doppler shape for mobile wireless channels but has no infinities or sharp discontinuities. Such a model was proposed by Parsons [6], and the PDF of the angle of arrival $p_\beta(\beta)$ is given by:

$$p_\beta(\beta) = \begin{cases} \dfrac{\pi}{4|\beta_m|}\cos\left(\dfrac{\pi}{2}\cdot\dfrac{\beta}{\beta_m}\right) & |\beta| \le |\beta_m| \le \dfrac{\pi}{2} \\ 0 & \text{elsewhere} \end{cases} . \qquad (1.45)$$

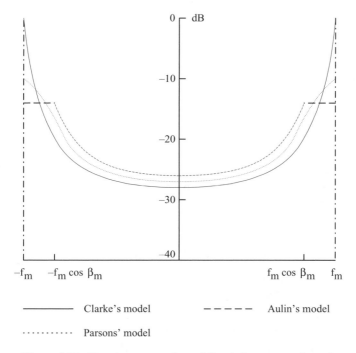

Figure 1.25 Doppler spectra of a mobile wireless access channel.

Using (1.45), the power spectral density can be expressed as:

$$S(f) = \text{FT}\left[\frac{E_0}{2}\int_{-\pi}^{+\pi} J_0\left(2\pi f_m \tau \cos\beta\right)p_\beta\left(\beta\right)\mathrm{d}\beta\right], \tag{1.46}$$

where FT is the Fourier transform and J_0 is the zero-order Bessel function of the first kind.

More information on the origin of these equations can be found in [7]. The Doppler spectra of (1.43), (1.44) and (1.46) are shown in Figure 1.25.

1.5.8 Simulating the Fixed Broadband Wireless Access Channel Model

There are different SUI(Stanford University Interim) models presented in [13]. The description of the SUI-3 model is explained here, and the model description and parameters are given in Table 1.2. A three-tap model is described here to represent the multipath scenario. Doppler PSD of fixed wireless is used to model the real environment.

Figure 1.26 shows the delay profile of each tap. The magnitude of the channel coefficients of all the taps, generated and plotted versus time, are shown below in Figure 1.27.

Table 1.2 SUI-3 channel parameters.

SUI-3 Channel				
	Tap 1	Tap 2	Tap 3	Units
Delay	0	0.5	1.0	μs
Power (Omni ant.)	0	-5	-10	dB
K Factor (Omni ant.)	1	0	0	
Power (30° ant.)	0	-11	-22	dB
K Factor (30° ant.)	3	0	0	
Doppler	0.4	0.4	0.4	Hz
Antenna Correlation:	$\rho_{ENV} = 0.4$			
Gain Reduction Factor:	GRF $= 3$ dB			
Normalization Factor:	$F_{omni} = -1.5113$ dB, $F_{30} = -0.3573$ dB			

1.5.9 Simulating the UMTS Mobile Radio Channel

The parameters for three different UMTS mobile channel models based on the standards put forward by ETSI, which model an *indoor*, a *pedestrian* and a *vehicular* environment, are given in this chapter. Tables 1.3–1.5 give the relative delays between taps and the average tap power for each scenario [14].

Simulation results evaluating the bit-error rate (BER) performance of QPSK modulation on the indoor, pedestrian and vehicular UMTS channels and BFWA channel are presented in Figure 1.28.

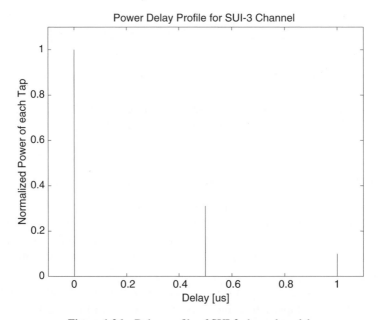

Figure 1.26 Delay profile of SUI-3 channel model.

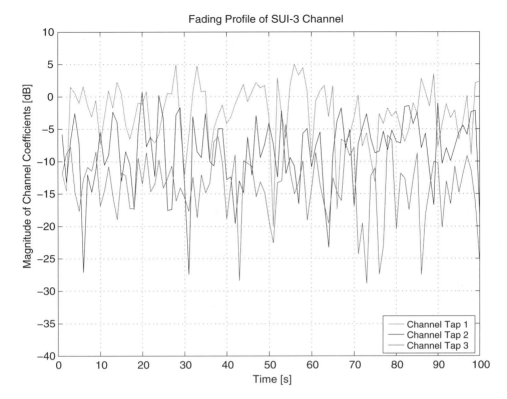

Figure 1.27 Fading profile of SUI-3 channel model.

Table 1.3 Channel parameters for the indoor UMTS channel model.

Tap	Relative delay (ns)	Average power (dB)
1	0	0
2	50	−3.0
3	110	−10.0
4	170	−18.0
5	290	−26.0
6	310	−32.0

Table 1.4 Channel parameters for the pedestrian UMTS channel model.

Tap	Relative delay (ns)	Average power (dB)
1	0	0
2	110	−9.7
3	190	−19.2
4	410	−22.8

Table 1.5 Channel parameters for the vehicular UMTS channel model.

Tap	Relative delay (ns)	Average power (dB)
1	0	0
2	310	−1.0
3	710	−9.0
4	1090	−10.0
5	1730	−15.0
6	2510	−20.0

1.6 (Multiple-Input–Multiple-Output) (MIMO) Channel

The performance analysis of the WiMAX system is extended from a SISO model to the MIMO model. Let us consider a multi-antenna system with N_T transmit and N_R receive antennas characterized by input/output in (1.47); the antenna configuration is shown in Figure 1.29.

$$y_m(t) = \sum_{n=1}^{N_T} \int_{-\infty}^{\infty} h_{mn}(t, \tau) x_n(t - \tau) \, d\tau, \quad m = 1, 2, \ldots, N_R, \qquad (1.47)$$

where $x_n(t)$ is the signal transmitted from the nth transmitted antenna, $y_m(t)$ is the signal received at the mth receive antenna and $h_{mn}(t, \tau)$ is the impulse response of the channel between the nth transmitted and mth receive antennas. The above equation

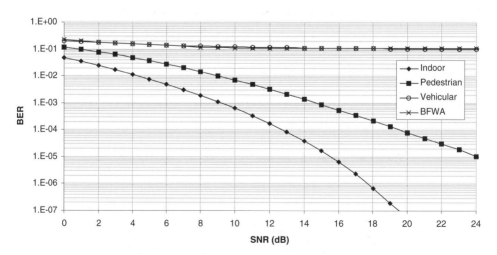

Figure 1.28 BER performance of QPSK modulation on the indoor, pedestrian and vehicular and BFWA channels.

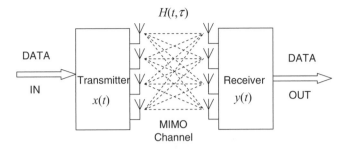

Figure 1.29 General MIMO channel with matrix H.

can be written in matrix form as follows:

$$y(t) = \int_{-\infty}^{\infty} H(t, \tau) x(t - \tau) \, d\tau,$$ (1.48)

where $\mathbf{H}(t, \tau)$ is the $N_R \times N_T$ channel matrix and is given as:

$$\mathbf{H}(t, \tau) = \begin{bmatrix} h_{11}(t, \tau) & h_{12}(t, \tau) & h_{13}(t, \tau) & \dots\dots\dots & h_{1N_T}(t, \tau) \\ h_{21}(t, \tau) & h_{22}(t, \tau) & h_{23}(t, \tau) & \dots\dots\dots & h_{2N_T}(t, \tau) \\ h_{31}(t, \tau) & h_{32}(t, \tau) & h_{33}(t, \tau) & \dots\dots\dots & h_{3N_T}(t, \tau) \\ \cdot & \cdot & \cdot & \dots\dots\dots & \cdot \\ \cdot & \cdot & \cdot & \dots\dots\dots & \cdot \\ h_{N_R1}(t, \tau) & h_{N_R2}(t, \tau) & h_{N_R3}(t, \tau) & \dots\dots\dots & h_{N_RN_T}(t, \tau) \end{bmatrix}, \quad (1.49)$$

$\mathbf{y}(t) = \begin{bmatrix} y_1(t), y_2(t), \dots, y_{N_R}(t) \end{bmatrix}$ is the N_R size row vector containing the signals received from N_R antenna and $\mathbf{x}(t) = \begin{bmatrix} x_1(t), x_2(t), \dots, x_{N_T}(t) \end{bmatrix}$ is the N_T size row vector containing signals transmitted from N_T antennas. The impulse response in the case of the MIMO channel can be explained using the same criteria as in the SISO case, but the distinguishing feature of MIMO systems is the spatial correlation among the impulse response composed of $\mathbf{H}(t, \tau)$.

The channel matrix $\mathbf{H}$ defines the input-output relations of the MIMO system and is known as the channel transfer function. Let us consider the two-transmit and two-receive antennae configuration, as shown in Figure 1.30. The system model comprises two transmitting and two receiving antennae and the medium between them is modelled as the MIMO channel.

The antenna correlations among the different paths are calculated and shown in Table 1.6, showing small correlations between the paths of the same channel. Also note that, for example, path 0 of channel A (A0) has correlation factor of 0.4 with path 0 of channels B, C and D (i.e. B0, C0 and D0). However, the same path 0 (A0) has a much lower correlation between other paths of the channels B, C and D.

Similarly, examples of correlation tables for the indoor, pedestrian and vehicular UMTS MIMO channels used in simulations are given in Tables 1.7–1.9.

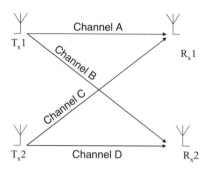

Figure 1.30 2×2 MIMO channel.

Table 1.6   Correlations between four paths, A, B, C and D, with subchannel paths 0, 1, 2 of a 2×2 MIMO system.

	A0	A1	A2	B0	B1	B2	C0	C1	C2	D0	D1	D2
A0	1	0.017	0.008	0.38	0.013	0.011	0.39	0.008	0.005	0.38	0.007	0.004
A1	0.017	1	0.021	0.003	0.4	0.02	0.003	0.39	0.01	0.003	0.403	0.015
A2	0.008	0.021	1	0.009	0.003	0.398	0.012	0.008	0.39	0.01	0.006	0.402
B0	0.38	0.003	0.009	1	0.003	0.008	0.39	0.004	0.006	0.39	0.008	0.002
B1	0.001	0.4	0.003	0.003	1	0.015	0.008	0.39	0.01	0.008	0.39	0.012
B2	0.01	0.02	0.39	0.008	0.015	1	0.016	0.01	0.39	0.01	0.008	0.39
C0	0.39	0.003	0.01	0.39	0.008	0.016	1	0.002	0.008	0.4	0.009	0.0006
C1	0.008	0.39	0.008	0.004	0.39	0.01	0.002	1	0.008	0.007	0.4	0.009
C2	0.005	0.011	0.39	0.006	0.01	0.39	0.008	0.008	1	0.009	0.007	0.401
D0	0.38	0.003	0.01	0.39	0.008	0.01	0.4	0.007	0.009	1	0.008	0.0039
D1	0.007	0.4	0.006	0.008	0.39	0.008	0.009	0.4	0.007	0.008	1	0.012
D2	0.004	0.001	0.4	0.002	0.01	0.39	0.006	0.009	0.4	0.003	0.012	1

Table 1.7   Mean correlation values between each path of the indoor MIMO channel.

	A	B	C	D
A	1	0.006874	0.006467	0.006548
B	0.006874	1	0.005903	0.006568
C	0.006467	0.005903	1	0.006312
D	0.006548	0.006568	0.006312	1

Table 1.8   Mean correlation values between each path of the pedestrian MIMO channel.

	A	B	C	D
A	1	0.049111	0.054736	0.050264
B	0.049111	1	0.056464	0.057746
C	0.054736	0.056464	1	0.062907
D	0.050264	0.057746	0.062907	1

Table 1.9 Mean correlation values between each path of the vehicular MIMO channel.

	A	B	C	D
A	1	0.057106	0.04603	0.052264
B	0.057106	1	0.040664	0.044029
C	0.04603	0.040664	1	0.061777
D	0.052264	0.044029	0.061777	1

In Chapter 7, the UMTS MIMO channels are used to evaluate the performance of space-time ring-TCM codes. The simplest space-time code uses a delay diversity code. Its performance on the indoor, pedestrian and vehicular MIMO channels is given in Figure 1.31.

1.7 Magnetic Storage Channel Modelling

Another application of error-correcting codes can be found in magnetic data storage devices. An error-correcting scheme in this situation must be able to correct long bursts of errors and have an efficient decoding algorithm, minimizing latency, to ensure high data rates. In this section we present a simple channel model for longitudinal magnetic recording, a method of writing data to a magnetic disc that is currently in use. Data is written to the magnetic disc by magnetizing microscopic areas on the disc in one of

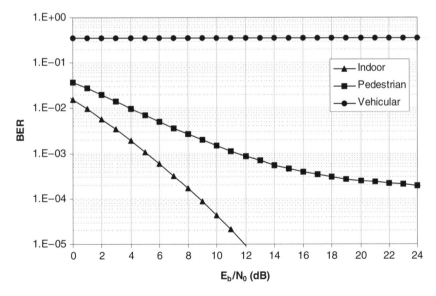

Figure 1.31 Simulation results evaluating the performance of the delay diversity code on the indoor, pedestrian and vehicular 2×2 MIMO channels.

two directions, representing either a '1' or a '0'. The data is recovered by passing the *read head* over the disc, which detects changes in the magnetic field corresponding to transitions from '0' to '1' or '1' to '0'. Traditionally, data is written on the plane of the magnetic disc; this is known as *longitudinal recording*. However, we have reached the limits of storage capacity for this particular technique, and now data is written perpendicular to the disc, known as *perpendicular recording*, which allows storage capacity to be further increased. In this chapter, modelling a longitudinal magnetic recording channel is explained, as this is still the most common recording method used for most hard drives.

1.7.1 Longitudinal Recording

A simple linear channel model for a longitudinal magnetic recording system is given in Figure 1.32 [15]. As stated previously, the read head measures the changes in the direction of magnetization, and this can be modelled as a differentiator. The differentiator with transfer function $1 - D$, where D is a memory element, subtracts the previous bit value from the current bit value.

The transition response for longitudinal magnetic recording can be modelled as a *Lorentzian* pulse, given by [15]:

$$h(t) = \frac{1}{1 + \left(\dfrac{2t}{PW50} \right)^2},\qquad (1.50)$$

where $PW50$ is the pulse width of the Lorentzian pulse at half its peak value. Some Lorentzian pulses and the effect of varying $PW50$ are shown in Figure 1.33. It will be shown that increasing $PW50$ increases the level of intersymbol interference (ISI).

A useful measure of ISI is the *recording linear density*, denoted by D_s and given by [15]:

$$D_s = \frac{PW50}{\tau},\qquad (1.51)$$

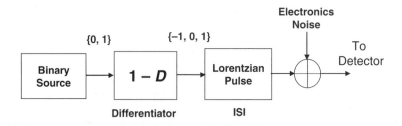

Figure 1.32 Simple channel model for a longitudinal magnetic recording channel.

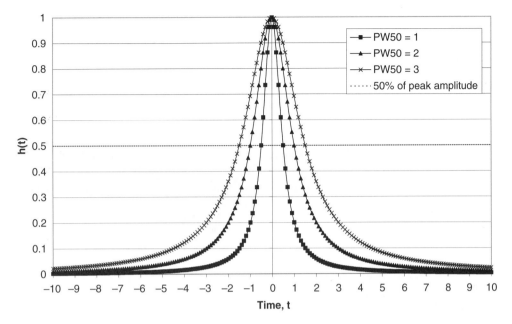

Figure 1.33 Lorentzian pulses with varying values of *PW50*.

where τ is the bit period. It gives the number of bits present within a time interval equal to *PW*50. A higher value of D_s corresponds to a higher level of ISI.

The response of a positive transition from '1' to a '0' is called a *dibit* response, expressed as:

$$d(t) = h(t) - h(t - \tau). \tag{1.52}$$

An example of a dibit response where the Lorentzian pulse has $PW50 = \tau = 1$ is given in Figure 1.34. A positive Lorentzian pulse is initiated at $t = 0$ and is followed by a negative Lorentzian pulse at $t = 1$. The sum of the two pulses gives the dibit response. The peak value of the dibit response is 80% of the peak value of the Lorentzian pulses since *PW*50 is wide enough that both pulses overlap. Increasing *PW*50 reduces the peak value of the dibit response further and this illustrates how ISI occurs in a longitudinal magnetic recording channel.

1.7.2 (Partial Response Maximum Likelihood) (PRML) Detection

Partial response maximum likelihood (PRML) detection is a two-stage process consisting of a transversal FIR filter concatenated with a Viterbi decoder (explained in Chapter 7), as shown in Figure 1.35.

The idea is to design a partial response (PR) equalizer with coefficients that shape the frequency response of the channel output to a predetermined *target response*

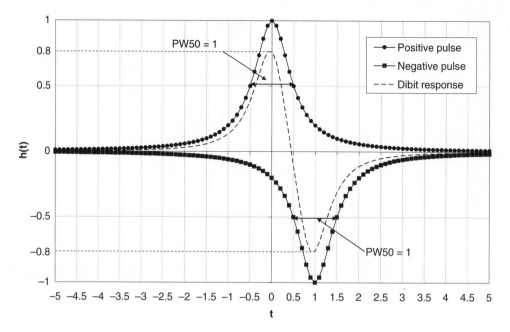

Figure 1.34 A positive Lorentzian pulse at $t = 0$ followed by a negative Lorentzian pulse at $t = 1$, both with *PW50* = 1. The sum of the two pulses results in the dibit response.

[15, 16]. It is well known that frequency response of the channel output for longitudinal recording with varying values of D_s matches closely with the frequency response of the polynomial of the form:

$$G(D) = (1 - D)(1 + D)^n, \tag{1.53}$$

where D is a delay element and n is a positive integer. For $n = 1$, 2 and 3 the polynomials are known as PR4, EPR4 and E^{2}PR4 respectively. For example, a longitudinal magnetic recording channel with $D_s = 2$ has a response that matches closely with the response of $G(D) = (1 - D)(1 + D) = 1 - D^2$, known as PR4. Given a binary input

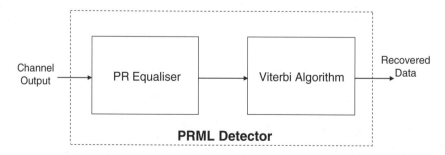

Figure 1.35 Block diagram showing the two processes in PRML detection.

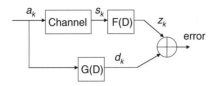

Figure 1.36 PR equalizer design criterion.

a_k, the output z_k of the PR equalizer should closely match the output d_k of the target polynomial $G(D)$, as shown in Figure 1.36 [16].

1.7.2.1 PR Equalizer Design

The PR equalizer polynomial $F(D)$ and target polynomial $G(D)$ are expressed as vectors **F** and **G** respectively [16]:

$$\mathbf{F} = \{-f_K \; -f_{K+1} \cdots f_0 \cdots f_{K-1} \; f_K\}, \tag{1.54}$$

where K is a positive integer and $\pm f_i$, $i = 0, 1, \ldots, K$, are the coefficients of $F(D)$.

$$\mathbf{G} = \{g_0 \; g_1 \cdots g_{L-1}\}, \tag{1.55}$$

where L is a positive integer and g_i, $i = 0, 1, \ldots, L-1$, are the coefficients of $G(D)$. Two further matrices **R** and **T** are defined as:

$$\mathbf{R} = \{R_{i,j}\} = E\{s_{k-i}s_{k-j}\}, \quad -K \leq i, j \leq K, \tag{1.56}$$

where $E\{\}$ is the expectation operator. **R** therefore contains autocorrelation values of the channel output:

$$\mathbf{T} = \{T_{i,j}\} = E\{s_{k-i}a_{k-j}\}, \quad -K \leq i \leq K, 0 \leq j \leq L-1, \tag{1.57}$$

T contains the cross correlation values of the channel output with the binary input. The coefficients of **F** can then be determined by [16]:

$$\mathbf{F} = \mathbf{R}^{-1}\mathbf{TG}. \tag{1.58}$$

After equalization the output should be similar to the desired output from $G(D)$. A diagram of the PR4 polynomial $G(D) = 1 - D^2$ is given in Figure 1.37.

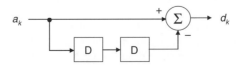

Figure 1.37 The PR4 target.

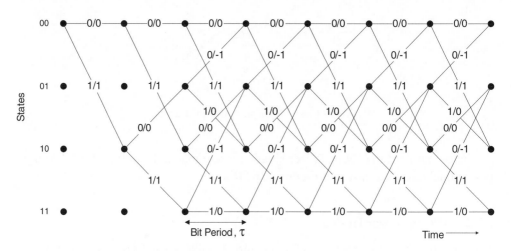

Figure 1.38   Trellis diagram of the PR4 target polyomial.

Since it has two memory elements, this can be expressed as a trellis, as shown in Figure 1.38, with four states and one input and one output for each branch.

The final part of PRML detection involves finding the maximum likelihood path through the trellis corresponding to the original binary input a_k. This can be achieved by using the soft-decision Viterbi algorithm.

1.8 Conclusions

In this chapter, an introduction to information theory and channel modelling was presented. In particular, the capacities of various channel models, such as the AWGN channel and SISO and MIMO fading channels modelling a fixed wireless access channel and UMTS cellular channels were examined. Simulation results were also presented, showing the performance of uncoded schemes on the FWA and UMTS channels for both MIMO and SISO situations. Finally, descriptions of different channel models were presented, allowing the performance for many coding schemes in wireless and magnetic storage channels to be evaluated by computer simulation.

References

[1] Shannon, C.E. (1948) A mathematical theory of communication. *Bell Systems Technical Journal*, **27**, 379–423, 623–56.
[2] Shannon, C.E. (1998) Communications in the presence of noise. *Proceedings of IEEE*, **86**, 447–58.
[3] McEliece, R.J. (1977) *The Theory of Information and Coding*, Addison-Wesley, Massachusetts.
[4] Abramson, N. (1962) *Information Theory and Coding*, McGraw-Hill, New York.
[5] Proakis, J.G. (1989) *Digital Communications*, 2nd edn, McGraw-Hill, New York.
[6] Parsons, J.D. (2000) *The Mobile Radio Propagation Channel*, 2nd edn, John Wiley & Sons Inc.

[7] Clarke, R.H. (1968) A statistical theory of mobile-radio reception. *Bell Systems Technical Journal*, **47**, 957–1000.

[8] Jakes, W.C. (1994) *Microwave Mobile Communications*, John Wiley & Sons, Inc., New York.

[9] Hata, M. (1980) Empirical formula for propagation loss in land mobile radio services. *IEEE Transactions on Vehicular Technology*, **VT-29** (3), 317–25.

[10] IEEE 802.16 (2001) *IEEE Standard for Local and Metropolitan Area Networks: part 1b, Air Interface for Fixed Broadband Wireless Access Systems*, April 8, 2002.

[11] Aulin, T. (1979) A modified model for the fading signal at a mobile radio channel. *IEEE Transactions on Vehicular Technology*, **VT-28** (3), 182–203.

[12] Greenstein, L.J., Erceg, V., Yeh, Y.S. and Clark, M.V. (1997) A new path-gain/delay-spread propagation model for digital cellular channels. *IEEE Transactions on Vehicular Technology*, **46** (2), 477–85.

[13] IEEE 802.11 (1994) *Wireless Access Method and Physical Layer Specification*, New York.

[14] Erceg, V. (1999) An empirically based path loss model for wireless channels in suburban environments. *IEEE Journal in Selected Areas in Communications*, **17** (7), 1205–11.

[15] Vasic, B. and Kurtas, E.M. (eds) (2005) *Coding and Signal Processing for Magnetic Recording Systems*, CRC Press.

[16] Moon, J. and Zeng, W. (1995) Equalization for maximum likelihood detectors. *IEEE Trans. Mag.*, **31** (2), 1083–8.

2

Basic Principles of Non-Binary Codes

2.1 Introduction to Algebraic Concepts

In this chapter, the basic mathematical concepts necessary for working with non-binary error-correcting codes are presented. The elements of the non-binary codes described in this book belong to a non-binary alphabet and so this chapter begins with a discussion of the definition and properties of a *Group*, which will lead to an introduction to *Rings* and *Fields*. Later in the book, knowledge of the properties of rings will be necessary to understanding the design of ring trellis coded modulation (ring-TCM) codes and ring block coded modulation (ring-BCM) codes. Similarly, a good understanding of finite fields is needed to be able to construct and decode Bose, Ray-Chaudhuri, Hocquenghem (BCH) codes, Reed–Solomon codes and Algebraic–Geometric codes.

2.1.1 Groups

A set contains any elements or objects with no conditions imposed on it and can be either finite or infinite. The number of elements or objects in a set is called the *cardinality*. There are two binary operations that can operate on a set: multiplication '·' and addition '+'. Certain conditions can be applied to the set under these binary operations. The most common conditions are [1]:

- *Commutativity:* For two elements a and b in the set, $a·b = b·a$ under multiplication or $a + b = b + a$ under addition.
- *Identity:* For any element a in the set there is an identity element b such that $a·b = a$ under multiplication or $a + b = a$ under addition.
- *Inverse:* For any element a in the set its inverse a^{-1} must also be in the set. This obeys $a·a^{-1} = a^{-1}·a = b$ (the identity element).

Non-Binary Error Control Coding for Wireless Communication and Data Storage   Rolando Antonio Carrasco and Martin Johnston
© 2008 John Wiley & Sons, Ltd

- *Associativity:* For three elements a, b and c in the set, $(a \cdot b) \cdot c = a \cdot (b \cdot c)$.
- *Distributivity:* For three elements a, b and c in the set, $a \cdot (b + c) = (a \cdot b) + (a \cdot c)$.

A *group* is defined as a set with the multiplication operation. Multiplying any two elements in the group must result in a third element that is also in the group. This is known as *closure*. A group also has the following conditions:

- associativity under multiplication,
- identity under multiplication,
- inverse.

The group is called a *commutative* or *abelian group* if it also has commutativity under multiplication.

2.1.2 Rings

If the two binary operations '$+$' and '$\cdot$' are allowed then a *ring* can be defined. A ring must have the following conditions:

1. associativity,
2. distributivity,
3. commutativity under addition.

The ring is called a *commutative ring* if it also has commutativity under multiplication. If the ring has a multiplicative identity 1 then it is called a *ring with identity*. An example of a ring is the ring of integers $\mathbb{Z}_q$ under modulo-q addition and multiplication, where q is the cardinality of the ring. For example, $\mathbb{Z}_4$ is defined as $\{0, 1, 2, 3\}$. It is easy to see that the elements obey the three definitions of a ring. Also, all the elements commute under multiplication and the multiplicative identity element 1 is present, meaning that $\mathbb{Z}_4$ is a *commutative ring with identity*. Tables 2.1 and 2.2 show the addition and multiplication tables respectively of the ring of integers $\mathbb{Z}_8 = \{0, 1, 2, 3, 4, 5, 6, 7\}$.

Table 2.1   Addition table for $\mathbb{Z}_8$.

$+$	0	1	2	3	4	5	6	7
0	0	1	2	3	4	5	6	7
1	1	2	3	4	5	6	7	0
2	2	3	4	5	6	7	0	1
3	3	4	5	6	7	0	1	2
4	4	5	6	7	0	1	2	3
5	5	6	7	0	1	2	3	4
6	6	7	0	1	2	3	4	5
7	7	0	1	2	3	4	5	6

Table 2.2 Multiplication table for $\mathbb{Z}_8$.

·	0	1	2	3	4	5	6	7
0	0	0	0	0	0	0	0	0
1	0	1	2	3	4	5	6	7
2	0	2	4	6	0	2	4	6
3	0	3	6	1	4	7	2	5
4	0	4	0	4	0	4	0	4
5	0	5	2	7	4	1	6	3
6	0	6	4	2	0	6	4	2
7	0	7	6	5	4	3	2	1

2.1.3 Ring of Polynomials

The set of all polynomials with coefficients defined in $\mathbb{Z}_q$ forms a ring under the addition and multiplication operations. If we define two polynomials $f(x)$ and $g(x)$ as $f(x) = f_0 + f_1 x + f_2 x^2 + \cdots + f_v x^v$ and $g(x) = g_0 + g_1 x + g_2 x^2 + \cdots + g_w x^w$, where f_i and $g_i \in \mathbb{Z}_q$, v and w are the degrees of both polynomials respectively with $v < w$, then the addition of both polynomials is:

$$f(x) + g(x) = \sum_i (f_i + g_i) \cdot x^i,$$

where $f_i + g_i \in \mathbb{Z}_q$.

Similarly, the product of the two polynomials is:

$$f(x) \cdot g(x) = \sum_i \left(\sum_{j=0}^{i} f_j \cdot g_{i-j} \right) \cdot x^i \quad \text{where} \quad \sum_i \left(\sum_{j=0}^{i} f_j \cdot g_{i-j} \right) \in \mathbb{Z}_q.$$

For example, if $f(x) = 1 + 2x + 3x^2$ and $g(x) = 3 + x + 2x^2 + x^3$ are defined in $\mathbb{Z}_4$, then the sum of both polynomials is:

$$
\begin{aligned}
f(x) + g(x) &= \left(1 + 2x + 3x^2\right) + \left(3 + x + 2x^2 + x^3\right) \\
&= (1+3) + (2+1)x + (3+2)x^2 + x^3 \\
&= 3x + x^2 + x^3.
\end{aligned}
$$

Similarly, the product of both polynomials is:

$$
\begin{aligned}
f(x) \cdot g(x) &= (1 + 2x + 3x^2) \cdot (3 + x + 2x^2 + x^3) \\
&= (1 \times 3) + (1 \times 1 + 2 \times 3)x + (1 \times 2 + 2 \times 1 + 3 \times 3)x^2 \\
&\quad + (1 \times 1 + 2 \times 2 + 3 \times 1)x^3 + (3 \times 2 + 2 \times 1)x^4 + (3 \times 1)x^3 \\
&= 3 + 7x + 13x^2 + 8x^3 + 8x^4 + 3x^3 \\
&= 3 + 3x + x^2 + 3x^3.
\end{aligned}
$$

Table 2.3 Multiplication table of nonzero elements in GF(5).

·	1	2	3	4
1	1	2	3	4
2	2	4	1	3
3	3	1	4	2
4	4	3	2	1

2.1.4 Fields

A *field* is similar to a ring as it also uses both binary operations. It has the following conditions:

1. *Commutativity* under addition,
2. *Distributivity*,
3. *Commutativity* under multiplication when the additive identity element 0 is removed,
4. *Identity*,
5. *Inverse*.

An example of a field is the set of real numbers. The set of integers is not a field since not all integers have a multiplicative inverse. A finite field has a finite number of elements and is known as a Galois Field, written as GF(q), where q is the cardinality of the field and is a prime number or a power of a prime greater than 1. To be a finite field the elements $\{1, 2, 3, \ldots, q - 1\}$ must form a group under multiplication modulo-q.

For example, GF(5) $= \{0, 1, 2, 3, 4\}$ is a finite field because the elements $\{1, 2, 3, 4\}$ are a group under modulo-q multiplication. The group has closure because no two elements multiplied together give 0, which is not in the group, as shown in Table 2.3.

However, let us say that the set $\{0, 1, 2, 3, 4, 5\}$ defines the finite field GF(6). To be a finite field the elements $\{1, 2, 3, 4, 5\}$ must form a group, but this is not the case. It can be seen in Table 2.4 that some elements produce a 0 when multiplied together,

Table 2.4 Multiplication table of nonzero elements in GF(6), proving it is not a finite field.

·	1	2	3	4	5
1	1	2	3	4	5
2	2	4	<u>0</u>	2	4
3	3	<u>0</u>	3	<u>0</u>	3
4	4	2	<u>0</u>	4	2
5	5	4	3	2	1

Table 2.5 The order of each nonzero element in GF(5).

Element in GF(5)	Order
1	1
2	4
3	4
4	2

which is not in the set $\{1, 2, 3, 4, 5\}$, and this proves that GF(6) is not a finite field. Hence, only sets with a prime or a power of a prime number of elements can be finite fields.

Every element β in a finite field has an *order*, denoted as $ord(\beta)$. The order of an element is the number of times the element is multiplied by itself until the produce equals 1, that is the value of m which gives $\beta^m = 1$. In a finite field GF(q) the order of an element divides $q - 1$. For example, the order of the element 2 in GF(5) is 4 because $2 \times 2 \times 2 \times 2 = 2^4 = 1$. The order of each element in GF(5) is shown in Table 2.5.

It can be seen that all the elements in GF(5) have an order which divides $q - 1 = 4$. An element which has an order equal to $q - 1$ is called a *primitive element*. Therefore, from Table 2.5, the elements 2 and 3 are primitive elements in GF(5) since they have an order of 4. This means that all the nonzero elements in a finite field can be expressed as powers of primitive elements. For example, in GF(5), 2 and 3 are the primitive elements, meaning that all the nonzero elements in GF(5) can be expressed using powers of 2 or 3, as shown in Table 2.6.

Finite fields of the form GF(p^m), where p is prime and $m > 0$, are called *extension fields*. They contain the elements $\{0, 1, \alpha, \alpha^2, \ldots, \alpha^{p^m-2}\}$, where α is the primitive element with order $p^m - 1$. This means that all the nonzero elements in GF(p^m) can

Table 2.6 Using primitive elements to define all nonzero elements in GF(5).

Powers of primitive element 2	Element in GF(5)
2^1	2
2^2	4
2^3	3
2^4	1

Powers of primitive element 3	Element in GF(5)
3^1	3
3^2	4
3^3	2
3^4	1

be represented as powers of α. For example, $\mathrm{GF}(2^2) = \{0, 1, \alpha, \alpha^2\}$. Each element in $\mathrm{GF}(p^m)$ can be expressed as an m-tuple vector with elements from $\mathrm{GF}(p)$ and addition between elements is done modulo-p.

2.1.5 Primitive Polynomials

A class of polynomials called primitive polynomials are used to define $\mathrm{GF}(p^m)$. An irreducible polynomial $f(x)$ of degree m defined over $\mathrm{GF}(p)$ is primitive if the smallest positive integer n for which $f(x)$ divides $x^n - 1$ is $n = p^m - 1$ [2]. As an example, take the irreducible polynomial

$$x^4 + x + 1$$

and $p = 2$.

This polynomial will be primitive if it divides $x^{2^4-1} + 1 = x^{16-1} + 1 = x^{15} + 1$, as $-1 \equiv 1$ when $p = 2$. It can be seen from Figure 2.1 that there is no remainder and so $x^4 + x + 1$ divides $1 + x^{15}$. Similar calculations will show that $x^4 + x + 1$ will not divide $x^n + 1$, $1 \leq n < 15$, which means that $x^4 + x + 1$ is a primitive polynomial. If a primitive element α is the root of a primitive polynomial then a higher power of α

$$
\begin{array}{l}
\qquad\qquad x^{11} + x^8 + x^7 + x^5 + x^3 + x^2 + x + 1 \\
x^4 + x + 1 \overline{)\, x^{15} + 0x^{14} + 0x^{13} + 0x^{12} + 0x^{11} + 0x^{10} + 0x^9 + 0x^8 + 0x^7 + 0x^6 + 0x^5 + 0x^4 + 0x^3 + 0x^2 + 0x + 1} \\
\qquad x^{15} \qquad x^{12} + x^{11} \\
\qquad 0 \qquad x^{12} + x^{11} \\
\qquad\quad x^{12} + \qquad x^9 + x^8 \\
\qquad\quad 0 \quad x^{11} \quad + x^9 + x^8 \\
\qquad\qquad x^{11} \qquad + x^8 + x^7 \\
\qquad\qquad 0 \qquad x^9 + 0 + x^7 \\
\qquad\qquad\quad x^9 \qquad + x^6 + x^5 \\
\qquad\qquad\quad 0 \qquad x^7 + x^6 + x^5 \\
\qquad\qquad\qquad x^7 \qquad + x^4 + x^3 \\
\qquad\qquad\qquad 0 \quad + x^6 + x^5 + x^4 + x^3 \\
\qquad\qquad\qquad\quad x^6 \qquad + x^3 + x^2 \\
\qquad\qquad\qquad\quad 0 + x^5 + x^4 + 0 + x^2 \\
\qquad\qquad\qquad\qquad x^5 \qquad + x^2 + x \\
\qquad\qquad\qquad\qquad 0 + x^4 \qquad + 0 + x + 1 \\
\qquad\qquad\qquad\qquad\quad x^4 \qquad + x + 1 \\
\qquad\qquad\qquad\qquad\qquad\qquad 0
\end{array}
$$

Figure 2.1 Long division showing that the primitive polynomial $x^4 + x + 1$ divides $x^{15} - 1$.

can be expressed as the sum of lower powers of α. For example, if α is a root of $x^4 + x + 1$, this implies $\alpha^4 + \alpha + 1 = 0$ or $\alpha^4 = -\alpha - 1$. So, $\alpha^5 = -\alpha^2 - \alpha$, $\alpha^6 = -\alpha^3 - \alpha^2$, $\alpha^7 = -\alpha^4 - \alpha^3 = (-\alpha - 1) - \alpha^3$ and so on. In this way an extension field $GF(p^m)$ can be defined with the elements expressed in the form $\alpha^{m-1} + \alpha^{m-2} + \ldots \alpha + 1$, or equivalently an m-dimensional vector space over $GF(p)$.

Example 2.1: Constructing the extension field $GF(2^3)$: The extension field $GF(2^3)$ can be constructed using the primitive polynomial $f(x) = x^3 + x + 1$. If α is the root of $f(x)$ then $\alpha^3 = \alpha + 1$. The elements of $GF(2^3)$ are shown in Table 2.7.

Table 2.7 Construction of $GF(2^3)$.

Element in $GF(2^3)$	Element expressed as the sum of lower powers of α	Element expressed as 3-tuple vector over $GF(2)$
0	**0**	000
1	**1**	001
α	α	010
α^2	α^2	100
α^3	$\alpha + \mathbf{1}$	011
α^4	$\alpha^2 + \alpha$	110
α^5	$\alpha^3 + \alpha^2 = \alpha + \mathbf{1} + \alpha^2$	111
α^6	$\alpha^4 + \alpha^3 = \alpha^2 + \alpha + \alpha + \mathbf{1} = \alpha^2 + \mathbf{1}$	101

To prove the field has closure, the element α^7 can be expressed as $\alpha^6 \cdot \alpha = (\alpha^2 + 1) \cdot \alpha = \alpha^3 + \alpha = \alpha + 1 + \alpha = 1$, that is $\alpha^{c(q-1)} = 1$ where c is a nonnegative integer. Addition in $GF(2^3)$ is accomplished by adding the vector representation of the elements modulo-2. For example, $\alpha^5 + \alpha^6 = (111) + (101) = (010) = \alpha$. Multiplication is accomplished by taking the powers of the elements and adding them modulo-7. For example, $\alpha^4 \cdot \alpha^5 = \alpha^{(4+5) \bmod 7} = \alpha^2$. The addition and multiplication tables for $GF(2^3)$ are shown in Tables 2.8 and 2.9, respectively.

Table 2.8 Addition table for $GF(2^3)$.

+	**0**	**1**	α	α^2	α^3	α^4	α^5	α^6
0	0	1	α	α^2	α^3	α^4	α^5	α^6
1	1	0	α^3	α^6	α	α^5	α^4	α^2
α	α	α^3	0	α^4	1	α^2	α^6	α^5
α^2	α^2	α^6	α^4	0	α^5	α	α^3	1
α^3	α^3	α	1	α^5	0	α^6	α^2	α^4
α^4	α^4	α^5	α^2	α	α^6	0	1	α^3
α^5	α^5	α^4	α^6	α^3	α^2	1	0	α
α^6	α^6	α^2	α^5	1	α^4	α^3	α	0

Table 2.9 Multiplication table for GF(2^3).

·	0	1	α	α^2	α^3	α^4	α^5	α^6
0	0	0	0	0	0	0	0	0
1	0	1	α	α^2	α^3	α^4	α^5	α^6
α	0	α	α^2	α^3	α^4	α^5	α^6	1
α^2	0	α^2	α^3	α^4	α^5	α^6	1	α
α^3	0	α^3	α^4	α^5	α^6	1	α	α^2
α^4	0	α^4	α^5	α^6	1	α	α^2	α^3
α^5	0	α^5	α^6	1	α	α^2	α^3	α^4
α^6	0	α^6	1	α	α^2	α^3	α^4	α^5

Example 2.2: Constructing the extension field GF(3^2): The extension field GF(3^2) can be constructed using the primitive polynomial $f(x) = x^2 - 2x - 1$. If the primitive element α is the root of $f(x)$ then $\alpha^2 = 2\alpha + 1$ and can be expressed as a 2-dimensional vector space over GF(3), as shown in Table 2.10.

Table 2.10 Construction of GF(3^2).

Element in GF(3^2)	Element expressed as the sum of lower powers of α	Element expressed as 2-tuple vector over GF(3)
0	**0**	00
1	**1**	01
α	α	10
α^2	$2\alpha + 1$	21
α^3	$2\alpha^2 + \alpha = 2(2\alpha + 1) + \alpha = 4\alpha + 2 + \alpha = 5\alpha + 2$ $= 2\alpha + 2$	22
α^4	$2\alpha^2 + 2\alpha = 2(2\alpha + 1) + 2\alpha = 4\alpha + 2 + 2\alpha = 6\alpha$ $+ 2 = 2$	02
α^5	2α	20
α^6	$2\alpha^2 = 2(2\alpha + 1) = 4\alpha + 2 = \alpha + 2$	12
α^7	$\alpha^2 + 2\alpha = 2\alpha + 1 + 2\alpha = 4\alpha + 1 = \alpha + 1$	11

To add two elements together, their corresponding vectors are added modulo-3, for example $\alpha^4 + \alpha^6 = (02) + (12) = (11) = \alpha^7$. To multiply two elements, their powers are added modulo-8, for example $\alpha^6 + \alpha^7 = \alpha^{(6+7)\bmod 8} = \alpha^5$.

2.1.6 Minimal Polynomials and Cyclotomic Cosets

It was stated that a primitive polynomial defined in GF(2) was irreducible, implying that it cannot be factorized, that is it has no roots in GF(2). However, that a polynomial is irreducible in one finite field usually does not mean that it will be irreducible in other finite fields. By substituting each element of GF(8) into the primitive polynomial $x^3 + x + 1$ it can be seen that this polynomial has three roots: $X_1 = \alpha$, $X_2 = \alpha^2$ and $X_3 = \alpha^4$. This means that $x^3 + x + 1$ can be factorized to $(x + \alpha)(x + \alpha^2)(x + \alpha^4)$.

In general, given one root $\beta \in \text{GF}(2^p)$, the remaining roots will be β^{2i}, $i = 1, 2, 3, \ldots, 2^{p-1} - 1$. These roots are called the *conjugates* of β and the polynomial

$$M(x) = (x + \beta)(x + \beta^2)(x + \beta^4) \ldots \left(x + \beta^{2^{p-1}-1}\right)$$

is called the *minimal polynomial* [2] of β, that is the degree of $M(x)$ is minimal. Therefore, the minimal polynomial of α, α^2 and α^4 in GF(8) is $M(x) = x^3 + x + 1$, which is also a primitive polynomial. The minimal polynomials for the other elements in GF(8) are calculated as follows.

For $\beta = 0$ the minimal polynomial is $M(x) = x$, and for $\beta = 1$, $M(x) = x + 1$. The conjugates for $\beta = \alpha^3$ are α^6 and α^5. The minimal polynomial for α^3 is then:

$$\begin{aligned}
M(x) &= (x + \alpha^3)(x + \alpha^6)(x + \alpha^5) \\
&= \left[x^2 + (\alpha^3 + \alpha^6)x + \alpha^9\right](x + \alpha^5) \\
&= (x^2 + \alpha^4 x + \alpha^2)(x + \alpha^5) \\
&= x^3 + (\alpha^5 + \alpha^4)x^2 + (\alpha^9 + \alpha^2)x + \alpha^7 \\
&= x^3 + x^2 + 1.
\end{aligned}$$

The powers of finite field elements associated with each minimal polynomial form *cyclotomic cosets*. The minimal polynomials for GF(8) and their corresponding cyclotomic cosets are summarized in Table 2.11.

2.1.7 Subfields

Some elements within a finite field GF(p^m) can also belong to a smaller finite field, known as a *subfield*. These elements will have a minimal polynomial that matches a known primitive polynomial that can define a smaller finite field. An example of a subfield can be found in GF(16), defined by the primitive polynomial $x^4 + x + 1$. The element α^5 has only one conjugate, namely α^{10}. The minimal polynomial for α^5 in

Table 2.11 Elements in GF(8) and their corresponding minimal polynomials.

Finite field element in GF(8)	Minimal polynomial	Cyclotomic coset
0	x	—
1	$x + 1$	$\{0\}$
$\alpha, \alpha^2, \alpha^4$	$x^3 + x + 1$	$\{1, 2, 4\}$
$\alpha^3, \alpha^6, \alpha^5$	$x^3 + x^2 + 1$	$\{3, 6, 5\}$

GF(16) is therefore:

$$\begin{aligned}
M(x) &= (x + \alpha^5)(x + \alpha^{10})\\
&= x^2 + (\alpha^5 + \alpha^{10})x + \alpha^{15}\\
&= x^2 + ((\alpha^2 + \alpha) + (\alpha^2 + \alpha + 1))x + 1\\
&= x^2 + x + 1.
\end{aligned}$$

This minimal polynomial is known to be the primitive polynomial that defines GF(4), and this implies that $\beta = \alpha^5$ is a primitive element for the subfield GF(4) within GF(16). This subfield then consists of the elements $\{0, 1, \beta, \beta^2\} = \{0, 1, \alpha^5, \alpha^{10}\}$.

More examples of subfields can be found in larger finite fields, such as GF(64), defined by the primitive polynomial $x^6 + x + 1$. A list of minimal polynomials for GF(64) and many others can be found in [1]. The element α^{21} has a minimal polynomial of $x^2 + x + 1$, implying that it is the primitive element of the subfield GF(4) with elements $\{0, 1, \alpha^{21}, \alpha^{42}\}$. Also, the element α^9 has a minimal polynomial of $x^3 + x^2 + 1$, implying that it is the primitive element of the subfield GF(8) with elements $\{0, 1, \beta, \beta^2, \beta^3, \beta^4, \beta^5, \beta^6\} = \{0, 1, \alpha^9, \alpha^{18}, \alpha^{36}, \alpha^8, \alpha^{16}, \alpha^{32}\}$.

2.2 Algebraic Geometry Principles

2.2.1 Projective and Affine Space

The construction of AG codes requires a set of points that satisfy an irreducible affine curve and a set of rational functions defined on the curve. *Projective space* is $(n + 1)$-dimensional and elements in this space defined over some finite field are called projective points, that is [3]:

$$(c_1, c_2, c_3, \ldots, c_n, c_{n+1}),$$

where c_i, $i = 0, 1, \ldots, n, n + 1$, are elements in some finite field.

Affine space is n-dimensional. Elements in this space defined over some finite field are called affine points and are of the form [3] $(c_1, c_2, c_3, \ldots, c_n, 1)$.

Finally, when points are of the form $(c_1, c_2, c_3, \ldots, c_n, 0)$, the space is called the *hyperplane at infinity* [3]. All points in the hyperplane at infinity are called *points at infinity*. For the construction of AG codes we are interested in finding all affine points and points at infinity that cause an irreducible smooth affine curve to vanish.

2.2.2 Projective and Affine Curves

A projective curve is an $(n + 1)$-dimensional curve defined by projective points. It is made up of n-dimensional affine curves in the $n + 1$ different coordinate systems.

For a 3-dimensional projective curve $C(x, y, z)$ there are three affine curves: $C(x, y, 1)$, $C(x, 1, z,)$ and $C(1, y, z)$.

Algebraic–geometric codes are constructed from irreducible affine smooth curves. An irreducible curve is a curve that cannot be expressed as the product of curves of lower degrees. For example, the curve $C(x, y) = x^2 + y^2$ is not irreducible over GF(2) because:

$$(x + y)(x + y)$$
$$= (x^2 + xy + xy + y^2) \quad \text{and} \quad xy + xy = 0.$$
$$= x^2 + y^2$$

Similarly, the curve $C(x, y) = x^3 + y^3$ is irreducible over GF(2) but is not irreducible over GF(3) because:

$$(x + y)(x + y)(x + y)$$
$$= (x^2 + 2xy + y^2)(x + y)$$
$$= x^3 + x^2y + 2x^2y + 2xy^2 + xy^2 + y^3.$$
$$= x^3 + 3x^2y + 3xy^2 + y^3 \quad \text{and} \quad 3 \equiv 0 \bmod 3$$
$$= x^3 + y^3$$

A point on a curve is said to be nonsingular if all partial derivatives of the curve do not vanish at this point. If all the points that satisfy the curve are nonsingular then the curve is said to be nonsingular or smooth [4]. An important class of curve is the Hermitian curve, which is defined over square finite fields $GF(w^2)$. The projective Hermitian curve is:

$$C(x, y, z) = x^{w+1} + y^w z + yz^w. \tag{2.1}$$

This is a smooth curve, and to prove it the three partial derivatives are calculated:

$$\frac{\partial C(x, y, z)}{\partial x} = (w + 1)x^w = x^w$$
$$\frac{\partial C(x, y, z)}{\partial y} = wy^{w-1}z + z^w = z^w.$$
$$\frac{\partial C(x, y, z)}{\partial z} = y^w + wyz^{w-1} = y^w$$

The only point that makes all three partial derivatives vanish is $(0, 0, 0)$, but this is not a point in projective space, and so all the points are nonsingular and the curve is smooth.

2.2.3 Finding Points on an Affine Curve

The projective points that satisfy the projective curve $C(x, y, z) = 0$ are of the form (α, β, δ), where α, β, δ are elements in a finite field. For the construction of an AG code, only the affine points of the form $(\alpha, \beta, 1)$, that is $z = 1$, and points at infinity of the form $(\alpha, \beta, 0)$, that is $z = 0$, are required. To find all the affine points the projective curve is broken down into its affine component curves. For each affine curve, the affine points $(\alpha, \beta, 1)$ and the points at infinity $(\alpha, \beta, 0)$ that cause the curve to vanish are kept. By using all the affine component curves more points can be found, resulting in longer codes.

The affine form of the Hermitian curve from (2.1) in the $(x - y)$ coordinate system (with $z = 1$) is:

$$C(x, y, 1) = x^{w+1} + y^w + y. \tag{2.2}$$

In the $(x - z)$ coordinate system (with $y = 1$) it is:

$$C(x, 1, z) = x^{w+1} + z + z^w. \tag{2.3}$$

In the $(y - z)$ coordinate system (with $x = 1$) it is:

$$C(1, y, z) = 1 + y^w z + yz^w. \tag{2.4}$$

The Hermitian curves are known to have $w^3 + 1$ points. For GF(2^2), the Hermitian curve will have $2^3 + 1 = 8 + 1 = 9$ points and is defined as:

$$C(x, y, z) = x^3 + y^2 z + yz^2, \tag{2.5}$$

with the affine curve in the $(x - y)$ system defined as:

$$C(x, y) = x^3 + y^2 + y. \tag{2.6}$$

The affine points that satisfy $C(x, y, 1) = 0$ are found by substituting all possible values of x and y (with $z = 1$) in GF(2^2), where $\alpha^2 = \alpha + 1$, into (2.6). This is shown below:

$$C(0, 0, 1) = 0^3 + 0^2 + 0 = 0$$
$$C(0, 1, 1) = 0^3 + 1^2 + 1 = 0$$
$$C(0, \alpha, 1) = 0^3 + \alpha^3 + \alpha = 1 + \alpha = \alpha^2 \neq 0$$
$$C(0, \alpha^2, 1) = 0^3 + \alpha^6 + \alpha^2 = 1 + \alpha^2 = \alpha \neq 0$$

$$C(1, 0, 1) = 1^3 + 0^2 + 0 = 1 \neq 0$$
$$C(1, 1, 1) = 1^3 + 1^2 + 1 = 1 \neq 0$$
$$C(1, \alpha, 1) = 1^3 + \alpha^2 + \alpha = 1 + 1 = 0$$
$$C(1, \alpha^2, 1) = 1^3 + \alpha^4 + \alpha^2 = 1 + \alpha + \alpha^2 = 1 + 1 = 0$$

$$C(\alpha, 0, 1) = \alpha^3 + 0^2 + 0 = \alpha^3 = 1 \neq 0$$
$$C(\alpha, 1, 1) = \alpha^3 + 1^2 + 1 = \alpha^3 = 1 \neq 0$$
$$C(\alpha, \alpha, 1) = \alpha^3 + \alpha^2 + \alpha = 1 + 1 = 0$$
$$C(\alpha, \alpha^2, 1) = \alpha^3 + \alpha^4 + \alpha^2 = 1 + \alpha + \alpha^2 = 1 + 1 = 0$$

Table 2.12 Eight projective points for the Hermitian curve in the $(x - y)$ system.

$P_1 = (0, 0, 1)$	$P_2 = (0, 1, 1)$	$P_3 = (1, \alpha, 1)$	$P_4 = (1, \alpha^2, 1)$
$P_5 = (\alpha, \alpha, 1)$	$P_6 = (\alpha, \alpha^2, 1)$	$P_7 = (\alpha^2, \alpha, 1)$	$P_8 = (\alpha^2, \alpha^2, 1)$

$$C(\alpha^2, 0, 1) = \alpha^6 + 0^2 + 0 = 1 \neq 0$$
$$C(\alpha^2, 1, 1) = \alpha^6 + 1^2 + 1 = \alpha^6 = 1 \neq 0$$
$$C(\alpha^2, \alpha, 1) = \alpha^6 + \alpha^2 + \alpha = \alpha^6 + 1 = 1 + 1 = 0.$$
$$C(\alpha^2, \alpha^2, 1) = \alpha^6 + \alpha^4 + \alpha^2 = \alpha^6 + \alpha + \alpha^2 = \alpha^6 + 1 = 1 + 1 = 0$$

Therefore, the affine points that satisfy $C(x, y, 1) = 0$ are given in Table 2.12.

We must also find the points of the other two affine curves, $C(x, 1, z) = x^3 + z + z^2$ and $C(1, y, z) = 1 + y^2z + yz^2$. There are eight projective points that satisfy $C(x, 1, z) = 0$, as shown in Table 2.13.

However, P_2 is the only affine point, since it is of the form $(x, y, 1)$, but it is also in Table 2.12. There is also a point at infinity, $P_1 = (0, 1, 0)$. Finally, there are four projective points that satisfy $C(1, y, z) = 0$, as shown in Table 2.14. In this case, only P_1 and P_3 are affine points, since they are of the form $(x, y, 1)$, and they are also present in Table 2.12. Therefore, the projective Hermitian curve in (2.5) has eight affine points and one point at infinity, $Q = (0, 1, 0)$.

All codes constructed from curves with one point at infinity are called one-point AG or Goppa codes [4] and these are the most commonly used codes in the literature. Elliptic and hyperelliptic curves are other examples of curves that have one point at infinity. An upper bound on the number of points N, including any points at infinity, that satisfy a curve over a field GF(q) is the Hasse-Weil bound, defined as [3]:

$$|N| \leq (m - 1)(m - 2)\sqrt{q} + 1 + q, \tag{2.7}$$

where m is the degree of the curve. Taking the Hermitian curve defined in (2.5) with degree $m = 3$ and $q = 4$, the upper bound from (2.7) is:

$$|N| \leq (3 - 1)(3 - 2)\sqrt{4} + 1 + 4$$
$$|N| \leq 9.$$

Table 2.13 Eight projective points for the Hermitian curve in the $(x - z)$ system.

$P_1 = (0, 1, 0)$	$P_2 = (0, 1, 1)$	$P_3 = (1, 1, \alpha)$	$P_4 = (1, 1, \alpha^2)$
$P_5 = (\alpha, 1, \alpha)$	$P_6 = (\alpha, 1, \alpha^2)$	$P_7 = (\alpha^2, 1, \alpha)$	$P_8 = (\alpha^2, 1, \alpha^2)$

Table 2.14 Four projective points for the Hermitian curve in the $(y - z)$ system.

$P_1 = (1, \alpha, 1)$	$P_2 = (1, \alpha, \alpha^2)$	$P_3 = (1, \alpha^2, 1)$	$P_4 = (1, \alpha^2, \alpha)$

Therefore, the maximum number of points a curve can have with $m = 3$ and defined over GF(2^2) is nine. For the Hermitian curve we have eight affine points and one point at infinity, giving a total of nine points. Therefore, the Hermitian curves meet the Hasse-Weil bound and are known as maximal curves. These curves are desirable because they produce long codes. Other types of maximal curve include the elliptic curves over certain finite fields. For example, the elliptic curve [3, 5]

$$C(x, y) = x^3 + x^2 + x + y^2 + y + 1$$

is a maximal curve over GF(2^4). It has degree $m = 3$ and from (2.7) it has 25 points. Compare this to the Hermitian curve defined over GF(2^4)

$$C(x, y) = x^5 + y^4 + y,$$

which has $m = 5$ and from (2.7) has 65 points. We can see that codes from elliptic curves will be much shorter than codes constructed from Hermitian curves, however they are still longer that Reed–Solomon codes, which are constructed from the affine line

$$y = 0$$

and have degree $m = 1$. For a finite field GF(q), the Hasse-Weil bound in (2.7) simplifies to:

$$|N| \leq q + 1,$$

so for GF(2^4) the longest possible Reed–Solomon code is $16 + 1 = 17$, which is shorter than a code constructed from an elliptic curve. However, it will be seen later that maximizing the degree of a curve is not the only design criterion for constructing good AG codes.

2.2.4 Rational Functions on Curves

To construct the generator matrix of an AG code, a basis of rational functions on the curve must first be defined. Each rational function is evaluated at each of the n affine points to form a row of the generator matrix, where n is the block length of the code. A rational function $f(x, y, z)$ is the quotient of two other functions, $g(x, y, z)$ and $h(x, y, z)$, that both have the same degree, that is $f(x, y, z) = \frac{g(x,y,z)}{h(x,y,z)}$. Defining a rational function on a curve changes its behaviour. For example, the function $f(x, y) = \frac{x}{y+1}$ defined over the cubic curve $C(x, y) = x^3 + y^3 + 1 = 0$ over GF(2^2) can be rewritten as:

$$x^3 = y^3 + 1$$
$$x = \frac{y^3 + 1}{x^2} = \frac{(y + 1)(y^2 + y + 1)}{x^2}.$$
$$\therefore \frac{x}{y + 1} = \frac{y^2 + y + 1}{x^2}$$

The cubic curve has six points: $P_1 = (0, 1)$, $P_2 = (0, \alpha)$, $P_3 = (0, \alpha^2)$, $P_4 = (1, 0)$, $P_5 = (\alpha, 0)$, $P_6 = (\alpha^2, 0)$. In this case the function $f(x, y) = \frac{x}{y+1}$ is defined for all the points on the cubic curve except for $P_1 = (0, 1)$, since $f(0, 1) = \frac{0}{1+1} = \frac{0}{0}$.

If the function was not defined on the curve it would have a zero of order 1 and a pole of order 1 at this point. However, since it is on the curve the function evaluated at $(0, 1)$ gives:

$$f(x, y) = \frac{y^2 + y + 1}{x^2},$$

$$f(0, 1) = \frac{1^2 + 1 + 1}{0^2} = \frac{1}{0}$$

which has no zeroes but has a pole of order 2 due to the x^2 in the denominator. The order of a function is denoted by $v(f(x, y, z))$ and is the sum of its zero order and pole order. To construct an AG code, the basis of the rational functions must have a pole at the points of infinity, but have no other poles at any of the affine points.

Example 2.3: Rational functions on the projective line $y = 0$: The points in the $(x, y, 1)$ system are of the form $(\alpha^i, 0, 1)$ for a finite field GF(q), where α^i is a primitive element, so there are q points in this system. In the $(x, 1, z)$ system there are no points because $y = 0$. In the $(1, y, z)$ system the points are of the form $(1, 0, \alpha^i)$, but only $(1, 0, 1)$ is of the form $(\alpha, \beta, 1)$. In this system there is also the point $(1, 0, 0)$, which is in the hyperplane at infinity (since $z = 0$) and so is a point at infinity. Therefore there are q affine points and 1 point at infinity. Next, a sequence of rational functions on the projective line that have a pole only at the point of infinity $Q = (1, 0, 0)$ must be found. The sequence of rational functions $\{\frac{x^i}{z^i}\}$, $i > 0$, has a pole of order i at Q but has no poles at any of the affine points because $z = 1$. From this example, we can see that the number of points on the projective line cannot exceed the cardinality of the finite field the line is defined over. The codes constructed from the line $y = 0$ are actually the well-known Reed–Solomon codes described in Chapter 3, which can be viewed as the simplest type of algebraic–geometric code. The small number of points on the projective line is the reason why Reed–Solomon codes only have short code lengths – no greater than the size of the finite field.

Example 2.4: Rational functions on the Hermitian curve: It is well known that a sequence of rational functions on the projective Hermitian curve in (2.1) can be formed from $\frac{x}{z}$ and $\frac{y}{z}$ [3]. The term $\frac{x}{z}$ can be rewritten as:

$$x^{w+1} = y^w z + y z^w$$

$$\frac{x}{z} = \frac{y^w + y z^{w-1}}{x^w},$$

which has an order of w at the point at infinity $Q = (0, 1, 0)$. Similarly, the term $\frac{y}{z}$ can be rewritten as:

$$x^{w+1} = y^w z + yz^w$$

$$\frac{y^w z + yz^w}{x^{w+1}} = 1$$

$$\frac{1}{z} = \frac{y^w + yz^{w-1}}{x^{w+1}} \quad ,$$

$$\frac{y}{z} = \frac{y^{w+1} + y^2 z^{w-1}}{x^{w+1}}$$

which has an order of $w + 1$ at the point at infinity $Q = (0, 1, 0)$. Therefore, for the Hermitian curve defined over GF(2^2) in (2.3) with $r = 2$, $v\left(\frac{x}{z}\right) = 2$ and $v\left(\frac{y}{z}\right) = 3$, other rational functions on the curve are formed by combining the products of different powers of $\frac{x}{z}$ and $\frac{y}{z}$ and adding their orders together. For example, $\frac{x}{z} \cdot \frac{x}{z} = \frac{x^2}{z^2}$ has an order of $2 + 2 = 4$, $\frac{x}{z} \cdot \frac{y}{z} = \frac{xy}{z^2}$ has an order of $2 + 3 = 5$, $\frac{y}{z} \cdot \frac{y}{z} = \frac{y^2}{z^2}$ has an order of $3 + 3 = 6$ and so on.

In general, the rational function $\frac{x^i y^j}{z^{i+j}}$ on the Hermitian curve has an order of [3]:

$$v\left(\frac{x^i y^j}{z^{i+j}}\right) = iw + j(w + 1). \tag{2.8}$$

This sequence of rational functions is denoted by $L(G)$, where G is a divisor on the curve. A divisor of a curve assigns an integer value to every point of the curve. An AG code is defined by two divisors, D and G. The divisor D assigns the value $D(P) = 1$ to every affine point and is the sum of all the affine points [3, 4]:

$$D = \sum_{i=1}^{n} D(P_i)P_i = \sum_{i=1}^{n} P_i. \tag{2.9}$$

The sum $\sum_{i=1}^{n} D(P)$ is called the degree of divisor D, $d(D)$.

Similarly, the divisor G assigns an integer value $G(Q)$ to each point at infinity Q and is defined as the sum of all points at infinity. For curves with one point at infinity, the divisor G is just the point at infinity multiplied by the degree of G, $d(G)$.

$$G = \sum_{i} G(Q_i)Q_i = d(G)Q \tag{2.10}$$

The space $L(G)$ contains rational functions of order up to $d(G)$. Continuing the example above, if $G = 7Q$ with $d(G) = 7$ then:

$$L(7Q) = \left\{1, \frac{x}{z}, \frac{y}{z}, \frac{x^2}{z^2}, \frac{xy}{z^2}, \frac{y^2}{z^2}, \frac{x^3}{z^3}, \frac{x^2 y}{z^3}\right\}.$$

Notice that the order of $\frac{x^3}{z^3}$ is 6, which is the same as the order of $\frac{y^2}{z^2}$. In this case all functions with a power of x greater than or equal to 3 are removed to avoid functions with the same order. Alternatively, all functions with a power of y greater than 1 can be removed. Therefore, there are seven rational functions in $L(G)$:

$$L(G) = \left\{ 1, \frac{x}{z}, \frac{y}{z}, \frac{x^2}{z^2}, \frac{xy}{z^2}, \frac{y^2}{z^2}, \frac{x^2 y}{z^3} \right\}.$$

2.2.5 Riemann–Roch Theorem

The Riemann–Roch theorem can be used to calculate the number of rational functions in $L(G)$ with order up to and including $d(G)$, and hence to determine the dimension and minimum distance of the code. The number of rational functions in $L(G)$ is called the dimension of G, $l(G)$. The theorem states that there exists a nonnegative integer γ such that [3, 4]:

$$l(G) - d(G) = 1 - \gamma, \tag{2.11}$$

provided that $d(G) > 2\gamma - 2$.

The nonnegative integer γ is called the **genus** and is defined as [3, 4]:

$$\gamma = \frac{(m-1)(m-2)}{2}, \tag{2.12}$$

where m is the degree of the curve, and plays an important part in the size of the code parameters.

Example 2.5: The number of rational functions on the Hermitian curve with pole order less than or equal to 21 at the point at infinity Q = (0, 1, 0): The rational function $\frac{x}{z}$ has an order of $w = 4$, and $\frac{y}{z}$ has an order of $w + 1 = 5$. Therefore, the sequence of functions of order up to and including 21 is:

$$L(21Q) = \left\{ 1, \frac{x}{z}, \frac{y}{z}, \frac{x^2}{z^2}, \frac{xy}{z^2}, \frac{y^2}{z^2}, \frac{x^3}{z^3}, \frac{x^2 y}{z^3}, \frac{xy^2}{z^3}, \frac{y^3}{z^3}, \frac{x^4}{z^4}, \frac{x^3 y}{z^4}, \frac{x^2 y^2}{z^4}, \frac{xy^3}{z^4}, \frac{y^4}{z^4}, \frac{x^4 y}{z^5} \right\}$$

All functions with a power of x greater than 4 have been removed to avoid duplicate orders, so there is no $\frac{x^5}{z^5}$ in $L(G)$ as its order is the same as $v(\frac{y^4}{z^4})$. The orders of these functions are:

$$\{1, 4, 5, 8, 9, 10, 12, 13, 14, 15, 16, 17, 18, 19, 20, 21\}$$

and there are 16 functions in $L(G)$. These numbers are also known as *nongaps*. Also, the orders 0, 2, 3, 6, 7 and 11 are not present, and these are known as *gaps*. The number of gaps is equal to γ.

The degree of the Hermitian curve is $m = 5$ and, from (2.12), the genus is $\gamma = 6$ and $d(G) = 21$. Next, $2\gamma - 2 = 10$, which is less than $d(G)$, so applying the Riemman–Roch theorem in (2.11) gives:

$$l(G) = d(G) + 1 - \gamma$$
$$= 21 + 1 - 6 \quad,$$
$$= 16$$

which is equal to the number of functions in $L(G)$.

2.2.6 The Zero Order of a Monomial

We can define a set Z of monomials $\psi_{p_i,u}(x, y)$ with a zero of order u at an affine point P_i as [6]:

$$\psi_{p_i,u}(x, y) = \psi_{p_i,\lambda+(w+1)\delta}(x, y) = (x - x_i)^\lambda[(y - y_i) - x_i^w(x - x_i)]^\delta, \quad (2.13)$$

where $\lambda, \delta \in \mathbb{N}$, $0 \leq \lambda \leq w$ and $\delta \geq 0$. To evaluate the zero order, $\psi_{p_i,\alpha}$ is divided by $(x - x_i)$ until a unit has been obtained. The zero order is equal to the number of divisions.

Example 2.6: Determining the zero order of a monomial at the affine point $(1, \alpha)$ on the Hermitian curve: The first eight monomials with respect to the point $p_i = (x_i, y_i) = (1, \alpha)$ from (2.13) are determined by:

$$\psi_{p_i,u}(x, y) = \psi_{(1,\alpha),\lambda+3\delta}(x, y) = (x - 1)^\lambda[(y - a) - 1^2(x - 1)]^\delta,$$

where $w = 2$ and $0 \leq \lambda \leq 2$.
   Therefore:

$\psi_{p_i,0}(x, y) = (x - 1)^0[(y - \alpha) - 1^2(x - 1)] = 1$
$\psi_{p_i,1}(x, y) = (x - 1)^1[(y - \alpha) - 1^2(x - 1)]^0 = 1 + x$
$\psi_{p_i,2}(x, y) = (x - 1)^2[(y - \alpha) - 1^2(x - 1)]^0 = 1 + x^2$
$\psi_{p_i,3}(x, y) = (x - 1)^0[(y - \alpha) - 1^2(x - 1)]^1 = y + \alpha + x + 1 = \alpha^2 + x + y$
$\psi_{p_i,4}(x, y) = (x - 1)1[(y - \alpha) - 1^2(x - 1)]^1 = (x + 1)(y + \alpha + x + 1)$
$\qquad = \alpha^2 + \alpha x + y + x^2 + xy$
$\psi_{p_i,5}(x, y) = (x - 1)^2[(y - \alpha) - 1^2(x - 1)]^1 = (x^2 + 1)(y + \alpha + x + 1)$
$\qquad = \alpha^2 + x + \alpha^2 x^2 + y^2 + x^2 y$
$\psi_{p_i,6}(x, y) = (x - 1)^0[(y - \alpha) - 1^2(x - 1)]^2 = (y + \alpha + x + 1)^2 = \alpha + x^2 + y^2$
$\psi_{p_i,7}(x, y) = (x - 1)^1[(y - \alpha) - 1^2(x - 1)]^2 = (x + 1)(y + \alpha + x + 1)^2$
$\qquad = \alpha + \alpha x + y + x^2 + xy^2.$

To evaluate a function's zero order at an affine point of the Hermitian curve $x^{w+1} + y^w + y = 0$, it is important to have the following equation associated with the curve [6]:

$$\frac{y - y_i}{x - x_i} = \frac{(x - x_i)^w + x_i(x - x_i)^{w-1} + x_i^w}{e(y)}, \tag{2.14}$$

where $e(y) = (y - y_i)^{w-1} + 1$. Notice that $e(y_i) = (y_i - y_i)^{w-1} + 1 = 1$.

It can be seen that $\psi_{p_i,3}(x, y) = (y - \alpha) - (x - 1)$ and $e(y) = (y - \alpha) + 1$.

Initialize $\psi^{(0)}(x, y) = \psi_{p_i,3}(x, y) = (y - \alpha) - (x - 1)$.

The first division:

$$\psi^{(1)}(x, y) = \frac{\psi^{(0)}(x, y)}{x - 1} = \frac{y - \alpha}{x - 1} - 1 = \frac{(x - 1)^2 + (x - 1) + 1}{e(y)} - 1$$

$$= \frac{(x - 1)^2 + (x - 1) + 1 - (y - \alpha) - 1}{e(y)}$$

$$= (x - 1)e(y)^{-1} + (y - \alpha)e(y)^{-1} + (x - 1)^2 e(y)^{-1}.$$

We have $\psi^{(1)}(p_i) = (1 - 1) \cdot 1 + (\alpha - \alpha) \cdot 1 + (1 - 1)^2 \cdot 1 = 0$.

The second division:

$$\psi^{(2)}(x, y) = \frac{\psi^{(1)}(x, y)}{x - 1} = e(y)^{-1} - \frac{y - \alpha}{x - 1}e(y)^{-1} + (x - 1)e(y)^{-1}$$

$$= e(y)^{-1} - [(x - 1)^2 + (x - 1) + 1]e(y)^{-2} + (x - 1)e(y)^{-1}$$

$$= (e(y)^{-1} - e(y)^{-2}) - (x - 1)(e(y)^{-2} - e(y)^{-1}) - (x - 1)^2 e(y)^{-2}.$$

We have $\psi^{(2)}(p_i) = (1 - 1) - (1 - 1) \cdot (1 - 1) - (1 - 1)^2 \cdot 1 = 0$.

The third division:

$$\psi^{(3)}(x, y) = \frac{\psi^{(2)}(x, y)}{x - 1} = \frac{e(y)^{-1} - e(y)^{-2}}{x - 1} - (e(y)^{-2} - e(y)^{-1}) - (x - 1)e(y)^{-2}$$

$$= \frac{e(y) - 1}{(x - 1)e(y)^2} - (e(y)^{-2} - e(y)^{-1}) - (x - 1)e(y)^{-2}.$$

$$= \frac{y - \alpha}{x - 1}e(y)^{-2} - (e(y)^{-2} - e(y)^{-1}) - (x - 1)e(y)^{-2}$$

$$= [(x - 1)^2 + (x - 1) + 1]e(y)^{-3} - (e(y)^{-2} - e(y)^{-1}) - (x - 1)e(y)^{-2}$$

$$= (e(y)^{-3} - e(y)^{-2} + e(y)^{-1}) + (x - 1)(e(y)^{-3}$$

$$-e(y)^{-2}) + (x - 1)^2 e(y)^{-3}.$$

We have $\psi^{(3)}(p_i) = (1 - 1 + 1) + (1 - 1) \cdot (1 - 1) + (1 - 1)^2 \cdot 1 = 1 \neq 0$.

There are 3 divisions in order obtain a unit. Therefore, the zero order of $\psi_{p_i,3}$ at p_i is 3 as: $v_{p_i}(\psi_{p_i,3}) = 3$. $\psi_{p_i,3}$ can also be written as: $\psi_{p_i,3} = (x-1)^3[(e(y)^{-3} - e(y)^{-2} + e(y)^{-1}) + (x-1)(e(y)^{-3} - e(y)^{-2}) + (x-1)^2 e(y)^{-3}]$.

2.2.7 AG Code Parameters

There are two types of AG code: Functional Goppa codes, C_L, and Residue Goppa codes, C_Ω, which are the dual of functional Goppa codes. In both cases the block length is the number of affine points, n.

For functional Goppa codes the message length is the number of rational functions in the divisor $L(G)$, as given by the Reimann–Roch theorem [3, 4]:

$$k = l(G) = d(G) + 1 - \gamma. \tag{2.15}$$

An accurate value of the Hamming distance of AG codes cannot always be calculated, so a lower bound called the *designed minimum distance d^** is used. The genus of the curve plays an important role in determining the distance of the code. The optimal Hamming distance occurs when the Singleton bound is met [1, 2]:

$$d = n - k + 1. \tag{2.16}$$

However, for AG codes the genus of the curve penalizes the minimum distance [4]:

$$d^* = n - k - \gamma + 1. \tag{2.17}$$

In this case a large genus will reduce the designed minimum distance of the code, but long codes such as the Hermitian codes have a large genus. Reed–Solomon codes are constructed from an affine line, which has degree of one and a genus equal to zero. Therefore they do not suffer any genus penalty and have optimal Hamming distances.

By substituting (2.15) into (2.16) the designed minimum distance of the code is [4]:

$$\begin{aligned} d^* &= n - (d(G) + 1 - \gamma) - \gamma + 1 \\ &= n - d(G) \end{aligned} \tag{2.18}$$

For residue Goppa codes the message length k of the code is simply the code length minus the message length of the functional Goppa codes, that is [4]:

$$k = n - l(G) = n - d(G) - 1 + \gamma. \tag{2.19}$$

Substituting (2.19) into (2.17) gives the designed minimum distance of the residue Goppa code [4]:

$$\begin{aligned} d^* &= n - (n - d(G) - 1 + \gamma) - \gamma + 1 \\ &= d(G) - 2\gamma + 2 \end{aligned} \tag{2.20}$$

2.3 Conclusions

This chapter has presented some of the more important mathematical concepts associated with non-binary error-correcting codes, explaining groups, rings and fields, minimal polynomials and cyclotomic cosets, and introducing algebraic geometry. These are all important tools for the design of binary and non-binary error-correcting codes and will be used often in the remaining chapters of this book.

Most of this chapter was dedicated to the principles of algebraic geometry since the construction and decoding of algebraic–geometric codes is more mathematically intensive than the other non-binary coding schemes covered in this book.

References

[1] Lin, S. and Costello, D.J. Jr. (2004) *Error Control Coding*, 2nd edn, Pearson Prentice Hall, ISBN 0-13-017973-6.
[2] Wicker, S.B. (1995) *Error Control Systems for Digital Communication and Storage*, Prentice Hall, Eaglewood Cliffs, NJ, ISBN 978-0132008099.
[3] Blake, I.F., Heegard, C., Hoholdt, T. and Wei, V. (1998) Algebraic geometry codes. *IEEE Trans. Inform. Theory*, **44** (6), 2596–618.
[4] Pretzel, O. (1998) *Codes and Algebraic Curves*, Oxford Science Publications, Oxford University Press (Oxford Lecture Series in Mathematics and its Applications 8), ISBN 0198500394.
[5] Driencourt, Y. (1985) Some properties of elliptic codes over a field of characteristic 2. *Proceedings of AAECC-3*, **229**, 185–93, Lecture Notes in Computer Science.
[6] Høholdt, T. and Nielsen, R.R. (1999) Decoding Hermitian codes with Sudan's algorithm, in *Applied Algebra, Algebraic Algorithms and Error-Correcting Codes*, Vol. **1719**, Springer-Verlag, Berlin, Germany, pp. 260–70 (Lecture Notes in Computer Science).

3

Non-Binary Block Codes

3.1 Introduction

The two previous chapters have covered the necessary mathematics to enable the reader to understand the theory of error-correcting codes. From now on, this chapter and succeeding chapters will present the different types and classes of binary and non-binary error-correcting code, focusing on their design, construction and decoding algorithms. In this chapter, we first introduce binary block codes, beginning with the Hamming code, one of the first block codes presented in 1953 by Hamming [1]. It will be seen that this coding scheme is not suitable for practical applications, but these codes are simple enough to demonstrate encoding and decoding block codes. Following this, a very important type of block code known as the *cyclic code* is presented. The cyclic code has improved error-correction and its encoder can be simply implemented using shift registers. The first important class of cyclic code is the Bose–Chaudhuri–Hocquengem (BCH) code [2, 3]. We describe the construction of binary BCH codes and then move on to the construction of non-binary BCH codes defined over finite fields. This will lead to the most important and commonly used class of non-binary BCH code, called the *Reed–Solomon code* [4]. Two different decoding algorithms are given to decode non-binary BCH codes: *Euclid's algorithm* [5] and the *Berlekamp–Massey algorithm* [6, 7].

We then explain the concept of *coded modulation* presented by Ungerboeck [8], whereby the encoding and modulation processes are treated as a single entity and the constellation set is increased to compensate for the redundancy in a code word, thus keeping the data rate constant without requiring the bandwidth use to be expanded. *Block Coded Modulation* (BCM) employs simple block codes and maps the symbols of the code word to an expanded constellation set, while at the same time maximizing the Euclidean distance between adjacent symbols. Binary BCM codes are introduced, covering their encoding procedure and their decoding by constructing the trellis diagram of the BCM codes and using the well-known Viterbi algorithm [9] to recover the original message. This work is then extended to non-binary BCM codes and we

Non-Binary Error Control Coding for Wireless Communication and Data Storage   Rolando Antonio Carrasco and Martin Johnston
© 2008 John Wiley & Sons, Ltd

demonstrate that the performance can be improved with only a small increase in complexity.

The chapter finishes by discussing the applications of non-binary block codes related to wireless communications, such as a cellular environment and fixed and wireless broadband access (WiMax), and to magnetic and optical data storage.

3.2 Fundamentals of Block Codes

In any error-correcting coding scheme a number of redundant bits must be included in the original message in order for it to be possible to recover that message at the receiver. In general, these redundant bits are determined by algebraic methods. All block codes are defined by their code word length n, their message length (or dimension) k and their minimum Hamming distance d (described later). When referring to a block code, it is written as a (n, k, d) block code.

One of the simplest types of block code is the *parity check code*, where for each binary message an extra bit, known as a *parity check bit*, is appended to the message to ensure that each code word has an even number of 1s. In this case the message is k bits in length and therefore the code word length $n = k + 1$. For example, if $k = 3$ then the code word length is $n = 4$ and the code is:

$$000 \rightarrow 0000$$
$$001 \rightarrow 0011$$
$$010 \rightarrow 0101$$
$$011 \rightarrow 0110.$$
$$100 \rightarrow 1001$$
$$101 \rightarrow 1010$$
$$110 \rightarrow 1100$$
$$111 \rightarrow 1111$$

It can be seen that if a single error is added to any of the code words it will result in an odd number of 1s in the received word and will be detected. Unfortunately, the code will not know the location of the error in the received word. Hence, parity check codes can only detect one error and can correct none. To obtain the parity check bit, one simply adds the three message bits modulo-2. If the message bits are denoted as u_1, u_2 and u_3 and the parity check bit is denoted as p_1 then the parity check equation for the $(4, 3)$ parity check code is:

$$p_1 = u_1 \oplus u_2 \oplus u_3. \tag{3.1}$$

A simple encoder for this is illustrated in Figure 3.1.

The number of errors that a block code can detect and correct is determined by its minimum Hamming distance d. This is defined as the minimum number of places where any two code words differ. For example, by observation the minimum Hamming

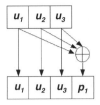

Figure 3.1   A simple encoder for the (4, 3) parity check code.

distance of the above parity check code is $d = 2$. Hence, if any two bits of a code word are flipped this will result in another valid code word, that is if two errors occurred they would not be detected. In general, the number of errors v that can be detected for a block code is:

$$v = d - 1. \tag{3.2}$$

The error detection and correction of a block code can be improved by increasing the number of parity check bits or equivalently increasing the number of parity check equations. For example, take the following three parity check equations:

$$p_1 = u_1 \oplus u_3 \oplus u_4$$
$$p_2 = u_1 \oplus u_2 \oplus u_3.$$
$$p_3 = u_2 \oplus u_3 \oplus u_4 \tag{3.3}$$

Each parity check bit in (3.3) is independent in order to ensure the minimum Hamming distance is maximized. The corresponding encoder is given in Figure 3.2. This is one of the first block codes known as *Hamming codes*. These have the following properties:

Codeword length $n = 2^m - 1$

Message length $k = 2^m - m - 1$

Minimum Hamming distance $d = 3$,

where $m \geq 3$ and is a positive integer.

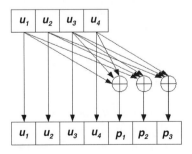

Figure 3.2   Encoder for the parity check equations of (3.3).

A list of the code words for the (7, 4, 3) Hamming code is shown below, with the parity bits highlighted:

$$0000 \rightarrow 0000\mathbf{000}$$
$$0001 \rightarrow 0001\mathbf{101}$$
$$0010 \rightarrow 0010\mathbf{111}$$
$$0011 \rightarrow 0011\mathbf{010}$$
$$0100 \rightarrow 0100\mathbf{011}$$
$$0101 \rightarrow 0101\mathbf{110}$$
$$0110 \rightarrow 0110\mathbf{100}$$
$$0111 \rightarrow 0111\mathbf{001}$$
$$1000 \rightarrow 1000\mathbf{110}$$
$$1001 \rightarrow 1001\mathbf{011}$$
$$1010 \rightarrow 1010\mathbf{001}.$$
$$1011 \rightarrow 1011\mathbf{100}$$
$$1100 \rightarrow 1100\mathbf{101}$$
$$1101 \rightarrow 1101\mathbf{000}$$
$$1110 \rightarrow 1110\mathbf{010}$$
$$1111 \rightarrow 1111\mathbf{111}$$

From (3.2), the Hamming code can detect two errors, improving on the parity check code. However, we do not know how many errors this code can correct. We know that changing a code word in d positions can result in another valid code word. Each code word can be interpreted as the centre of a circle of a certain radius t that contains all other vectors that differ from the code word in t places or less. Therefore, if $t = d$ then two neighbouring code words will overlap, as shown in Figure 3.3. If $t = d/2$ then the two circles will meet at their circumferences. This implies that for $t < d/2$ the circle will contain unique n-dimensional vectors that do not occur in any other circle. Hence, if the received word matches any of these vectors we know that it can only be one code word. So the number of errors t that a block code can correct up to is:

$$t < \left\lfloor \frac{d}{2} \right\rfloor \quad \text{or} \quad t \leq \left\lfloor \frac{d-1}{2} \right\rfloor, \tag{3.4}$$

where $\lfloor \rfloor$ is the floor function and any value inside is rounded down to the nearest integer. This is important for the situation where d is even and the number of errors that can be corrected would not be an integer. Since the Hamming code has a minimum Hamming distance $d = 3$ it can correct one error.

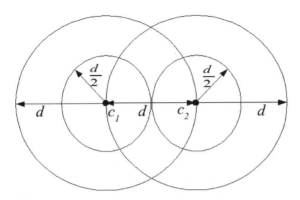

Figure 3.3 Representation of two code words that differ in d places. Any vector that differs from a code word in less than $d/2$ places is unique and can therefore be corrected.

3.2.1 Generator and Parity Check Matrices

The encoders of Figures 3.1 and 3.2 can be represented as a $(k \times n)$ matrix for convenience. Each column in the matrix corresponds to the connections from the message and the code word, that is a connection is denoted by a 1 and no connection by a 0. For example, the matrix for the parity check code is written as:

$$\mathbf{H} = \begin{bmatrix} 1 & 0 & 0 & 1 \\ 0 & 1 & 0 & 1 \\ 0 & 0 & 1 & 1 \end{bmatrix}.$$

Taking the first column, there is only one connection in Figure 3.1 between the first element of the code vector and the first element of the message vector, and this is denoted as 1. There are no connections between the first elements of the code vector and the second and third elements in the message vector so these are both denoted as 0. Similarly, for the last column there are connections between the final element of the code vector and all three elements of the message vector so this column contains all 1s. Multiplying a 3-bit message by this matrix will generate a code word, so the matrix is called a *generator matrix*, denoted as $\mathbf{G}$. The generator matrix of the Hamming code encoder in Figure 3.2 is:

$$\mathbf{G} = \begin{bmatrix} 1 & 0 & 0 & 0 & 1 & 1 & 0 \\ 0 & 1 & 0 & 0 & 0 & 1 & 1 \\ 0 & 0 & 1 & 0 & 1 & 1 & 1 \\ 0 & 0 & 0 & 1 & 1 & 0 & 1 \end{bmatrix}. \tag{3.5}$$

It can be seen from (3.5) that every code word can be obtained by adding different combinations of the rows in $\mathbf{G}$. For example, encoding the message

$\mathbf{u} = [1\ 1\ 1\ 1]$ gives:

$$\mathbf{c} = \mathbf{u} \cdot \mathbf{G} = 1 \cdot [1\ 0\ 0\ 0\ 1\ 1\ 0] \oplus 1 \cdot [0\ 1\ 0\ 0\ 0\ 1\ 1] \oplus 1 \cdot [0\ 0\ 1\ 0\ 1\ 1\ 1] \oplus$$
$$1 \cdot [0\ 0\ 0\ 1\ 1\ 0\ 1] = [1\ 1\ 1\ 1\ 1\ 1\ 1].$$

A second important matrix in error-correction coding is the *parity check matrix*, denoted by $\mathbf{H}$. It has the property that any row in $\mathbf{H}$ is *orthogonal* to the rows of $\mathbf{G}$, that is the inner product of a row in $\mathbf{G}$ and a row in $\mathbf{H}$ will be zero. This is important as it allows us to validate any code word in a block code since multiplying a code word by $\mathbf{H}$ should give an all-zero vector, indicating that the code word is valid, as given by [10]:

$$\mathbf{H} \cdot \mathbf{c}^{\mathbf{T}} = \mathbf{0}, \tag{3.6}$$

where $\mathbf{c}^{\mathbf{T}}$ is the transpose of a code word and $\mathbf{0}$ is the $(n - k) \times 1$ column vector containing all zeros. It can be seen that $\mathbf{G}$ is of the form:

$$\mathbf{G} = [\mathbf{I}_k | \mathbf{P}], \tag{3.7}$$

where $\mathbf{I}_k$ is the $k \times k$ identity matrix and $\mathbf{P}$ is a parity matrix. This is called a *systematic* generator matrix and generates a code word which contains the original message with parity bits appended to it. The $(n - k) \times n$ systematic parity check matrix is related to $\mathbf{G}$ by:

$$\mathbf{H} = [\mathbf{P}^{\mathbf{T}} | \mathbf{I}_{\mathbf{n-k}}], \tag{3.8}$$

where $\mathbf{P}^{\mathbf{T}}$ is the transpose of $\mathbf{P}$ and $\mathbf{I}_{\mathbf{n-k}}$ is the $(n-k) \times (n-k)$ identity matrix. The parity check matrix of the (4, 3, 2) parity check code is, by (3.8), simply $\mathbf{H} = [1\ 1\ 1\ 1]$. Similarly, the parity check matrix for the (7, 4, 3) Hamming code is:

$$\mathbf{H} = \begin{bmatrix} 1 & 0 & 1 & 1 & 1 & 0 & 0 \\ 1 & 1 & 1 & 0 & 0 & 1 & 0 \\ 0 & 1 & 1 & 1 & 0 & 0 & 1 \end{bmatrix}. \tag{3.9}$$

As an example, multiplying $\mathbf{H}$ by the transpose of the code word $\mathbf{c} = [1\ 0\ 0\ 0\ 1\ 1\ 0]$ results in:

$$\mathbf{H} \cdot \mathbf{c}^{T} = \begin{bmatrix} 1 & 0 & 1 & 1 & 1 & 0 & 0 \\ 1 & 1 & 1 & 0 & 0 & 1 & 0 \\ 0 & 1 & 1 & 1 & 0 & 0 & 1 \end{bmatrix} \cdot \begin{bmatrix} 1 \\ 0 \\ 0 \\ 0 \\ 1 \\ 1 \\ 0 \end{bmatrix} = \begin{bmatrix} 0 \\ 0 \\ 0 \end{bmatrix}.$$

3.2.2 Decoding Block Codes

As well as validating code words, the parity check matrix can also be used to decode a received word containing errors. It has been seen that multiplying the parity check matrix by any code word results in an all-zero column vector. Therefore, if the parity check matrix is multiplied by a received word containing errors, the column vector will contain nonzero elements. These values are known as *syndromes* and the column vector is a syndrome vector, $\mathbf{s}$ [10].

$$\mathbf{s} = \mathbf{H} \cdot \mathbf{r}^{\mathrm{T}}, \tag{3.10}$$

where $\mathbf{r}$ is the received vector defined as $\mathbf{r} = \mathbf{c} + \mathbf{e}$ and $\mathbf{e}$ is an error vector. Therefore, (3.10) can be written as:

$$\mathbf{s} = \mathbf{H} \cdot (\mathbf{c} + \mathbf{e})^{\mathrm{T}} = \mathbf{H} \cdot \mathbf{c}^{\mathrm{T}} + \mathbf{H} \cdot \mathbf{e}^{\mathrm{T}} = \mathbf{H} \cdot \mathbf{e}^{\mathrm{T}}. \tag{3.11}$$

We can see that the syndromes are only dependent on the error pattern. The (7, 4, 3) Hamming code can correct a single error and so there are seven possible error patterns:

$$\begin{bmatrix} 1 & 0 & 0 & 0 & 0 & 0 & 0 \end{bmatrix}, \begin{bmatrix} 0 & 1 & 0 & 0 & 0 & 0 & 0 \end{bmatrix}, \begin{bmatrix} 0 & 0 & 1 & 0 & 0 & 0 & 0 \end{bmatrix},$$
$$\begin{bmatrix} 0 & 0 & 0 & 1 & 0 & 0 & 0 \end{bmatrix}, \begin{bmatrix} 0 & 0 & 0 & 0 & 1 & 0 & 0 \end{bmatrix}, \begin{bmatrix} 0 & 0 & 0 & 0 & 0 & 1 & 0 \end{bmatrix}$$
$$\text{and} \quad \begin{bmatrix} 0 & 0 & 0 & 0 & 0 & 0 & 1 \end{bmatrix}.$$

Multiplying $\mathbf{H}$ by the transpose of each of these error patterns gives the seven unique syndrome vectors.

Now suppose the code word $\mathbf{c} = \begin{bmatrix} 1 & 0 & 0 & 0 & 1 & 1 & 0 \end{bmatrix}$ has been transmitted and the received vector $\mathbf{r} = \begin{bmatrix} 1 & 1 & 0 & 0 & 1 & 1 & 0 \end{bmatrix}$ has a single error in its second position as highlighted. From (3.10) the syndrome vector is:

$$\begin{bmatrix} 1 & 0 & 1 & 1 & 1 & 0 & 0 \\ 1 & 1 & 1 & 0 & 0 & 1 & 0 \\ 0 & 1 & 1 & 1 & 0 & 0 & 1 \end{bmatrix} \cdot \begin{bmatrix} 1 \\ 1 \\ 0 \\ 0 \\ 1 \\ 1 \\ 0 \end{bmatrix} = \begin{bmatrix} 0 \\ 1 \\ 1 \end{bmatrix}.$$

Looking up this syndrome vector in Table 3.1 shows that the associated error pattern is $\mathbf{e} = \begin{bmatrix} 0 & 1 & 0 & 0 & 0 & 0 & 0 \end{bmatrix}$. Therefore, adding this error pattern to the received word gives the decoded code word $\hat{\mathbf{c}}$:

$$\begin{array}{r} 1100110 \\ \oplus \underline{0100000} \\ 1000110 \end{array} .$$

Table 3.1 Error patterns and their associated syndrome
vectors for the (7, 4, 3) Hamming code.

Error pattern **e**	Syndrome vector **s**
1 0 0 0 0 0 0	1 1 0
0 1 0 0 0 0 0	0 1 1
0 0 1 0 0 0 0	1 1 1
0 0 0 1 0 0 0	1 0 1
0 0 0 0 1 0 0	1 0 0
0 0 0 0 0 1 0	0 1 0
0 0 0 0 0 0 1	0 0 1

However, for block codes that can correct more than one error this method becomes
prohibitive.

3.3 Cyclic Codes

A cyclic code is a block code which has the property that for any code word, a
cyclic shift of this code word results in another code word [11]. Cyclic codes have
the advantage that simple encoders can be constructed using shift registers, and low-
complexity decoding algorithms exist to decode them. A cyclic code is constructed
by first choosing a generator polynomial $g(x)$ and multiplying this by a message
polynomial $m(x)$ to generate a code word polynomial $c(x)$.

3.3.1 Polynomials

There are two conventions for representing polynomials. A polynomial $f(x)$ can be
of the form $f_0 + f_1x + f_2x^2 + \cdots f_nx^n$, where $f_i \in$ GF(q) and $i = 0, 1, \ldots, n$. In
this case the corresponding vector would be $[f_0 f_1 f_2 \ldots f_n]$. Alternatively, $f(x)$ can
be of the form $f_nx^n + f_{n-1}x^{n-1} + \cdots + f_0$ and the corresponding vector is $[f_n f_{n-1}$
$f_{n-2} \ldots f_0]$. Both conventions are acceptable but in this book the latter representation
is chosen. So for the vector [1 1 0 1] the polynomial would be $x^3 + x^2 + 1$.

The (7, 4, 3) Hamming code is actually also a cyclic code and can be constructed
using the generator polynomial $g(x) = x^3 + x^2 + 1$.

For example, to encode the binary message 1010 we first write it as the message
polynomial $m(x) = x^3 + x$ and then multiply it with $g(x)$:

$$
\begin{aligned}
c(x) &= m(x)g(x) \\
&= (x^3 + x)(x^3 + x^2 + 1) \\
&= x^6 + x^5 + x^3 + x^4 + x^3 + x. \\
&= x^6 + x^5 + x^4 + x
\end{aligned}
$$

This code word polynomial corresponds to 1 1 1 0 0 1 0. The complete list of code
words is shown in Table 3.2.

Table 3.2 Code words for the $(7, 4, 3)$ cyclic Hamming code.

Binary message	Message polynomial $m(x)$	Code word polynomial $c(x)$	Binary code word
0000	0	0	0000000
0001	1	$x^3 + x^2 + 1$	0001101
0010	x	$x^4 + x^3 + x$	0011010
0011	$x + 1$	$x^4 + x^2 + x + 1$	0010111
0100	x^2	$x^5 + x^4 + x^2$	0110100
0101	$x^2 + 1$	$x^5 + x^4 + x^3 + 1$	0111001
0110	$x^2 + x$	$x^5 + x^3 + x^2 + x$	0101110
0111	$x^2 + x + 1$	$x^5 + x + 1$	0100011
1000	x^3	$x^6 + x^5 + x^3$	1101000
1001	$x^3 + 1$	$x^6 + x^5 + x^2 + 1$	1100101
1010	$x^3 + x$	$x^6 + x^5 + x^4 + x$	1110010
1011	$x^3 + x + 1$	$x^6 + x^5 + x^4 + x^3 + x^2 + x + 1$	1111111
1100	$x^3 + x^2$	$x^6 + x^4 + x^3 + x^2$	1011100
1101	$x^3 + x^2 + 1$	$x^6 + x^4 + 1$	1010001
1110	$x^3 + x^2 + x$	$x^6 + x^2 + x$	1000110
1111	$x^3 + x^2 + x + 1$	$x^6 + x^3 + x + 1$	1001011

Taking a code word from Table 3.2 and shifting each element results in another code word. For example, the code word 0110100 shifted one place to the left gives the vector 1101000, which is another valid code word in Table 3.2.

Multiplying a message polynomial by the generator polynomial is equivalent to multiplying the binary message by a generator matrix $\mathbf{G}$ where each row contains the coefficients of $g(x)$ shifted one place to the right with respect to the previous row.

For the $(7, 4)$ cyclic Hamming code the generator matrix would be:

$$\mathbf{G} = \begin{bmatrix} 1 & 1 & 0 & 1 & 0 & 0 & 0 \\ 0 & 1 & 1 & 0 & 1 & 0 & 0 \\ 0 & 0 & 1 & 1 & 0 & 1 & 0 \\ 0 & 0 & 0 & 1 & 1 & 0 & 1 \end{bmatrix}.$$

Multiplying the message $\mathbf{m} = \begin{bmatrix} 1 & 0 & 1 & 0 \end{bmatrix}$ by $\mathbf{G}$ gives the code word:

$$\begin{bmatrix} 1 & 0 & 1 & 0 \end{bmatrix} \begin{bmatrix} 1 & 1 & 0 & 1 & 0 & 0 & 0 \\ 0 & 1 & 1 & 0 & 1 & 0 & 0 \\ 0 & 0 & 1 & 1 & 0 & 1 & 0 \\ 0 & 0 & 0 & 1 & 1 & 0 & 1 \end{bmatrix} = \begin{bmatrix} 1 & 1 & 1 & 0 & 0 & 1 & 0 \end{bmatrix},$$

which matches the code word in Table 3.2.

It can be seen that the code words of the cyclic code are nonsystematic and so recovering messages is more difficult. One method would be to divide the code word polynomial by the generator polynomial to obtain the message polynomial, but this is an unfeasible method for larger cyclic codes. However, it is desirable for the code word to be in systematic form so that it is simpler to recover the transmitted message.

3.3.2 Systematic Cyclic Codes

A systematic cyclic code word is made up of two polynomials: a message polynomial $m(x)$ and a parity polynomial $p(x)$, where the degree of $p(x)$ is less than $(n - k)$. The message polynomial is of the form:

$$m(x) = m_k x^k + m_{k-1} x^{k-1} + \cdots + m_1 x + m_0,$$

where k is the message length. It is first multiplied by x^{n-k}, resulting in:

$$x^{n-k} m(x) = m_n x^n + m_{n-1} x^{n-1} + \cdots + m_{n-k+1} x^{n-k+1} + m_{n-k} x^{n-k}.$$

It is then divided by the generator polynomial $g(x)$, giving [10]:

$$
\frac{x^{n-k} m(x)}{g(x)} = \underbrace{q(x)}_{\text{quotient}} + \underbrace{\frac{p(x)}{g(x)}}_{\text{remainder}} ,
$$

$$
\Rightarrow \frac{x^{n-k} m(x) + p(x)}{g(x)} = q(x)
$$

(3.12)

where $q(x)$ is a quotient polynomial. From (3.12) it can be seen that the term $x^{n-k} m(x) + p(x)$ is a valid code word polynomial since it is a factor of $g(x)$.

Example 3.1:   Generating a systematic code word using the (7, 4, 3) cyclic Hamming code: First multiply $m(x)$ by $x^{n-k} = x^3$:

$$x^3 m(x) = x^6 + x^4.$$

Now divide this by the generator polynomial $g(x) = x^3 + x^2 + 1$:

$$
\require{enclose}
\begin{array}{r}
x^3 + x^2 + 1 \\
x^3 + x^2 + 1 \enclose{longdiv}{\,x^6 \qquad\quad + x^4} \\
\underline{x^6 + x^5 \qquad\quad + x^3} \\
x^5 + x^4 + x^3 \\
\underline{x^5 + x^4 \qquad\qquad + x^2} \\
x^3 + x^2 \\
\underline{x^3 + x^2 + 1} \\
1
\end{array}
$$

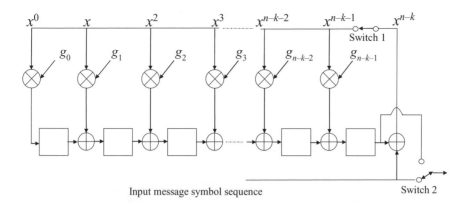

Figure 3.4 A general systematic cyclic code encoder.

In this case the quotient is $q(x) = x^3 + x^2 + 1$ and the remainder of parity polynomial is $p(x) = 1$. Hence the systematic code word polynomial $c(x) = m(x) + p(x) = x^6 + x^4 + x^2 + 1$ and the code word is $\mathbf{c} = \begin{bmatrix} 1 & 0 & 1 & 0 & 0 & 0 & 1 \end{bmatrix}$.

This operation can be achieved using an $(n-k)$-stage shift register, as shown in Figure 3.4. Initially, the shift register is initialized to zero and each one of the k input bits is fed into the $(n-k)$-stage shift register. At the same time switch 2 ensures that the output also has those k input bits. After the final bit is fed in, switch 1 is open and switch 2 moves to its second position, emptying the contents of the shift registers, which are the redundant parity bits. An example of the $(7,4,3)$ cyclic Hamming code systematic encoder is shown in Figure 3.5. This is a $(n-k) = 7 - 4 = 3$-stage shift register in which each delay element corresponds to a power of x in the parity polynomial $p(x)$.

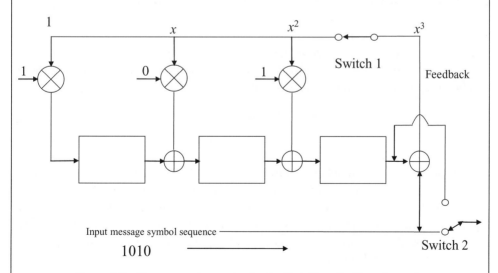

Figure 3.5 The systematic encoder for the $(7, 4, 3)$ cyclic Hamming code.

Table 3.3 Contents of the shift register for each cycle in the systematic encoder.

		Shift register contents		
Input	Feedback	1	x	x^2
1	1	1	0	1
0	1	1	1	1
1	0	0	1	1
0	1	1	0	0

If the input message is $\mathbf{m} = [1\ 0\ 1\ 0]$ then Table 3.3 shows the contents of the shift register at each cycle in the systematic encoder.

The contents of the shift register for the final information bit show that the parity polynomial is $p(x) = 0x^2 + 0x + 1 = 1$ and the systematic code word is $\mathbf{c} = \begin{bmatrix} 1\ 0\ 1\ 0\ 0\ 0\ 1 \end{bmatrix}$.

3.4 Bose–Chaudhuri–Hocquenghem (BCH) Codes

One of the most important classes of cyclic code is the Bose–Chaudhuri–Hocquenghem (BCH) code [2, 3]. When constructing a cyclic code the actual value of the minimum Hamming distance is unknown, but BCH codes have a lower bound on the minimum Hamming distance, called the *designed minimum distance*, denoted as d^*. This is due to the BCH bound, defined as follows [12].

The BCH Bound: For any q-ary (n, k) cyclic code, choose a value of m that ensures the extension field $GF(q^m)$ is minimal and contains a primitive nth root of unity α. Choose a generator polynomial $g(x)$ of minimal degree in $GF(q)x$ so that $g(\alpha) = g(\alpha^2) = g(\alpha^4) = \cdots = g(\alpha^{d^*-2})$.

From the above, the cyclic code has minimum distance $d \geq d^*$ [12].

The value of d^* is determined by deciding how many errors the BCH code should be able to correct. This can be achieved by rearranging (3.4) to get:

$$d^* = 2t + 1. \tag{3.13}$$

Example 3.2: Constructing a binary BCH code of length $n = 15$ bits: The BCH bound tells us that we must choose the smallest extension field $GF(2^m)$ with an nth root of unity, which in this case is a 15th root of unity. The smallest extension field that satisfies this is $GF(16)$. Next, we must decide how many errors the BCH code will be able to correct; for this example it can correct $t = 2$ errors, giving a designed minimum distance of $d^* = 5$. Finally, we need a generator polynomial $g(x)$ where $g(\alpha) = g(\alpha^2) = g(\alpha^3) = g(\alpha^4) = 0$. To obtain this the minimal polynomials of GF(16) are required, which are given in Table 2.11, but are repeated here for convenience:

Cyclotomic Cosets	Minimal Polynomials $M(x)$
$\{1, 2, 4, 8\}$	$M_1(x) = x^4 + x + 1$
$\{3, 6, 12, 9\}$	$M_3(x) = x^4 + x^3 + x^2 + x + 1$
$\{5, 10\}$	$M_5(x) = x^2 + x + 1$
$(7, 14, 13, 11\}$	$M_7(x) = x^4 + x^3 + 1$

The minimal polynomial for α and α^2 is $M_1(x)$, the minimal polynomial for α^3 is $M_3(x)$, and for α^4 it is also $M_1(x)$, so to satisfy the BCH bound the generator polynomial must be $g(x) = M_1(x)M_3(x)$.

$$g(x) = (x^4 + x + 1)(x^4 + x^3 + x^2 + x + 1)$$
$$= x^8 + x^7 + x^6 + (1+1)x^5 + (1+1+1)x^4 + (1+1)x^3 + (1+1)x^2 + (1+1)x + 1.$$
$$= x^8 + x^7 + x^6 + x^4 + 1$$

This means that the message length of the code is $k = n - \deg(g(x)) = 15 - 8 = 7$, resulting in the (15, 7, 5) binary BCH code.

The design procedure for constructing non-binary BCH codes is the same as for the binary case.

Example 3.3:   Constructing a non-binary BCH code over GF(4) of length $n = 15$ symbols: In this case we must now find the smallest extension field GF(4^m) with a 15th root of unity. Again this is the extension field GF(16) with $m = 2$. The elements $\{0, 1, \alpha^5, \alpha^{10}\}$ in GF(16) form a subfield of GF(4), so let $\beta = \alpha^5$ be a primitive element in GF(4). The cyclotomic cosets and minimal polynomials of GF(4^2) are given in Table 3.4.

Table 3.4 Cyclotomic cosets and minimal polynomials for GF(4^2).

Cyclotomic cosets	Minimal polynomials $M(x)$
$\{1, 4\}$	$M_1(x) = x^2 + x + \beta$
$\{2, 8\}$	$M_2(x) = x^2 + x + \beta^2$
$\{3, 12\}$	$M_3(x) = x^2 + \beta^2 x + 1$
$\{5\}$	$M_5(x) = x + \beta$
$\{6, 9\}$	$M_6(x) = x^2 + \beta x + 1$
$\{7, 13\}$	$M_7(x) = x^2 + \beta x + \beta$
$\{10\}$	$M_{10}(x) = x + \beta^2$
$\{11, 14\}$	$M_{11}(x) = x^2 + \beta^2 x + \beta^2$

To obtain Table 3.4, it must first be observed that the conjugates of α^i in $GF(4^2)$ are α^{4i}, $i = 1, 2, \ldots$, that is, the conjugates of α are α^4, α^{16}, $\alpha^{64}, \ldots$, but $\alpha^{16} = \alpha$ and $\alpha^{64} = \alpha^4$ so the cyclotomic coset would be $\{1, 4\}$. The associate minimal polynomial is:

$$M_1(x) = (x + \alpha)(x + \alpha^4) = x^2 + (\alpha + \alpha^4)x + \alpha^5$$
$$= x^2 + x + \beta.$$

To correct two errors, the generator polynomial will be $g(x) = M_1(x)M_2(x)M_3(x)$:

$$\begin{aligned}
g(x) &= (x^2 + x + \beta)(x^2 + x + \beta^2)(x^2 + \beta^2 x + 1) \\
&= [x^4 + (1 + 1)x^3 + (\beta^2 + 1 + \beta)x^2 + (\beta^2 + \beta)x + 1](x^2 + \beta^2 x + 1) \\
&= (x^4 + x + 1)(x^2 + \beta^2 x + 1) \\
&= x^6 + \beta^2 x^5 + x^4 + x^3 + (\beta^2 + 1)x^2 + (\beta^2 + 1)x + 1 \\
&= x^6 + \beta^2 x^5 + x^4 + x^3 + \beta x^2 + \beta x + 1.
\end{aligned}$$

There are two important features of this non-binary BCH code. Firstly, the message length of this code is $k = n - \deg(g(x)) = 15 - 6 = 9$, which means there is less redundancy in the code than in the binary BCH code in Example 3.2 (the binary BCH code has code rate of 0.47, compared with the 4-ary BCH code which has a code rate of 0.6). Secondly, although both the $(15, 7, 5)$ binary BCH code and the $(15, 9, 5)$ 4-ary BCH code correct up to two errors it should be noted that the 4-ary BCH code corrects two *symbol* errors. In $GF(4)$ each element is represented by two bits, which implies that it can correct up to 4 bit errors, provided the bit errors span two symbols, whereas the binary BCH code can only correct up to 2 bit errors. Of course it must be remembered that the actual minimum Hamming distance of these codes could be greater than the designed minimum distance and so they may be able to correct more than two errors.

From these two examples, we now begin to see the advantages of using non-binary block codes.

3.5 Reed–Solomon Codes

A BCH code defined over $GF(q)$ of length n can always be constructed provided that there is an nth root of unity in some extension field $GF(q^m)$. A special case occurs if we construct a BCH code of length $n = q^m - 1$ over $GF(q^m)$. Obviously, $GF(q^m)$ is the smallest extension field with a $(q^m - 1)$th root of unity. In this case, each cyclotomic coset will contain only one element. For example, the conjugates of the primitive element α will be α^{q^m}, $\alpha^{2q^m}, \ldots$, which is equal to α. Therefore, each minimal polynomial of α^i will be $(x - \alpha^i)$, $i = 1, 2, 3 \ldots, q^m - 2$.

Example 3.4:   Constructing a non-binary BCH code over GF(16) of length
$n = 15$ **symbols:** The 15th root of unity in GF(16) is the primitive element α
and the cyclotomic cosets and their associated minimal polynomials are given in
Table 3.5.

Table 3.5   Cyclotomic cosets and minimal polynomials for GF(16).

Cyclotomic cosets	Minimal polynomials $M(x)$
$\{1\}$	$M_1(x) = x + \alpha$
$\{2\}$	$M_2(x) = x + \alpha^2$
$\{3\}$	$M_3(x) = x + \alpha^3$
$\{4\}$	$M_4(x) = x + \alpha^4$
$\{5\}$	$M_5(x) = x + \alpha^5$
$\{6\}$	$M_6(x) = x + \alpha^6$
$\{7\}$	$M_7(x) = x + \alpha^7$
$\{8\}$	$M_8(x) = x + \alpha^8$
$\{9\}$	$M_9(x) = x + \alpha^9$
$\{10\}$	$M_{10}(x) = x + \alpha^{10}$
$\{11\}$	$M_{11}(x) = x + \alpha^{11}$
$\{12\}$	$M_{12}(x) = x + \alpha^{12}$
$\{13\}$	$M_{13}(x) = x + \alpha^{13}$
$\{14\}$	$M_{14}(x) = x + \alpha^{14}$

To correct up to $t = 2$ symbol errors the generator matrix will be $g(x) = M_1(x)M_2(x)M_3(x)M_4(x)$.

$$
\begin{aligned}
g(x) &= (x + \alpha)(x + \alpha^2)(x + \alpha^3)(x + \alpha^4) \\
&= (x^2 + \alpha^5 x + \alpha^3)(x^2 + \alpha^7 x + \alpha^7) \\
&= x^4 + (\alpha^7 + \alpha^5)x^3 + (\alpha^7 + \alpha^{12} + \alpha^3)x^2 + (\alpha^{12} + \alpha^{10})x + \alpha^{10} \\
&= x^4 + \alpha^{13}x^3 + \alpha^6 x^2 + \alpha^3 x + \alpha^{10}.
\end{aligned}
$$

The message length of the code is $k = n - \deg(g(x)) = 15 - 4 = 11$, resulting in
a $(15, 11, 5)$ 16-ary BCH code. Once again it can be seen that the code rate of this
code is even higher than the 4-ary BCH code of Example 3.3, and it can correct
two symbol errors, where one symbol now represents 4 bits. This means that it can
correct up to 8 bit errors, provided they span two symbols.

Non-binary BCH codes defined over GF(q^m) of length $n = q^m - 1$ are called
Reed–Solomon codes [4] and are the best-performing BCH codes. Hence, the
non-binary BCH code of Example 3.4 is the $(15, 11, 5)$ Reed–Solomon code.
Reed–Solomon codes also differ from other BCH codes in that their minimum
Hamming distance is equal to the designed minimum distance.

An important upper bound on the minimum Hamming distance in error-correction is the *Singleton Bound*, defined as [10]:

$$d \leq n - k + 1. \tag{3.14}$$

The Reed–Solomon code of Example 3.4 meets this bound, and this is the case for all Reed–Solomon codes. Any code with a minimum Hamming distance $d = n - k + 1$ is known as *maximum distance separable* (MDS) and has optimal minimum Hamming distance. Therefore, Reed–Solomon codes can correct:

$$t = \left\lfloor \frac{d-1}{2} \right\rfloor = \left\lfloor \frac{n-k+1-1}{2} \right\rfloor = \left\lfloor \frac{n-k}{2} \right\rfloor. \tag{3.15}$$

Rearranging (3.15) gives a message length of $k = n - 2t$. In summary, Reed–Solomon codes defined over GF(q^m) have the following parameters [12]:

Codeword length $n = q^m - 1$

Message length $k = n - 2t$

Minimum Hamming distance $d = n - k + 1$,

and the generator matrix $g(x)$ is of the form [12]:

$$g(x) = \prod_{i=1}^{2t} (x + \alpha^i). \tag{3.16}$$

The parity check matrix of a Reed–Solomon code is [12]:

$$\mathbf{H} = \begin{bmatrix} 1 & \alpha & \alpha^2 & \cdots & \alpha^{q^m-2} \\ 1 & \alpha^2 & \alpha^4 & \cdots & \alpha^{q^m-3} \\ 1 & \alpha^3 & \alpha^6 & \cdots & \alpha^{q^m-4} \\ \vdots & \vdots & \vdots & \ddots & \vdots \\ 1 & \alpha^{2t} & \alpha^{4t} & \cdots & \alpha^{q^m-2t-1} \end{bmatrix}. \tag{3.17}$$

To obtain the $2t$ syndromes the parity check matrix is multiplied by the transpose of the received vector, as in (3.10). This is equivalent to substituting successive powers of α into the received vector.

3.6 Decoding Reed–Solomon Codes

In Section 3.2.2 it was shown how syndromes could be determined from the received word and used to identify the correct error pattern and obtain the originally transmitted code word. This method becomes too complex for larger codes and alternative decoding algorithms are needed. To decode binary codes only the locations of the errors in the received word are required, since the value at these location can be flipped, that is a '1' becomes a '0' and vice versa. However, for a non-binary code an error value can be many different values and a secondary process is needed to evaluate the error value. So, two algorithms are required to decode a non-binary block code: an error-locating algorithm and an error-evaluation algorithm.

For decoding we will make use of two polynomials: an error-locating polynomial $\Lambda(x)$ and an error-magnitude polynomial $\Omega(x)$. These two polynomials are related to each other by the *key equation*, given as [12]:

$$\Lambda(x)[1 + S(x)] \equiv \Omega(x) \bmod x^{2t+1}, \tag{3.18}$$

where $S(x)$ is the syndrome polynomial.

3.6.1 Euclid's Algorithm

Euclid's algorithm [5] is used to determine the *greatest common divisor* (GCD) between two elements a and b, with $a > b$. If a and b are elements then a is divided by b to obtain a remainder r. If the remainder is zero then GCD$(a, b) = b$, otherwise we let $a = b$, $b = r$ and repeat the division to obtain another remainder. This is continued until the remainder $r = 0$. Euclid's algorithm can be extended, by using the quotient q from the division process, to also determine two further elements, u and v, that satisfy the relationship:

$$ua + vb = \gcd(a, b). \tag{3.19}$$

3.6.1.1 Euclid's Extended Algorithm

1. Initialization: $r^{(-1)} = a$, $r^{(0)} = b$, $u^{(-1)} = 1$, $u^{(0)} = 0$, $v^{(-1)} = 0$, $v^{(0)} = 1$.
2. $k = 1$.
3. $r^{(k)} = r^{(k-2)}/r^{(k-1)}$. Store the quotient $q^{(k)}$ from this division.
4. If $r^{(k)} = 0$ terminate the algorithm, otherwise $u^{(k)} = u^{(k-2)} - q^{(k)}u^{(k-1)}$ and $v^{(k)} = v^{(k-2)} - q^{(k)}v^{(k-1)}$.
5. $k = k + 1$.
6. Go to step 3.

As an example, take the elements $a = 121$ and $b = 33$ and determine their greater common divisor. The values for each iteration are given in Table 3.6.

It can be seen that each row in Table 3.6 satisfies (3.19).

We can also use Euclid's extended algorithm to decode Reed–Solomon codes. The key equation of (3.17) and (3.18) can be rewritten as:

$$\Theta(x)x^{2t+1} + \Lambda(x)[1 + S(x)] = \Omega(x), \tag{3.20}$$

Table 3.6 Euclid's algorithm to determine GCD(121, 33).

k	$r^{(k)}$	$q^{(k)}$	$u^{(k)}$	$v^{(k)}$
-1	121	—	1	0
0	33	—	0	1
1	22	3	1	-3
2	11	1	-1	4
3	0	2	3	-11

where $\Theta(x)$ is a polynomial. It is now in the form of (3.19), with $u = \Theta(x)$, $a = x^{2t+1}$, $v = \Lambda(x)$, $b = 1 + S(x)$ and $\text{GCD}(x^{2t+1}, 1 + S(x)) = \Omega(x)$.

3.6.1.2 Euclid's Algorithm to decode Reed–Solomon Codes [5, 12]

1. Initialization: $\Omega^{(-1)}(x) = x^{2t+1}$, $\Omega^{(0)} = 1 + S(x)$, $\Theta^{(-1)}(x) = 1$, $\Theta^{(0)}(x) = 0$, $\Lambda^{(-1)}(x) = 0$, $\Lambda^{(0)}(x) = 1$.
2. $k = 1$.
3. Find the remainder $\Omega^{(k)}(x)$ from the division of $\Omega^{(k-2)}$ by $\Omega^{(k-1)}$ and store the quotient $q^{(k)}$.
4. If $\Omega^{(k)}(x) = 0$ terminate the algorithm otherwise $\Theta^{(k)}(x) = \Theta^{(k-2)}(x) - q^{(k)} \Theta^{(k-1)}(x)$ and $\Lambda^{(k)}(x) = \Lambda^{(k-2)}(x) - q^{(k)} \Lambda^{(k-1)}(x)$.
5. $k = k + 1$.
6. Go to step 3.

The error locations are then determined by first finding the roots of the error-locating polynomial. The locations are the inverses of these roots.

Example 3.5:   Determining $\Lambda(x)$ and $\Omega(x)$ for the (7, 3, 5) Reed–Solomon code using Euclid's algorithm: The generator polynomial for the (7, 3, 5) Reed–Solomon code defined over GF(8) is $g(x) = x^4 + \alpha^3 x^3 + x^2 + \alpha x + \alpha^3$. Let the message polynomial be $m(x) = x^2 + \alpha x + \alpha^2$; the code word polynomial is $c(x) = x^6 + x^5 + \alpha^3 x^4 + \alpha^5 x^3 + \alpha^3 x^2 + \alpha^6 x + \alpha^5$. At the receiver it is assumed that two errors have occurred in the received word polynomial $r(x)$, in the third and sixth positions, so that $r(x) = x^6 + x^5 + \alpha x^4 + \alpha^5 x^3 + \alpha^3 x^2 + \alpha^2 x + \alpha^5$.

The $2t = 4$ syndromes are found by substituting α, α^2, α^3 and α^4 into $r(x)$ as follows:

$$
\begin{aligned}
S_1 = r(\alpha) &= \alpha^6 + \alpha^5 + \alpha \cdot \alpha^4 + \alpha^5 \cdot \alpha^3 + \alpha^3 \cdot \alpha^2 + \alpha^2 \cdot \alpha + \alpha^5 \\
&= \alpha^6 + \alpha^5 + \alpha^5 + \alpha + \alpha^5 + \alpha^3 + \alpha^5 \\
&= \alpha^6 + \alpha + \alpha^3 \\
&= (\alpha^2 + 1) + \alpha + (\alpha + 1) \\
&= \alpha^2
\end{aligned}
$$

$$S_2 = r(\alpha^2) = \alpha^{12} + \alpha^{10} + \alpha \cdot \alpha^8 + \alpha^5 \cdot \alpha^6 + \alpha^3 \cdot \alpha^4 + \alpha^2 \cdot \alpha^2 + \alpha^5$$
$$= \alpha^5 + \alpha^3 + \alpha^2 + \alpha^4 + 1 + \alpha^4 + \alpha^5$$
$$= \alpha^3 + \alpha^2 + 1$$
$$= (\alpha + 1) + \alpha^2 + 1$$
$$= \alpha^4$$

$$S_3 = r(\alpha^3) = \alpha^{18} + \alpha^{15} + \alpha \cdot \alpha^{12} + \alpha^5 \cdot \alpha^9 + \alpha^3 \cdot \alpha^6 + \alpha^2 \cdot \alpha^3 + \alpha^5$$
$$= \alpha^4 + \alpha + \alpha^6 + 1 + \alpha^2 + \alpha^5 + \alpha^5$$
$$= \alpha^4 + \alpha + \alpha^6 + 1 + \alpha^2$$
$$= (\alpha^2 + \alpha) + \alpha + (\alpha^2 + 1) + 1 + \alpha^2$$
$$= \alpha^2$$

$$S_4 = r(\alpha^4) = \alpha^{24} + \alpha^{20} + \alpha \cdot \alpha^{16} + \alpha^5 \cdot \alpha^{12} + \alpha^3 \cdot \alpha^8 + \alpha^2 \cdot \alpha^4 + \alpha^5$$
$$= \alpha^3 + \alpha^6 + \alpha^3 + \alpha^3 + \alpha^4 + \alpha^6 + \alpha^5$$
$$= \alpha^3 + \alpha^4 + \alpha^5$$
$$= (\alpha + 1) + (\alpha^2 + \alpha) + (\alpha^2 + \alpha + 1)$$
$$= \alpha$$

Therefore, $1 + S(x) = 1 + \alpha^2 x + \alpha^4 x^2 + \alpha^2 x^3 + \alpha x^4$.
Euclid's algorithm is initialized as follows:

$$\Omega^{(-1)}(x) = x^5, \ \Omega^{(0)}(x) = 1 + \alpha^2 x + \alpha^4 x^2 + \alpha^2 x^3 + \alpha x^4, \ \Lambda^{(-1)}(x) = 0,$$
$$\Lambda(0)(x) = 1, \ \Theta^{(-1)}(x) = 1, \ \Theta^{(0)}(x) = 0.$$

We must now divide $\Omega^{(-1)}(x)$ by $\Omega^{(0)}(x)$ to determine the remainder $\Omega^{(1)}(x)$:

$$
\begin{array}{r}
\alpha^6 x + 1 \\
\hline
\alpha x^4 + \alpha^2 x^3 + \alpha^4 x^2 + \alpha^2 x + 1 \overline{)\, x^5 } \\
x^5 + \alpha x^4 + \alpha^3 x^3 + \alpha x^2 + \alpha^6 x \\
\hline
\alpha x^4 + \alpha^3 x^3 + \alpha x^2 + \alpha^6 x \\
\alpha x^4 + \alpha^2 x^3 + \alpha^4 x^2 + \alpha^2 x + 1 \\
\hline
\alpha^5 x^3 + \alpha^2 x^2 + x + 1
\end{array}
$$

So, $\Omega^{(1)}(x) = \alpha^5 x^3 + \alpha^2 x^2 + x + 1$ and the quotient $q^{(1)} = \alpha^6 x + 1$. Hence:

$$\Theta^{(1)}(x) = \Theta^{(-1)}(x) + q^{(1)} \Theta^{(0)}(x) = 1 + (\alpha^6 x + 1) \cdot 0 = 1$$
$$\text{and} \quad \Lambda^{(1)}(x) = \Lambda^{(-1)}(x) + q^{(1)} \Lambda^{(0)}(x) = 0 + (\alpha^6 x + 1) \cdot 1 = \alpha^6 x + 1.$$

For the second iteration we must now divide $\Omega^{(0)}(x)$ by $\Omega^{(1)}(x)$ to determine the remainder $\Omega^{(2)}(x)$ and $q^{(2)}$:

$$
\begin{array}{r}
\alpha^3 x + \alpha^5 \\[2pt]
\alpha^5 x^3 + \alpha^2 x^2 + x + 1 \overline{\smash{\big)}\ \alpha x^4 + \alpha^2 x^3 + \alpha^4 x^2 + \alpha^2 x + 1} \\
\alpha x^4 + \alpha^5 x^3 + \alpha^3 x^2 + \alpha^3 x \\
\hline
\alpha^3 x^3 + \alpha^6 x^2 + \alpha^5 x + 1 \\
\alpha^3 x^3 + x^2 + \alpha^5 x + \alpha^5 \\
\hline
\alpha^2 x^2 + \alpha^4
\end{array}
$$

So $\Omega^{(2)}(x) = \alpha^2 x^2 + \alpha^4$ and the quotient $q^{(2)} = \alpha^3 x + \alpha^5$. Hence $\Theta^{(2)}(x) = \Theta^{(0)}(x) + q^{(2)} \Theta^{(1)}(x) = 0 + (\alpha^3 x + \alpha^5).1 = \alpha^3 x + \alpha^5$ and $\Lambda^{(2)}(x) = \Lambda^{(0)}(x) + q^{(2)} \Lambda^{(1)}(x) = 1 + (\alpha^3 x + \alpha^5)(\alpha^6 x + 1) = \alpha^2 x^2 + \alpha^6 x + \alpha^4$. The algorithm is now terminated, since the iteration step $k = t$. The example is summarized in Table 3.7.

Table 3.7   Euclid's algorithm for Example 3.6.

k	$\Omega^{(k)}(x)$	$q^{(k)}$	$\Theta^{(k)}(x)$	$\Lambda^{(k)}(x)$
-1	x^5	—	1	0
0	$\alpha x^4 + \alpha^2 x^3 + \alpha^4 x^2 + \alpha^2 x + 1$	—	0	1
1	$\alpha^5 x^3 + \alpha^2 x^2 + x + 1$	$\alpha^6 x + 1$	1	$\alpha^6 x + 1$
2	$\alpha^2 x^2 + \alpha^4$	$\alpha^3 x + \alpha^5$	$\alpha^3 x + \alpha^5$	$\alpha^2 x^2 + \alpha^6 x + \alpha^4$

The error-locating polynomial is therefore $\Lambda(x) = \Lambda^{(2)}(x) = \alpha^2 x^2 + \alpha^6 x + \alpha^4$ and the error-magnitude polynomial is $\Omega(x) = \Omega^{(2)}(x) = \alpha^5 x^3 + \alpha^2 x^2 + x + 1$. The error-locating polynomial can be factorized to $\Lambda(x) = (\alpha x + \alpha^4)(\alpha x + 1)$ and the two roots are $x_1 = \alpha^3$ and $x_2 = \alpha^6$. The error locations are the inverses of these roots and are $X_1 = \alpha^4$ and $X_2 = \alpha$, corresponding to the x^4 and x terms in the received word. It can be seen that generated polynomials from Euclid's algorithm satisfy (3.20).

$$
\begin{aligned}
\Theta(x)x^5 + \Lambda(x)(1 + S(x)) &= (\alpha^3 x + \alpha^5)x^5 + (\alpha^2 x^2 + \alpha^6 x + \alpha^4) \\
&\quad \times (1 + \alpha^2 x + \alpha^4 x^2 + \alpha^2 x^3 + \alpha x^4) \\
&= \alpha^3 x^6 + \alpha^5 x^5 + \alpha^3 x^6 + (\alpha^4 + 1)x^5 \\
&\quad + (\alpha^6 + \alpha + \alpha^5)x^4 + (\alpha^4 + \alpha^3 + \alpha^6)x^3 \\
&\quad + (\alpha^2 + \alpha + \alpha)x^2 + (\alpha^6 + \alpha^6)x + \alpha^4 \\
&= \alpha^2 x^2 + \alpha^4 = \Omega(x).
\end{aligned}
$$

Since this is equivalent to the key equation of (3.19) we actually do not need to generate $\Theta(x)$ to be able to obtain the error-locating and error-magnitude polynomials.

3.6.2 Berlekamp–Massey's Algorithm

Berlekamp–Massey's algorithm [6, 7] is more difficult to understand than Euclid's algorithm but has a more efficient implementation.

3.6.2.1 Berlekamp–Massey's Algorithm

1. Initialization: $k = 0$, $\Lambda^{(0)}(x) = 1$, $L = 0$, $T(x) = x$.
2. $k = k + 1$.
3. Compute the discrepancy $\Delta^{(k)} = S_k - \sum_{i=1}^{L} \Lambda_i^{(k-1)} S_{k-i}$.

   If $\Delta^{(k)} = 0$, go to step 8.
4. Modify the connection polynomial: $\Lambda^{(k)}(x) = \Lambda^{(k-1)}(x) - \Delta^{(k)} T(x)$.
5. If $2L \geq k$, go to step 8.
6. Set $L = k - L$ and $T(x) = \frac{\Lambda^{(k-1)}(x)}{\Delta^{(k)}}$.
7. Set $T(x) = xT(x)$.
8. If $k < 2t$, go to step 3.
9. Determine the roots of $\Lambda(x) = \Lambda^{(2t)}(x)$.

Example 3.6: Determining $\Lambda(x)$ and $\Omega(x)$ for the $(7, 3, 5)$ Reed–Solomon code using Berlekamp–Massey's algorithm: Assuming the same received word as in Example 3.5, $r(x) = x^6 + x^5 + \alpha x^4 + \alpha^5 x^3 + \alpha^3 x^2 + \alpha^2 x + \alpha^5$, the syndromes were found to be $S_1 = \alpha^2$, $S_2 = \alpha^4$, $S_3 = \alpha^2$ and $S_4 = \alpha$. The algorithm proceeds as follows:

First, initialize the algorithm variables: $k = 0$, $\Lambda^{(0)}(x) = 1$, $L = 0$ and $T(x) = x$.

$$\Delta^{(1)} = S_1 - \sum_{i=1}^{0} \Lambda_i^{(0)} S_{1-i} = S_1 = \alpha^2$$

$$\Lambda^{(1)}(x) = \Lambda^{(0)}(x) - \Delta^{(1)} T(x)$$
$$= 1 - \alpha^2 x.$$

At step 5, $2L = 2 \times 0 = 0$, which is less than k, so go to step 6: $L = k - L = 1 - 0 = 1$.

$$T(x) = \frac{\Lambda^{(0)}(x)}{\Delta^{(1)}} = \frac{1}{\alpha^2} = \frac{\alpha^7}{\alpha^2} = \alpha^5.$$

At step 7, $T(x) = xT(x) = \alpha^5 x$. At step 8, $k < 2t = 4$, so go to step 3.
For $k = 2$, the discrepancy $\Delta^{(2)}$ and error-locating polynomial $\Lambda^{(2)}(x)$ are:

$$\Delta^{(2)} = S_2 - \sum_{i=1}^{1} \Lambda_i^{(1)} S_{2-i} = \alpha^4 - \alpha^2 \alpha^2 = 0$$

$$\Lambda^{(2)}(x) = \Lambda^{(1)}(x) - \Delta^{(2)} T(x) = \Lambda^{(1)}(x) = 1 - \alpha^2 x.$$

$\Delta^{(2)} = 0$, meaning there is no discrepancy and the error-locating polynomial does not need to be modified, so go to step 7: $T(x) = xT(x) = \alpha^5 x^2$.

At step 8, $k < 2t$, so go to step 3.

For $k = 3$, the discrepancy $\Delta^{(3)}$ and error-locating polynomial $\Lambda^{(3)}(x)$ are:

$$\Delta^{(3)} = S_3 - \sum_{i=1}^{1} \Lambda_i^{(1)} S_{3-i} = \alpha^2 - \alpha^2 \alpha^4 = \alpha^2 + \alpha^2 + 1 = 1$$

$$\Lambda^{(3)}(x) = \Lambda^{(2)}(x) - \Delta^{(3)} T(x) = 1 - a^2 x - a^5 x^2.$$

At step 5, $2L = 2$, which is less than k, so go to step 6.

$$L = k - L = 3 - 1 = 2.$$
$$T(x) = \frac{\Lambda^{(2)}(x)}{\Delta^{(3)}} = \frac{1 - \alpha^2 x}{1} = 1 - \alpha^2 x.$$

At step 7, $T(x) = xT(x) = x - \alpha^2 x^2$. At step 8, $k < 2t$, so go to step 3.

Finally, for $k = 4$, the discrepancy $\Delta^{(4)}$ and error-locating polynomial $\Lambda^{(4)}(x)$ are:

$$\Delta^{(4)} = S_4 - \sum_{i=1}^{2} \Lambda_i^{(3)} S_{4-i} = S_4 - \left[\Lambda_1^{(3)} S_3 + \Lambda_2^{(3)} S_2 \right]$$
$$= \alpha + \alpha^2 \alpha^2 + \alpha^5 \alpha^4 = \alpha + \alpha^4 + \alpha^2$$
$$= \alpha + \alpha^2 + \alpha + \alpha^2 = 0$$
$$\Lambda^{(4)}(x) = \Lambda^{(3)}(x) - \Delta^{(4)} T(x) = \Lambda^{(3)}(x) - 0 = 1 - \alpha^2 x - \alpha^5 x^2$$

At step 3, $\Delta^{(4)} = 0$, meaning there is no modification to the error-locating polynomial, so go to step 7: $T(x) = xT(x) = x^2 - \alpha^2 x^3$.

At step 8, $k = 2t$, so go to step 9 and end the algorithm.

The error-locating polynomial $\Lambda(x) = \Lambda^{(4)}(x) = 1 - \alpha^2 x - \alpha^5 x^2$ can be factorized to $\Lambda(x) = (1 - \alpha^4 x)(1 - \alpha x)$, which has the roots $X_1 = \alpha^3$ and $X_2 = \alpha^6$. This completes the Berlekamp–Massey algorithm. A summary of the above example is shown in Table 3.8.

Table 3.8 Summary of the Berlekamp–Massey algorithm steps.

k	S_k	$\Lambda^{(k)}(x)$	$\Delta^{(k)}$	L	$T(x)$
0	—	1	—	0	x
1	α^2	$1 - \alpha^2 x$	α^2	1	$\alpha^5 x$
2	α^4	$1 - \alpha^2 x$	0	1	$\alpha^5 x^2$
3	α^2	$1 - \alpha^2 x - \alpha^5 x^2$	1	2	$x - \alpha^2 x^2$
4	α	$1 - \alpha^2 x - \alpha^5 x^2$	0	2	$x^2 - \alpha^2 x^3$

However, the decoding process is still unfinished as the error magnitudes at these locations need to be calculated.

3.6.3 Determining the Error Magnitudes Using Forney's Formula

Forney's formula [13] is a very efficient method for finding the error magnitudes and only requires the error locations, the error-locating polynomial and the error-magnitude polynomial $\Omega(x)$ given by (3.19). If errors have occurred at the positions $i_1, i_2, i_3, \ldots i_{2t}$, the error magnitudes are determined by Forney's formula as [12]:

$$e_{i_k} = \frac{-X_k \Omega\left(X_k^{-1}\right)}{\Lambda'\left(X_k^{-1}\right)}, \tag{3.21}$$

where $\Lambda'(x)$ is the *formal derivative* of the error-locating polynomial. The formal derivative of a function $f(x) = f_0 + f_1 x + f_2 x^2 + \cdots + f_n x^n$ whose coefficients belong to some finite field GF(q) is given as [12] :

$$f'(x) = f_1 + 2f_2 x + 3f_3 x^2 + \cdots + n f_{n-1} x^{n-1}. \tag{3.22}$$

Completing the previous example, the error locations were found to be at $X_1 = \alpha^4$ and $X_2 = \alpha$. The error magnitude $\Omega(x)$ is:

$$
\begin{aligned}
\Omega\left(x\right) &= \Lambda\left(x\right)\left[1 + S\left(x\right)\right] \bmod x^{2t+1} \\
&= (1 - \alpha^2 x - \alpha^5 x^2)(1 + \alpha^2 x + \alpha^4 x^2 + \alpha^2 x^3 + \alpha x^4) \bmod x^5 \\
&= 1 + (\alpha^2 + \alpha^2)x + (\alpha^4 + \alpha^4 + \alpha^5)x^2 + (\alpha^2 + \alpha^6 + 1)x^3 + (\alpha + \alpha^4 + \alpha^2)x^4 \\
&= 1 + \alpha^5 x^2 + (\alpha^2 + \alpha^2 + 1 + 1)x^3 + (\alpha + \alpha^2 + \alpha + \alpha^2)x^4 \\
&= 1 + \alpha^5 x^2.
\end{aligned}
$$

Applying (3.22) to $\Lambda(x)$, the formal derivative of the error-locating polynomial is:

$$
\begin{aligned}
f'(x) &= f_1 + 2f_2 x + 3f_3 x^2 + \cdots + n f_{n-1} x^{n-1} \\
\Lambda'(x) &= \alpha^2 + 2\alpha^5 x \\
&= \alpha^2 + \left(\alpha^5 + \alpha^5\right) x \\
&= \alpha^2.
\end{aligned}
$$

Therefore, the error magnitude at the location $X_1 = \alpha^4$ is:

$$e_{i_1} = \frac{-X_1 \Omega\left(X_1^{-1}\right)}{\Lambda'\left(X_1^{-1}\right)} = \frac{\alpha^4 \left(1 + \alpha^5 \left(\alpha^4\right)^{-2}\right)}{\alpha^2} = \frac{\alpha^4(1 + \alpha^5 \alpha^6)}{\alpha^2} = \frac{\alpha^4(1 + \alpha^4)}{\alpha^2} = \frac{\alpha^9}{\alpha^2} = \alpha^7 = 1.$$

Similarly, the error magnitude at the location $X_2 = \alpha$ is:

$$e_{i_2} = \frac{-X_1 \Omega\left(X_1^{-1}\right)}{\Lambda'\left(X_1^{-1}\right)} = \frac{\alpha(1 + \alpha^5(\alpha)^{-2})}{\alpha^2} = \frac{\alpha(1 + \alpha^5 \alpha^5)}{\alpha^2} = \frac{\alpha \left(1 + \alpha^3\right)}{\alpha^2} = \frac{\alpha^2}{\alpha^2} = 1.$$

This results in an error polynomial of $e(x) = x^4 + x$. Therefore, the decoded code word is $r(x) + e(x)$, giving:

$$
\begin{aligned}
r(x) + e(x) &= \left(x^6 + x^5 + \alpha x^4 + \alpha^5 x^3 + \alpha^3 x^2 + \alpha^2 x + \alpha^5\right) + \left(x^4 + x\right) \\
&= x^6 + x^5 + (\alpha + 1)x^4 + \alpha^5 x^3 + \alpha^3 x^2 + \left(\alpha^2 + 1\right)x + \alpha^5 \\
&= x^6 + x^5 + \alpha^3 x^4 + \alpha^5 x^3 + \alpha^3 x^2 + \alpha^6 x + \alpha^5,
\end{aligned}
$$

which matches the transmitted code word polynomial $c(x)$.

3.7 Coded Modulation

When conventional coding techniques are introduced in a transmission system, the bandwidth of the coded signal after modulation is wider than that of the uncoded signal for the same information rate and the same modulation scheme. In fact, the encoding process requires a bandwidth expansion that is inversely proportional to the code rate, being traded for a coding gain. This is the reason why, in the past, conventional coding schemes have been very popular on power-limited channels (where bandwidth is readily available) but not on bandwidth-limited channels.

The first important contribution to coding on bandwidth-limited channels was made by Ungerboeck in 1982 [8]. He presented trellis coded modulation (*TCM*) as a coded modulation scheme that accommodates the redundancy of a code on an expanded signal set, and showed that excellent coding gains over uncoded modulation can be achieved with no bandwidth expansion required. TCM codes are explained in Chapter 7 of this book.

As well as combining convolutional codes with modulation, it is also possible to combine block codes with modulation; this is known as *block coded modulation* (*BCM*). BCM codes have the advantage of being simpler to design than TCM codes, but the decoding complexity can become too high for larger BCM codes.

3.7.1 Block Coded Modulation (BCM) Codes

The encoding procedure involves adding redundant bits or symbols to the message. This redundancy does not contain any message bits and transmitting it results in the overall data rate being reduced. The Quadrature Phase Shift Keying (QPSK) modulation scheme carries two information bits per QPSK symbol without coding. However, if the information is encoded with encoder of code rate $R = 0.5$ then half the bits in the code word are redundant bits and so each QPSK symbol only carries one information bit. One solution is to increase the size of the constellation to 16-PSK (or alternatively 16-QAM), where each 16-PSK symbol carries four information bits without coding. Therefore, using an encoder with $R = 0.5$ and 16-PSK means that each 16-PSK symbol carries two information bits, which is the same as uncoded QPSK.

One problem with this solution is that as the constellation size is increased the Euclidean distance between constellation points decreases, resulting in poorer performance. However, combining code design with the modulation scheme can ensure that code words have a minimum Euclidean distance greater than the Euclidean distance between neighbouring constellation points. The combination of convolutional codes and modulation is known as *trellis coded modulation (TCM)*, while the combination of block codes with modulation is known as *block coded modulation (BCM)*.

3.7.2 Multi-Level Block Coding

A general block diagram of a binary BCM encoder is shown in Figure 3.6. The encoding process takes m parallel messages $\mathbf{u}^{(1)}$, $\mathbf{u}^{(2)}$, ..., $\mathbf{u}^{(m)}$ of lengths k_1, k_2, ..., k_m bits and each message is encoded by one of m parallel block encoders, resulting in m code words all of length n bits with minimum Hamming distances d_1, d_2, ..., d_m. This arrangement of block encoders is called *multi-level block coding* and was proposed by Imai and Hirakawa [14] . Finally, each bit from the m encoders is mapped onto a 2^m-ary constellation [15].

From Figure 3.6, we can see that the code rate of the multi-level block code is:

$$R = \frac{\sum\limits_{i=1}^{m} k_i}{mn}. \tag{3.23}$$

An example of a BCM code using $m = 3$ block encoders which are mapped onto a $2^3 = 8$-PSK constellation is given in Figure 3.7. The top block encoder is an (8, 1, 8) repetition code, which takes a message of $k_1 = 1$ bit and repeats that value eight times, for example $0 \rightarrow 00000000$ and $1 \rightarrow 11111111$. The second block code is the (8, 7, 2) parity check code, as explained in Section 3.2. It takes a 7 bit message and appends a parity check bit, which is either zero if the number of '1's in the message is even or one if the number of '1's in the message is odd. Finally, the last encoder

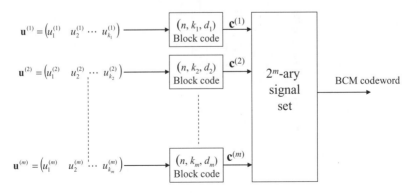

Figure 3.6 General multi-level BCM encoder.

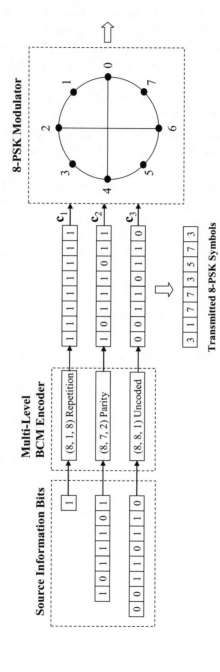

Figure 3.7 A 3-level BCM encoder for 8-PSK.

is not an encoder at all since $k_3 = n = 8$ and so there is no redundancy added. From (3.23) the code rate of this BCM code is $R = \frac{k_1 + k_2 + k_3}{3n} = \frac{1+7+8}{24} = \frac{16}{24} = \frac{2}{3}$. All that remains is to map each bit from the three encoders to the 8-PSK constellation, which is explained next.

3.7.3 Set Partitioning

The BCM encoder output is a constrained sequence of constellation symbols and the mapping of the multi-level block encoder to these symbols is very important for ensuring that the minimum Euclidean distance, or *free distance*, of the BCM code is maximized. To achieve this, Ungerboeck proposed the *Set Partitioning* [8] of the constellation, whereby the points in the constellation are recursively divided into subsets, with the Euclidean distance between neighbouring points in each subset being increased. Figure 3.8 shows set partitioning of the 8-PSK constellation as an example. This is called a *partition tree*.

Ungerboeck's rules for set partitioning any constellation are [8]:

- The signals in the bottom level of the partition tree are assigned parallel transitions.
- Parallel transitions must appear in subsets separated by the largest Euclidean distance.
- All signal points should be used equally often.

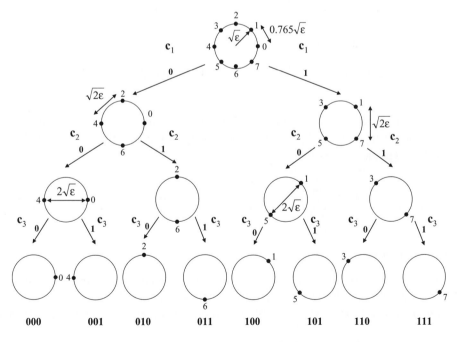

Figure 3.8   The set partitioning of the 8-PSK constellation.

From Figure 3.8 we can see that the coded bit from the repetition code c_1 in Figure 3.7 selects one of two QPSK subsets with constellation points separated by a Euclidean distance of $0.765\sqrt{\varepsilon}$. The coded bit from the parity check code c_2 then selects one of two subsets with constellation points separated at a greater distance of $2\sqrt{\varepsilon}$. Finally, the uncoded bits c_3 select signals in the bottom level of the partition tree.

It can be shown that the minimum squared Euclidean distance between any two sequences of the BCM code, otherwise known as the *free distance* d^2_{free}, is [15]:

$$d^2_{\text{free}} = \min \left(\Delta^2_1 d_1, \, \Delta^2_2 d_2, \, \Delta^2_3 d_3, \, \ldots, \, \Delta^2_m d_m \right), \tag{3.24}$$

where Δ^2_i, $i = 0, 1, 2, \ldots, m$ is the squared Euclidean distance between neighbouring points in a subset in the ith level of the partition tree and d_i is the minimum Hamming distance of the ith block encoder of the BCM code. A measure of the performance of the BCM code over an uncoded constellation can be determined by finding the asymptotic coding gain (ACG) γ [10]:

$$\gamma = \frac{d^2_{\text{free}}/\varepsilon}{d^2_{\text{min}}/\varepsilon'}, \tag{3.25}$$

where d^2_{min} is the squared Euclidian distance between points in the uncoded constellation, ε is the energy of the coded constellation and ε' is the energy of the uncoded constellation. For the BCM code in Figure 3.7 the minimum Hamming distances are $d_1 = 8$, $d_2 = 2$ and $d_3 = 1$. The corresponding Euclidean distances from the subsets of 8-PSK in Figure 3.8 are $\Delta^2_1 = 0.585\varepsilon'$, $\Delta^2_2 = 2\varepsilon'$ and $\Delta^2_3 = 4\varepsilon'$. Therefore, from (3.24) the free distance of the BCM code is:

$$d^2_{\text{free}} = \min \left(0.585\varepsilon' \times 8, \, 2\varepsilon' \times 2, \, 4\varepsilon' \times 1 \right) = \min(4.68\varepsilon', 4\varepsilon', 4\varepsilon') = 4\varepsilon'.$$

Since the uncoded and coded modulation schemes are QPSK and 8-PSK respectively, the energy of both constellations is the same; that is, $\varepsilon = \varepsilon'$. The minimum squared Euclidean distance between neighbouring points of QPSK constellation is $d^2_{\text{min}} = 2\varepsilon$, giving an ACG of:

$$\gamma = \frac{d^2_{\text{free}}/\varepsilon}{d^2_{\text{min}}/\varepsilon'} = \frac{4\varepsilon'/\varepsilon}{2\varepsilon/\varepsilon'} = \frac{4}{2} = 2 \quad \text{or} \quad 3 \text{ dB}.$$

3.7.4 Construction of a Block Code Trellis

The decoding of a BCM code can be achieved efficiently by representing it as a trellis diagram and applying the Viterbi algorithm. Wolf [16] showed how a trellis diagram of a linear block code could be constructed using its parity check matrix, and this method was extended by Paravalos and Fleisher [17] to construct BCM trellis diagrams.

Since the BCM encoder is made up of a number of block codes, its trellis is a combination of the trellis diagrams of the component codes. The state transitions in the trellis are determined by [17]:

$$s_\xi(\lambda + 1) = s_l(\lambda) + \sum_{i=1}^{m} \delta_i \mathbf{h}_{i,\lambda+1}, \qquad (3.26)$$

where λ is an index denoting the depth of the trellis, $l \in \mathbf{S}_\lambda$ (a set containing the index of all states created up to depth λ), $\xi \in \mathbf{S}_{\lambda+1}$ (a set containing the index of all states in $\mathbf{S}_\lambda$ plus all new states formed between λ and $\lambda + 1$), $\mathbf{h}_{i,\lambda+1}$ is a vector representing column $\lambda + 1$ of the parity check matrix of the ith block code and $\delta_i = \{0, 1\}$ is a binary input.

The trellis construction consists of four steps:

1. Each trellis begins and terminates at the all-zero state.
2. The depth of each trellis spans from $\lambda = 0$ to $\lambda = n$.
3. The number of states depends on the number and type of block codes used and the constellation size. It grows while proceeding further into the trellis. To begin the construction, a number of 2^m states is assumed. If a number of information bits are presented to the mapper uncoded, the initial number of states reduces to 2^{m-r}, where r is the number of absent encoders.
4. Each transition is determined by (3.26).

The following should be observed:

1. When the construction is complete all trellis paths that do not terminate at the all-zero state are removed as they do not represent valid code sequences.
2. Transitions in a trellis are assigned a symbol from the signal constellation. Each path within the trellis corresponds to a possible BCM sequence.
3. Due to the variation in component code sizes, the parity check matrix vectors used in (3.26) have different dimensions. When the size of one parity check vector is less than the size of the state binary vector s, an appropriate number of zeroes is appended at the top of the vector. Alternatively, a vector with more elements than the state binary vector s is reduced from top to bottom using modulo-2 addition.

The given procedure yields the final BCM trellis only in cases where the top encoder employs the repetition code and the remaining encoders are simple codes.

Example 3.7: BCM code trellis construction: Using the BCM code from Figure 3.7, an example of constructing a trellis is now given. The BCM encoder consists of three block encoders: the (8, 1, 8) Repetition code, the (8, 7, 2) Parity Check code and 8 bits of uncoded information. The parity check matrices of the Repetition code, $\mathbf{H}_R$, and the Parity Check code, $\mathbf{H}_P$, are:

$$\mathbf{H}_R = \begin{bmatrix} 1 & 1 & 0 & 0 & 0 & 0 & 0 & 0 \\ 1 & 0 & 1 & 0 & 0 & 0 & 0 & 0 \\ 1 & 0 & 0 & 1 & 0 & 0 & 0 & 0 \\ 1 & 0 & 0 & 0 & 1 & 0 & 0 & 0 \\ 1 & 0 & 0 & 0 & 0 & 1 & 0 & 0 \\ 1 & 0 & 0 & 0 & 0 & 0 & 1 & 0 \\ 1 & 0 & 0 & 0 & 0 & 0 & 0 & 1 \end{bmatrix} \quad \text{and} \quad H_P = [1\ 1\ 1\ 1\ 1\ 1\ 1\ 1].$$

The trellis will consist of $2^{m-r} = 2^{3-1} = 4$ states with two subtrellises each of two states. To accommodate two states per subtrellis, the parity check matrices are adjusted to attain a row size equal to the size of the binary state vector. Thus, the reduced vectors of $\mathbf{H}_R$ are obtain by top-to-bottom modulo-2 addition and the result is:

$$H_{R'} = \begin{bmatrix} 1\ 1\ 1\ 1\ 1\ 1\ 1\ 1 \end{bmatrix} \quad H_{P'} = \begin{bmatrix} 1\ 1\ 1\ 1\ 1\ 1\ 1\ 1 \end{bmatrix}.$$

Now the state transitions will be determined by (3.26):

1. At depth $\lambda = 1$, (3.26) will become:

$$s_\xi(1) = s_l(0) + \delta_1 \mathbf{h}_{1,1} + \delta_2 \mathbf{h}_{2,1}.$$

In the first subtrellis, the values of δ_i for the first path are $\delta_1 = 0$, $\delta_2 = 0$ and for the second path $\delta_1 = 0$, $\delta_2 = 1$.

$$s_0(1) = [0] + 0 \times [1] + 0 \times [1] = [0]$$
$$s_1(1) = [0] + 0 \times [1] + 1 \times [1] = [1].$$

In the second subtrellis, the values of δ_i for the first path are $\delta_1 = 1$, $\delta_2 = 0$ and for the second path $\delta_1 = 1$, $\delta_2 = 1$.

$$s_2(1) = [0] + 1 \times [1] + 0 \times [1] = [1]$$
$$s_3(1) = [0] + 1 \times [1] + 1 \times [1] = [0].$$

Note that the uncoded information is transmitted on the third BCM encoder level. This means that the parity check matrix does not exist and hence all corresponding transitions lead to the same state, causing parallel transitions in the trellis diagram.

2. At depth $\lambda = 2$, (3.26) will become:

$$s_\xi(2) = s_l(1) + \delta_1 \mathbf{h}_{1,2} + \delta_2 \mathbf{h}_{2,2}.$$

First Subtrellis

$$s_0(2) = [0] + 0 \times [1] + 0 \times [1] = [0]$$
$$s_1(2) = [0] + 0 \times [1] + 1 \times [1] = [1]$$

$$s_2(2) = [1] + 0 \times [1] + 0 \times [1] = [1]$$
$$s_3(2) = [1] + 0 \times [1] + 1 \times [1] = [0].$$

Second Subtrellis

$$s_4(2) = [0] + 1 \times [1] + 0 \times [1] = [1]$$
$$s_5(2) = [0] + 1 \times [1] + 1 \times [1] = [0]$$

$$s_6(2) = [1] + 1 \times [1] + 0 \times [1] = [0]$$
$$s_7(2) = [1] + 1 \times [1] + 1 \times [1] = [1].$$

3. At depth $\lambda = 3$, (3.26) will become:

$$s_\xi(3) = s_l(2) + \delta_1 \mathbf{h}_{1,3} + \delta_2 \mathbf{h}_{2,3}.$$

First Subtrellis

$$s_0(3) = [0] + 0 \times [1] + 0 \times [1] = [0]$$
$$s_1(3) = [0] + 0 \times [1] + 1 \times [1] = [1]$$

$$s_2(3) = [1] + 0 \times [1] + 0 \times [1] = [1]$$
$$s_3(3) = [1] + 0 \times [1] + 1 \times [1] = [0].$$

Second Subtrellis

$$s_4(3) = [0] + 1 \times [1] + 0 \times [1] = [1]$$
$$s_5(3) = [0] + 1 \times [1] + 1 \times [1] = [0]$$

$$s_6(3) = [1] + 1 \times [1] + 0 \times [1] = [0]$$
$$s_7(3) = [1] + 1 \times [1] + 1 \times [1] = [1].$$

The construction continues until $\lambda = n = 8$. The final trellis is shown in Figure 3.9.

3.7.5 Non-Binary BCM Codes

In 1994, Baldini and Farrell [18] introduced a class of non-binary BCM code over rings of integers suitable for M-PSK and M-QAM. The idea of non-binary BCM codes is to transmit m bits per channel symbol by using a modulator with $q > 2^m$ waveforms

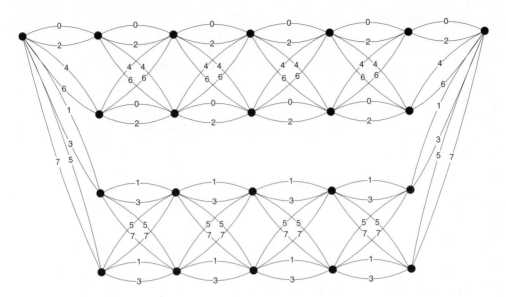

Figure 3.9 Trellis for a BCM code using 8-PSK modulation.

to accommodate the extra redundancy. The non-binary BCM encoder structure is shown in Figure 3.10. The binary source generates $m + 1$ parallel bits, which are Gray mapped onto one of 2^{m+1} channel symbols $a_i \in \mathbb{Z}_q$, $i = 0, 1, \ldots, k - 1$. These are then fed to the multi-level encoder to generate the BCM coded symbols $x_i \in \mathbb{Z}_q$, $i = 0, 1, \ldots n$, which will increase the minimum Euclidean distance.

One class of block code that can be used in the BCM encoder is the *systematic linear circulant* block code. Its generator matrix is of the form given in (3.7), where its parity block **P** is [18]:

$$
\mathbf{P} = \begin{bmatrix}
P_{1,1} & P_{k,1} & P_{k-1,1} & \cdots & P_{3,1} & P_{2,1} \\
P_{2,1} & P_{1,1} & P_{k,1} & \cdots & P_{4,1} & P_{3,1} \\
\vdots & \vdots & \vdots & \ddots & \vdots & \vdots \\
P_{k-1,1} & P_{k-2,1} & P_{k-3,1} & \cdots & P_{1,1} & P_{k,1} \\
P_{k,1} & P_{k-1,1} & P_{k-2,1} & \cdots & P_{2,1} & P_{1,1}
\end{bmatrix}.
\tag{3.27}
$$

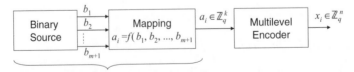

Figure 3.10 Non-binary BCM encoder structure.

Each column of $\mathbf{P}$ in (3.27) is a cyclic shift upwards of the preceding column. To identify this type of code the values of the first column of $\mathbf{P}$ are used, that is $P_{1,1}$, $P_{2,1}, \ldots, P_{k-1,1}, P_{k,1}$. An example of a systematic linear circulant block code over $\mathbb{Z}_4 = \{0, 1, 2, 3\}$ is the $(10, 5)$ block code denoted by 12233 [18]. Its generator matrix is:

$$\mathbf{G} = \begin{bmatrix} 1 & 0 & 0 & 0 & 0 & 1 & 3 & 3 & 2 & 2 \\ 0 & 1 & 0 & 0 & 0 & 2 & 1 & 3 & 3 & 2 \\ 0 & 0 & 1 & 0 & 0 & 2 & 2 & 1 & 3 & 3 \\ 0 & 0 & 0 & 1 & 0 & 3 & 2 & 2 & 1 & 3 \\ 0 & 0 & 0 & 0 & 1 & 3 & 3 & 2 & 2 & 1 \end{bmatrix}$$

and it has a minimum squared Euclidean distance of $d_{\text{free}}^2 = 12$. From (3.25) its asymptotic coding gain over uncoded BPSK is 4.77 dB.

Another class of multi-level block code is the *pseudocyclic multi-level code*, which has a generator matrix of the form [18]:

$$\mathbf{G} = \begin{bmatrix} g_{1,1} & g_{1,2} & \cdots & \cdots & g_{1,r} & 0 & 0 & \cdots & 0 \\ 0 & g_{1,1} & g_{1,2} & \cdots & \cdots & g_{1,r} & 0 & \cdots & 0 \\ \vdots & \vdots & \vdots & \ddots & \ddots & \vdots & \vdots & \ddots & \vdots \\ 0 & 0 & \cdots & g_{1,1} & g_{1,2} & \cdots & \cdots & \cdots & g_{1,r} \end{bmatrix}, \tag{3.28}$$

where $g_{i,j} \in \mathbb{Z}_q$ and r depends on the coded q-PSK modulation scheme and the code rate.

It is also possible to construct non-binary BCM codes that are invariant to phase rotations, that is a phase shift of the received symbols results in another valid code word. Codes that are *phase invariant* to rotations of $(360/q)°$ are called *transparent codes*. A code is transparent if it contains the all-one sequence $(1, 1, 1, \ldots, 1)$ as a valid code word. Since the BCM code is linear, adding any two code word modulo-q will result in another code word. Adding the all-one code word to another code word is equivalent to rotating all the coded symbols by 90°, which produces yet another valid code word.

To ensure that the all-one code word is present in the non-binary BCM code, the generator matrix from (3.28) is modified so that the bottom row is all ones [18].

$$\mathbf{G} = \begin{bmatrix} g_{1,1} & g_{1,2} & \cdots & \cdots & g_{1,r} & 0 & 0 & \cdots & 0 \\ 0 & g_{1,1} & g_{1,2} & \cdots & \cdots & g_{1,r} & 0 & \cdots & 0 \\ \vdots & \vdots & \vdots & \ddots & \ddots & \vdots & \vdots & \ddots & \vdots \\ 1 & 1 & \cdots & 1 & 1 & \cdots & \cdots & \cdots & 1 \end{bmatrix}. \tag{3.29}$$

Now the pseudocyclic multi-level code is transparent. To remove phase rotations from the received symbols a differential encoder can be added before the transparent

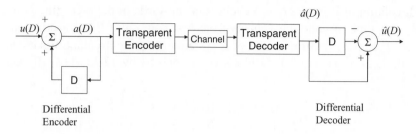

Figure 3.11 A differential encoder and decoder added to the non-binary BCM system to remove phase rotations.

encoder [18], as shown in Figure 3.11. Similarly, a differential decoder is added after the transparent decoder to obtain the decoded message.

3.8 Conclusions

In this chapter an introduction to block coding was given, beginning with the simplest binary block codes, such as the parity check codes. A class of block codes known as cyclic codes was then presented, with particular attention given to BCH codes. These codes have good parameters and, using BCH codes, it was demonstrated how extending to a non-binary alphabet can improve these parameters, for example giving large Hamming distance and higher code rates. A special class of non-binary BCH codes is the Reed–Solomon codes, which are the most commonly used error-correcting codes today and can be found in optical and magnetic storage, high-speed modems and wireless communications. Finally, the concept of block coded modulation was introduced, whereby the block encoder and modulator are treated as a single entity to avoid increasing bandwidth use. Proper mapping of the coded symbols to channel symbols ensures that the loss in performance due to the expansion of the signal set is more than compensated for by the coding gain of the BCM code. Further performance gains can be achieved by using non-binary block codes in the coded modulation scheme, but with an increase in the decoding complexity.

References

[1] Hamming, R. (1950) Error detecting and error correcting codes. *Bell Systems Technical Journal*, **29**, 41–56.

[2] Hocquenghem, A. (1959) Codes correcteurs d'erruers. *Chiffres*, **2**, 147–56.

[3] Bose, R. and Ray-Chaudhuri, D. (1960) On a class of error-correcting binary codes. *Information and Control*, **3**, 68–79.

[4] Reed, I. and Solomon, G. (1960) Polynomial codes over certain finite fields. *Journal of the Society of Industrial Mathematics*, **8**, 300–4.

[5] Sugiyama, Y., Kasahara, M., Hirasawa, S. and Namekawa, T. (1975) A method for solving key equation for Goppa codes. *Information and Control*, **27**, 87–99.

[6] Berlekamp, E. (1967) Nonbinary BCH decoding. Proceedings of the International Symposium on Information Theory, San Remo, Italy.

[7] Massey, J.L. (1969) Shift-register synthesis and BCH decoding. *IEEE Transactions on Information Theory*, **IT-15** (1), 122–7.

[8] Ungerboeck, G. (1982) Channel coding with multilevel/phase signals. *IEEE Transactions on Information Theory*, **28** (1), 55–67.

[9] Viterbi, A.J. (1971) Convolutional codes and their performance in communication systems. *IEEE Transactions on Communications Technology*, **19** (5), 751–72.

[10] Moon, T.K. (2005) *Error Correction Coding. Mathematical Methods and Algorithms*, Wiley Interscience, ISBN 0-471-64800-0.

[11] Prange, E. (1958) *Cyclic Error-Correcting Codes in Two Symbols*, Air Force Cambridge Research Center, Cambridge, MA, Technical Report, TN-58-156.

[12] Wicker, S.B. (1995) *Error Control Systems for Digital Communications and Storage*, Prentice Hall, Eaglewood Cliffs, NJ.

[13] Forney, G.D. (1965) On decoding BCH codes. *IEEE Transactions on Information Theory*, **11** (4), 549–57.

[14] Imai, H. and Hirakawa, S. (1977) A new multilevel coding method using error-correcting codes. *IEEE Transactions on Information Theory*, **23** (3), 371–7.

[15] Lin, S. and Costello, D.J. Jr. (2004) *Error Control Coding*, 2nd edn, Pearson Prentice Hall, ISBN 0-13-017973-6.

[16] Wolf, J.K. (1978) Efficient maximum likelihood decoding of linear block codes using a trellis. *IEEE Transactions on Information Theory*, **24** (1), 76–80.

[17] Paravalos, E. and Fleisher, S. (1991) Block coded modulation: an application to frequency/phase modulation and a procedure for the construction of trellis diagram. IEEE Proceedings of Pacific Rim Conference on Communications, Computers and Signal Processing, Victoria, USA, pp. 79–82.

[18] Baldini, F.R. and Farrell, P.G. (1994) Coded modulation based on rings of integers modulo-q. Part 1: block codes. *IEE Proceedings Communications*, **141** (3), 129–36.

4

Algebraic–Geometric Codes

4.1 Introduction

Algebraic geometry is a powerful mathematical tool for constructing very long non-binary block codes with excellent parameters, such as high code rate and large Hamming distance. In 1981, Goppa showed how algebraic geometry could be used to construct non-binary block codes called Goppa codes or algebraic–geometric (AG) codes [1]. These codes are constructed from the affine points of an irreducible projective curve and a set of rational functions defined on that curve. The simplest type of AG code is the well-known Reed–Solomon code, which can be constructed from the affine points of a projective line. The length of an AG code is equal to the number of affine points, which in the case of a line can be no greater than the cardinality of the chosen finite field. Hence, constructing codes from a line will result in short code lengths, as is the case for Reed–Solomon codes. More affine points can be obtained by instead choosing a projective curve, which can result in much longer codes without increasing the size of the finite field. The most desirable curves are those that have the maximum possible number of affine points, known as maximal curves, since these produce the longest possible codes. However, constructing AG codes requires an in-depth knowledge of the theory of algebraic geometry, which is very difficult to understand, and this could be one of the reasons why AG codes have yet to be implemented.

In 1989, Justesen *et al.* [2] presented a construction method that only requires a basic understanding of algebraic geometry, but this method can still produce AG codes. This method has the disadvantage that less AG codes can be constructed than through more complicated algebraic–geometric approach, but many can still be produced. The notation used by Justesen for the construction of AG codes discussed in this chapter is also used in the hard-decision decoding algorithm explained later in this chapter.

In this chapter, the construction of AG codes is presented using the theory of algebraic geometry discussed in Chapter 2. This is followed by Justesen's simplified construction method, which is described in detail. Comparisons are made between

Non-Binary Error Control Coding for Wireless Communication and Data Storage   Rolando Antonio Carrasco and Martin Johnston
© 2008 John Wiley & Sons, Ltd

Reed–Solomon and Hermitian codes, illustrating the limitations of Reed–Solomon codes in terms of code length and the number of codes that can be constructed.

The next section presents the construction of systematic AG codes. Unfortunately, Hermitian codes are not cyclic like Reed–Solomon codes, and so it is not possible to use a multi-stage shift register to produce systematic AG codes. Traditionally, systematic block codes are created by performing Gauss–Jordan elimination on the nonsystematic generator matrix.

4.2 Construction of Algebraic–Geometric Codes

The construction of AG codes can be accomplished using the theory described in Chapter 2. Firstly, an irreducible affine smooth curve over a finite field must be chosen. Curves that can be used to construct good AG codes are Hermitian curves, elliptic curves, hyperelliptic curves and so on, and all have a single point at infinity. Next, all affine points and the point at infinity must be found, as explained in Chapter 2. To determine the message length k and designed minimum distance d^* of the code, the degree a of the divisor aQ must be chosen, where Q is the point at infinity, using [3]:

$$2\gamma - 2 < a < n, \tag{4.1}$$

where γ is the genus of the curve and n is the number of affine points (code length). The code parameters can then be calculated using (2.13) and (2.16) for a functional Goppa code or (2.17) and (2.18) for a residue Goppa code.

To construct the generator matrix of the code, a set of rational functions with pole order up to and including a at Q must be determined, but these functions must not have poles at any of the affine points. From Chapter 2 we know the set of rational functions

$$L(aQ) = \left\{ \frac{x^i y^j}{z^{i+j}} \right\}, 0 \le i < m, \quad j \ge 0$$

(where m is the degree of the curve and the pole order of each rational function is $4i + 5j$) satisfy this requirement for the Hermitian curve. There are k rational function in $L(aQ)$, which are evaluated at each of the n affine points to generate the k rows of the $(k \times n)$ generator matrix **G**. Now an example of AG code construction is given.

Example 4.1: Construction of the (8, 5, 3) Hermitian code using algebraic geometry: In this example, the Hermitian curve over GF(2^2) defined by (2.1) with $\gamma = 1$ is chosen to construct a residue Goppa code. The curve has eight affine points, shown in Table 2.12, and one point at infinity, $Q = (0, 1, 0)$. The degree a of the divisor aQ is chosen using (4.1), which can vary between zero and eight.

In this example we choose $a = 3$ and from (2.19) the message length is:

$$k = n - l(G) = n - d(G) - 1 + \gamma$$
$$= 8 - 3 - 1 + 1 = 5$$

and from (2.18) the designed minimum distance d^* is:

$$d^* = n - (n - d(G) - 1 + \gamma) + 1 - \gamma$$
$$= d(G) - 2\gamma + 2$$
$$= 3 - 2 + 2 = 3$$

So we have the $(8, 5, 3)$ Hermitian code. The $k = 5$ rational functions that have pole orders up to and including $a = 5$ are:

$$L(5Q) = \left\{1, \frac{x}{z}, \frac{y}{z}, \frac{x^2}{z^2}, \frac{xy}{z^2}\right\}.$$

Since all the affine points have $z = 1$, $L(5Q)$ can be written as:

$$L(5Q) = \left\{1, x, y, x^2, xy\right\}.$$

The code word is then obtained by evaluating a message polynomial at each of the eight affine points of the Hermitian curve. From Chapter 2, the eight affine points of the Hermitian curve over GF(4) are:

$$P_0 = (0, 0), \, P_1 = (0, 1), \, P_2 = (1, \alpha), \, P_3 = (1, \alpha^2), \, P_4 = (\alpha, \alpha),$$
$$P_5 = (\alpha, \alpha^2), \, P_6 = (\alpha^2, \alpha) \text{ and } P_7 = (\alpha^2, \alpha^2).$$

Let the message polynomial be $f(x, y) = 1 + y + x^2$. Then each coded symbol $c_i = f(P_i)$, $i = 1, 2, \ldots, 8$.

$$c_1 = f(P_1) = f(0, 0) = 1 + 1 + 1 = 1,$$
$$c_2 = f(P_2) = f(0, 1) = 1 + 1 + 0 = 0,$$
$$c_3 = f(P_3) = f(1, \alpha) = 1 + 1 + \alpha^2 = \alpha^2,$$
$$c_4 = f(P_4) = f(1, \alpha^2) = 1 + 1 + \alpha^4 = \alpha,$$
$$c_5 = f(P_5) = f(\alpha, \alpha) = 1 + \alpha + \alpha^2 = 0,$$
$$c_6 = f(P_6) = f(\alpha, \alpha^2) = 1 + \alpha^2 + \alpha^2 = 1,$$
$$c_7 = f(P_7) = f(\alpha^2, \alpha) = 1 + \alpha + \alpha^4 = 1,$$
$$c_8 = f(P_8) = f(\alpha^2, \alpha^2) = 1 + \alpha^2 + \alpha^4 = 0,$$

Hence the code word is $\mathbf{c} = 1, 0, \alpha^2, \alpha, 0, 1, 1, 0$.

This is also equivalent to multiplying a message vector $\mathbf{m}$ by a generator matrix $\mathbf{G}$. To construct the rows of the generator matrix of an AG code, each monomial

ϕ_i, $i = 0, 1, \ldots, k$, in $L(5Q)$ is evaluated at each affine point, as shown in (4.2).

$$G = \begin{bmatrix} \phi_1(P_1) & \phi_1(P_2) & \phi_1(P_3) & \cdots & \phi_1(P_n) \\ \phi_2(P_1) & \phi_2(P_2) & \phi_2(P_3) & \cdots & \phi_2(P_n) \\ \phi_3(P_1) & \phi_3(P_2) & \phi_3(P_3) & \cdots & \phi_3(P_n) \\ \vdots & \vdots & \vdots & \ddots & \vdots \\ \phi_k(P_1) & \phi_k(P_2) & \phi_k(P_3) & \cdots & \phi_k(P_n) \end{bmatrix}. \tag{4.2}$$

If the first monomial is written as $\phi_1 = x^0 y^0$ then evaluating it at each point will give a row of all 1s due to powers of 0 in the x and y term. The remaining monomials are evaluated at each of the eight points using Table 2.12.

The second row of the generator matrix is obtained by evaluating the monomial $f(x, y) = x$ at all eight affine points. In this case each of the eight elements of this row is just the x-coordinate of one of the eight affine points.

$$f(0, 0) = 0, \ f(0, 1) = 0, \ f(1, \alpha) = 1, \ f(1, \alpha^2) = 1, \ f(\alpha, \alpha) = \alpha,$$
$$f(\alpha, \alpha^2) = \alpha, \ f(\alpha^2, \alpha) = \alpha^2, \ f(\alpha^2, \alpha^2) = \alpha^2.$$

The third row of the generator matrix is obtained by evaluating the monomial $f(x, y) = y$ at all eight affine points. In this case each of the eight elements of this row is just the y-coordinate of one of the eight affine points.

$$f(0, 0) = 0, \ f(0, 1) = 1, \ f(1, \alpha) = \alpha, \ f(1, \alpha^2) = \alpha^2, \ f(\alpha, \alpha) = \alpha,$$
$$f(\alpha, \alpha^2) = \alpha^2, \ f(\alpha^2, \alpha) = \alpha, \ f(\alpha^2, \alpha^2) = \alpha^2.$$

The fourth row of the generator matrix is obtained by evaluating the monomial $f(x, y) = x^2$ at all eight affine points. In this case each of the eight elements of this row is the square of the x-coordinate of one of the eight affine points.

$$f(0, 0) = 0, \ f(0, 1) = 0, \ f(1, \alpha) = 1, \ f(1, \alpha^2) = 1, \ f(\alpha, \alpha) = \alpha^2,$$
$$f(\alpha, \alpha^2) = \alpha^2, \ f(\alpha^2, \alpha) = \alpha, \ f(\alpha^2, \alpha^2) = \alpha.$$

The fifth row of the generator matrix is obtained by evaluating the monomial $f(x, y) = xy$ at all eight affine points. In this case each of the eight elements of this row is the product of the x-coordinate and y-coordinate of one of the eight affine points.

$$f(0, 0) = 0, \ f(0, 1) = 0, \ f(1, \alpha) = \alpha, \ f(1, \alpha^2) = \alpha^2, \ f(\alpha, \alpha) = \alpha^2,$$
$$f(\alpha, \alpha^2) = 1, \ f(\alpha^2, \alpha) = 1, \ f(\alpha^2, \alpha^2) = \alpha.$$

Therefore, the generator matrix is:

$$G = \begin{bmatrix} 1 & 1 & 1 & 1 & 1 & 1 & 1 & 1 \\ 0 & 0 & 1 & 1 & \alpha & \alpha & \alpha^2 & \alpha^2 \\ 0 & 1 & \alpha & \alpha^2 & \alpha & \alpha^2 & \alpha & \alpha^2 \\ 0 & 0 & 1 & 1 & \alpha^2 & \alpha^2 & \alpha & \alpha \\ 0 & 0 & \alpha & \alpha^2 & \alpha^2 & 1 & 1 & \alpha \end{bmatrix}.$$

The construction method for one-point AG codes is summarized below [3]:

1. Select an irreducible affine smooth curve and a finite field.
2. Find all the affine points of the form $(\alpha, \beta, 1)$ and point at infinity $Q = (\alpha, \beta, 0)$, where α and β are elements in the finite field, that make the curve vanish.
3. Choose the degree a of the divisor aQ using (4.1) and calculate the message length k and designed minimum distance d^* of the code using (2.13) and (3.16) for a functional Goppa code or (2.17) and (2.18) for a residue Goppa code.
4. Determine a set of rational functions $L(aQ)$ that have pole orders up to and including a at the point at infinity Q, but do not have poles at any of the affine points.
5. Evaluate each rational function in $L(aQ)$ at each affine point to obtain the nonsystematic generator matrix $\mathbf{G}$ of the AG code.

4.2.1 Simplified Construction of AG Codes

In this section, a simplified construction method attributed to Justesen [2] is described, which only requires a basic knowledge of algebraic geometry to construct an AG code. The first difference of this method is that projective curves are not considered. Only affine curves in the $(x - y)$ coordinate system are used and the point at infinity is excluded. Another difference is the use of a set of monomials in two dimensions instead of a set of rational functions in three dimensions to obtain the generator matrix.

To construct an algebraic–geometric code using Justesen's construction, a nonnegative integer j is first chosen that is bounded by [2]:

$$m - 2 \leq j \leq \left\lfloor \frac{n-1}{m} \right\rfloor. \tag{4.3}$$

(4.3) is derived from (4.1). In the simplified construction the degree of the divisor G is a multiple of the degree of the curve $C(x, y)$, that is $\deg(G) = mj$, where j is a nonnegative integer. Substituting $\deg(G) = mj$ and (2.12) into (4.1) gives:

$$2\frac{(m - 1)(m - 2)}{2} - 2 < mj < n$$

$$\Rightarrow m^2 - 3m + 2 - 2 < mj < n$$

$$\Rightarrow \frac{m(m - 3)}{m} < j < \frac{n}{m}$$

$$\therefore m - 2 \leq j \leq \left\lfloor \frac{n-1}{m} \right\rfloor$$

The codes obtained using Justesen's simplified construction are residue Goppa codes. Therefore, since $\deg(G) = mj$, the code parameters from (2.17) and (2.18) can be written as:

$$k = n - mj + \gamma - 1$$
$$d^* = mj - 2\gamma + 2 \qquad . \tag{4.4}$$

However, since $\deg(G)$ is limited to being a multiple of the degree of the curve, less AG codes can be constructed than with the algebraic–geometric construction. This is the only disadvantage of the simplified method, but many AG codes can still be constructed using it.

A monomial basis is defined as:

$$f = \{x^i y^j\}, \quad 0 \leq i < m, \qquad b \geq 0, \tag{4.5}$$

which contains k monomials. Hence, as before, the monomials are evaluated at each point to obtain the generator matrix of the code.

Example 4.2: Construction of the (8, 5, 3) Hermitian code over GF(2^2) using the simplified method: In this example, the same Hermitian code is constructed as in the previous example, but using Justesen's simplified construction method. Again, the Hermitian curve defined in (2.1) is used. From the previous example, we know that the Hermitian code over GF(2^2) has $n = 8$ points, but these are not projective points and so there is no z component. The points are given in Table 4.1.

Table 4.1 Points of the Hermitian curve over GF(2^2).

$P_1 = (0, 0)$	$P_2 = (0, 1)$	$P_3 = (1, \alpha)$	$P_4 = (1, \alpha^2)$
$P_5 = (\alpha, \alpha)$	$P_6 = (\alpha, \alpha^2)$	$P_7 = (\alpha^2, \alpha)$	$P_8 = (\alpha^2, \alpha^2)$

The degree of the Hermitian curve is $m = 3$, so from (4.3) the parameter j can have values of:

$$3 - 2 \leq j \leq \left\lfloor \tfrac{8-1}{3} \right\rfloor$$
$$= 1 \leq j \leq 2 \qquad .$$

The genus of the curve is $\gamma = 1$, and taking $j = 1$ the parameters of the code from (4.4) are:

$$k = 8 - 3 \times 1 + 1 - 1 = 5$$
$$d^* = 3 \times 1 - 2 \times 1 + 2 = 3 \qquad .$$

Since $k = 5$ there will be five monomials in the monomial basis. These are $\{1, x, y, x^2, xy\}$. To construct the generator matrix **G**, each monomial is evaluated at each of the eight points, and thus we obtain the same generator matrix as in the previous example. This example shows that, using Justesen's construction method, all that is required is the degree and genus of the curve and the points on the curve. It does not require an in-depth knowledge of algebraic geometry to construct AG codes.

Table 4.2   Comparison of Reed–Solomon and Hermitian code lengths for different finite fields (figures in parentheses are for doubly-extended Reed–Solomon codes).

Finite field size q	Reed–Solomon code length	Hermitian code length
4	3 (5)	8
16	15 (17)	64
64	63 (65)	512
256	255 (257)	4096

4.2.2 Comparison of AG Codes with Reed–Solomon Codes

Reed–Solomon codes are the most commonly used codes due to their efficient decoding algorithms and good performance over channels with burst errors. However, they do have limitations, such as short code length and the fact that there are not many of them. Table 4.2 shows how the lengths of Reed–Solomon codes and Hermitian codes vary for increasing finite field size.

It can be seen that the length of the Hermitian codes increases dramatically for increasing finite fields since for a finite field size of q the length $n = q^{3/2}$. However, the length of the Reed–Solomon code is restricted to the size of the finite field and has length $n = q - 1$, or $q + 1$ if doubly extended.

The number of possible AG codes for a given finite field can be found from (4.1). If $2\gamma - 2 < \deg(G) < n$ then $2\gamma - 1 \leq \deg(G) \leq n - 1$. This means that there are $(n - 1) - (2\gamma - 1) = n - 2\gamma$ possible AG codes for a given finite field. As stated previously, a Reed–Solomon code is constructed from an affine straight line which has a genus $\gamma = 0$. This implies there are n possible Reed–Solomon codes for a given finite field. A comparison of the numbers of Reed–Solomon and Hermitian codes for increasing finite field size is given in Figure 4.1.

It can be seen that the number of possible Hermitian codes increases exponentially with increasing field size, with a possible 3855 Hermitian codes defined over GF(2^8). In contrast, the number of Reed–Solomon codes is again restricted by the size of the finite field.

4.2.3 Systematic Algebraic–Geometric Codes

Evaluating the elements in the monomial basis at each point on the curve produces a generator matrix in nonsystematic form. Traditionally, systematic codes are achieved by applying Gauss–Jordan elimination [4] to the nonsystematic generator matrix. However, this approach had not been used to construct systematic AG codes since for large generator matrices the complexity could be too high. In the literature, only Heegard *et al.* [5] have considered the construction of systematic AG codes. They show how systematic encoding of algebraic–geometric codes can be achieved in a method analogous to the systematic encoding of Reed–Solomon codes by using the cyclic properties of the automorphisms of a curve. This method requires fewer operations

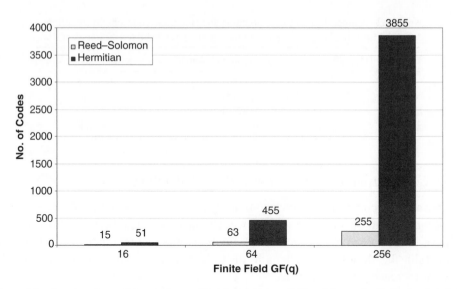

Figure 4.1   A comparison of the numbers of Reed–Solomon and Hermitian codes for increasing finite field size.

than applying Gauss–Jordan elimination on the nonsystematic generator matrix, but it necessitates a good knowledge of algebraic geometry. However, since their paper was written, computers have become much more powerful and it has been found that Gauss–Jordan elimination can be performed quickly even on large generator matrices from Hermitian codes defined over $GF(2^6)$. Gauss–Jordan elimination is made up of two algorithms: Gaussian elimination and Jordan elimination. Gaussian elimination converts the matrix into echelon form and Jordan elimination then ensures the matrix is in row-reduced echelon form. For the case of a nonsquare matrix, such as a $k \times n$ generator matrix, a $k \times k$ submatrix of the generator matrix in row-reduced echelon form is the $k \times k$ identity matrix $\mathbf{I_k}$. The remainder of the generator matrix is a $k \times (n - k)$ parity matrix $\mathbf{P}$. The generator matrix is now in systematic form, $\mathbf{G} = [\mathbf{I_k} \,|\, \mathbf{P}]$. The algorithms for Gaussian elimination and Jordan elimination are described below.

4.2.3.1 Gaussian Elimination

The Gaussian elimination algorithm uses the following notation: $\mathbf{G_j}$ is used to denote all the elements in row j of $\mathbf{G}$, $G_{i,j}$ denotes a single element located at (i, j) in $\mathbf{G}$, and k is the message length of the AG code. When Gaussian elimination is used on a matrix containing real numbers, elimination of a column occurs by multiplying previous rows by a factor and subtracting them from the row that is being operated on. In the case of a matrix containing finite field elements, Gaussian elimination is performed in the same way, using modulo-q addition instead of subtraction. Modulo-q addition can be used if the row to be subtracted is multiplied by the additional factor $q - 1$.

Initialize: $i = 0, j = 0$
1. Check that column i does not contain all zeros. If it does then $i = i + 1$ and go to step 1. Else go to step 2.
2. Find the row p with largest first element.
3. Swap row p with largest first element with row j.
4. Eliminate elements in column i from row $j + 1$ to row $k - 1$.
   For $row = j + 1$ to $k - 1$:

$$\mathbf{G}_{row} \rightarrow \mathbf{G}_{row} + (q - 1)\left(\frac{G_{i,row}}{G_{i,j}}\mathbf{G}_j\right).$$

5. $i = i + 1$ and $j = j + 1$. If $j = k$ go to step 6. Else go to step 2.
6. For $row = 0$ to $k - 1$:
   Divide $\mathbf{G}_{row}$ by the first nonzero element in $\mathbf{G}_{row}$ so that the main diagonal of $\mathbf{G}$ contains all 1s.

4.2.3.2 Jordan Elimination

1. Initialize: $i = 0, j = 0$.
2. If $G_{i,j} = 0$, swap column i with the nearest column that has a 1 in row j.
3. Eliminate elements in column i from row $j - 1$ to row 0.
   For $row = j - 1$ to 0:

$$\mathbf{G}_{row} \rightarrow \mathbf{G}_{row} + (q - 1)\left(\frac{G_{i,row}}{G_{i,j}}G_j\right).$$

4. $i = i + 1$ and $j = j + 1$. If $j = k$ STOP. Else go to step 2.

Example 4.3: Construction of the systematic (27, 13, 12) Hermitian code over GF(3^2): Using the primitive polynomial $x^2 = 2x + 1$, the finite field GF(3^2) is shown in Table 2.10. The Hermitian code defined over GF(3^2) is $C(x, y) = x^4 + y^3 + y$ and has 27 points that satisfy $C(x, y) = 0$, given in Table 4.3.

Table 4.3   The 27 points of the Hermitian curve over GF(3^2).

$P_1 = (0, 0)$	$P_2 = (0, \alpha^6)$	$P_3 = (0, \alpha^2)$	$P_4 = (1, 1)$	$P_5 = (1, \alpha)$	$P_6 = (1, \alpha^3)$
$P_7 = (\alpha^4, 1)$	$P_8 = (\alpha^4, \alpha)$	$P_9 = (\alpha^4, \alpha^3)$	$P_{10} = (\alpha, \alpha^4)$	$P_{11} = (\alpha, \alpha^7)$	$P_{12} = (\alpha, \alpha^5)$
$P_{13} = (\alpha^7, \alpha^4)$	$P_{14} = (\alpha^7, \alpha^7)$	$P_{15} = (\alpha^7, \alpha^5)$	$P_{16} = (\alpha^6, 1)$	$P_{17} = (\alpha^6, \alpha)$	$P_{18} = (\alpha^6, \alpha^3)$
$P_{19} = (\alpha^5, \alpha^4)$	$P_{20} = (\alpha^5, \alpha^7)$	$P_{21} = (\alpha^5, \alpha^5)$	$P_{22} = (\alpha^2, 1)$	$P_{23} = (\alpha^2, \alpha)$	$P_{24} = (\alpha^2, \alpha^3)$
$P_{25} = (\alpha^3, \alpha^4)$	$P_{26} = (\alpha^3, \alpha^7)$	$P_{27} = (\alpha^3, \alpha^5)$			

The genus of the curve from (2.12) is:

$$\gamma = \frac{(4-1)(4-2)}{2} = 3.$$

From (4.3), the parameter j can vary between 2 and 6. Taking $j = 4$, we get a message length $k = 13$ and designed minimum distance $d^* = 12$. The monomial basis from (4.5) therefore has the 13 elements $\{1, x, y, x^2, xy, y^2, x^3, x^2y, xy^2, y^3, x^3y, x^2y^2, xy^3\}$ and the nonsystematic generator matrix is:

$\mathbf{G} =$

$$
\begin{bmatrix}
1 & 1 & 1 & 1 & 1 & 1 & 1 & 1 & 1 & 1 & 1 & 1 & 1 & 1 & 1 & 1 & 1 & 1 & 1 & 1 & 1 & 1 & 1 & 1 & 1 & 1 & 1 \\
0 & 0 & 0 & 1 & 1 & 1 & \alpha^4 & \alpha^4 & \alpha^4 & \alpha & \alpha & \alpha & \alpha^7 & \alpha^7 & \alpha^7 & \alpha^6 & \alpha^6 & \alpha^6 & \alpha^5 & \alpha^5 & \alpha^5 & \alpha^2 & \alpha^2 & \alpha^2 & \alpha^3 & \alpha^3 & \alpha^3 \\
0 & \alpha^6 & \alpha^2 & 1 & \alpha & \alpha^3 & 1 & \alpha & \alpha^3 & \alpha^4 & \alpha^7 & \alpha^5 & \alpha^4 & \alpha^7 & \alpha^5 & 1 & \alpha & \alpha^3 & \alpha^4 & \alpha^7 & \alpha^5 & 1 & \alpha & \alpha^3 & \alpha^4 & \alpha^7 & \alpha^5 \\
0 & 0 & 0 & 1 & 1 & 1 & 1 & 1 & 1 & \alpha^2 & \alpha^2 & \alpha^2 & \alpha^6 & \alpha^6 & \alpha^6 & \alpha^4 & \alpha^4 & \alpha^4 & \alpha^2 & \alpha^2 & \alpha^2 & \alpha^4 & \alpha^4 & \alpha^4 & \alpha^6 & \alpha^6 & \alpha^6 \\
0 & 0 & 0 & 1 & \alpha & \alpha^3 & \alpha^4 & \alpha^5 & \alpha^7 & \alpha^5 & 1 & \alpha^6 & \alpha^3 & \alpha^6 & \alpha^4 & \alpha^6 & \alpha^7 & \alpha & \alpha & \alpha^4 & \alpha^2 & \alpha^2 & \alpha^3 & \alpha^5 & \alpha^7 & \alpha^2 & 1 \\
0 & \alpha^4 & \alpha^4 & 1 & \alpha^2 & \alpha^6 & 1 & \alpha^2 & \alpha^6 & 1 & \alpha^6 & \alpha^2 & 1 & \alpha^6 & \alpha^2 & 1 & \alpha^2 & \alpha^6 & 1 & \alpha^6 & \alpha^2 & 1 & \alpha^2 & \alpha^6 & 1 & \alpha^6 & \alpha^2 \\
0 & 0 & 0 & 1 & 1 & 1 & \alpha^4 & \alpha^4 & \alpha^4 & \alpha^3 & \alpha^3 & \alpha^3 & \alpha^5 & \alpha^5 & \alpha^5 & \alpha^2 & \alpha^2 & \alpha^2 & \alpha^7 & \alpha^7 & \alpha^7 & \alpha^6 & \alpha^6 & \alpha^6 & \alpha & \alpha & \alpha \\
0 & 0 & 0 & 1 & \alpha & \alpha^3 & 1 & \alpha & \alpha^3 & \alpha^6 & \alpha & \alpha^7 & \alpha^2 & \alpha^5 & \alpha^3 & \alpha^4 & \alpha^5 & \alpha^7 & \alpha^6 & \alpha & \alpha^7 & \alpha^4 & \alpha^5 & \alpha^7 & \alpha^2 & \alpha^5 & \alpha^3 \\
0 & 0 & 0 & 1 & \alpha^2 & \alpha^6 & \alpha^4 & \alpha^6 & \alpha^2 & \alpha & \alpha^7 & \alpha^3 & \alpha^7 & \alpha^5 & \alpha & \alpha^6 & 1 & \alpha^4 & \alpha^5 & \alpha^3 & \alpha^7 & \alpha^2 & \alpha^4 & 1 & \alpha^3 & \alpha & \alpha^5 \\
0 & \alpha^2 & \alpha^6 & 1 & \alpha^3 & \alpha & 1 & \alpha^3 & \alpha & \alpha^4 & \alpha^5 & \alpha^7 & \alpha^4 & \alpha^5 & \alpha^7 & 1 & \alpha^3 & \alpha & \alpha^4 & \alpha^5 & \alpha^7 & 1 & \alpha^3 & \alpha & \alpha^4 & \alpha^5 & \alpha^7 \\
0 & 0 & 0 & 1 & \alpha & \alpha^3 & \alpha^4 & \alpha^5 & \alpha^7 & \alpha^7 & \alpha^2 & 1 & \alpha & \alpha^4 & \alpha^2 & \alpha^2 & \alpha^3 & \alpha^5 & \alpha^3 & \alpha^6 & \alpha^4 & \alpha^6 & \alpha^7 & \alpha & \alpha^5 & 1 & \alpha^6 \\
0 & 0 & 0 & 1 & \alpha^2 & \alpha^6 & 1 & \alpha^2 & \alpha^6 & \alpha^2 & 1 & \alpha^4 & \alpha^6 & \alpha^4 & 1 & \alpha^4 & \alpha^6 & \alpha^2 & \alpha^2 & 1 & \alpha^4 & \alpha^4 & \alpha^6 & \alpha^2 & \alpha^6 & \alpha^4 & 1 \\
0 & 0 & 0 & 1 & \alpha^3 & \alpha & \alpha^4 & \alpha^7 & \alpha^5 & \alpha^5 & \alpha^6 & 1 & \alpha^3 & \alpha^4 & \alpha^6 & \alpha^6 & \alpha & \alpha^7 & \alpha & \alpha^2 & \alpha^4 & \alpha^2 & \alpha^5 & \alpha^3 & \alpha^7 & 1 & \alpha^2 \\
\end{bmatrix}.
$$

Applying Gaussian elimination to $\mathbf{G}$ gives:

$\mathbf{G} =$

$$
\begin{bmatrix}
1 & 1 & 1 & 1 & 1 & 1 & 1 & 1 & 1 & 1 & 1 & 1 & 1 & 1 & 1 & 1 & 1 & 1 & 1 & 1 & 1 & 1 & 1 & 1 & 1 & 1 & 1 \\
0 & 1 & \alpha^4 & \alpha^6 & \alpha & \alpha^7 & \alpha^6 & \alpha & \alpha^7 & \alpha^2 & \alpha^3 & \alpha^5 & \alpha^2 & \alpha^3 & \alpha^5 & \alpha^6 & \alpha & \alpha^7 & \alpha^2 & \alpha^3 & \alpha^5 & \alpha^6 & \alpha & \alpha^7 & \alpha^4 & \alpha^3 & \alpha^5 \\
0 & 0 & 1 & \alpha & 1 & \alpha^5 & \alpha & 1 & \alpha^5 & \alpha^3 & 1 & \alpha^7 & \alpha^3 & 1 & \alpha^7 & \alpha & 1 & \alpha^5 & \alpha^3 & 1 & \alpha^7 & \alpha & 1 & \alpha^5 & \alpha^3 & 1 & \alpha^7 \\
0 & 0 & 0 & 1 & 1 & 1 & 1 & 1 & 1 & \alpha^4 & \alpha^4 & \alpha^4 & \alpha^4 & \alpha^4 & \alpha^4 & 1 & 1 & 1 & \alpha^4 & \alpha^4 & \alpha^4 & 1 & 1 & 1 & \alpha^4 & \alpha^4 & \alpha^4 \\
0 & 0 & 0 & 0 & 1 & \alpha^4 & \alpha^6 & \alpha^7 & \alpha & 1 & \alpha^7 & \alpha^2 & \alpha^3 & 0 & \alpha^7 & \alpha^5 & \alpha^4 & \alpha^7 & \alpha^5 & \alpha & 0 & \alpha^3 & \alpha & 1 & \alpha^4 & \alpha^2 & \alpha \\
0 & 0 & 0 & 0 & 0 & 1 & \alpha^6 & \alpha^6 & \alpha^7 & \alpha^2 & \alpha & \alpha^6 & \alpha^7 & \alpha^2 & \alpha^6 & \alpha^5 & \alpha^5 & \alpha^4 & 0 & \alpha^7 & \alpha^6 & \alpha^3 & \alpha^3 & \alpha & \alpha & 0 & \alpha^6 \\
0 & 0 & 0 & 0 & 0 & 0 & 1 & 1 & 1 & 0 & 1 & \alpha^4 & \alpha^4 & 0 & 1 & \alpha^4 & 0 & 1 & \alpha^4 & 1 & 0 & \alpha^4 & 1 & 0 & 0 & \alpha^4 & 1 \\
0 & 0 & 0 & 0 & 0 & 0 & 0 & 1 & \alpha^4 & \alpha^6 & \alpha^3 & \alpha^4 & \alpha^7 & 1 & \alpha^2 & \alpha^7 & \alpha^6 & \alpha & \alpha & \alpha^2 & \alpha^4 & \alpha & \alpha^7 & \alpha^6 & \alpha^6 & 1 & \alpha^5 \\
0 & 0 & 0 & 0 & 0 & 0 & 0 & 0 & 1 & 0 & \alpha^7 & \alpha^7 & 1 & 1 & 0 & \alpha^6 & 0 & 1 & \alpha^2 & \alpha^6 & \alpha^4 & \alpha^7 & \alpha^5 & \alpha^2 & \alpha^2 & \alpha^3 & 0 \\
0 & 0 & 0 & 0 & 0 & 0 & 0 & 0 & 0 & 1 & 1 & 1 & \alpha^6 & \alpha^6 & \alpha^6 & \alpha^5 & \alpha^5 & \alpha^5 & 1 & 1 & 1 & \alpha^5 & \alpha^5 & \alpha^5 & \alpha^6 & \alpha^6 & \alpha^6 \\
0 & 0 & 0 & 0 & 0 & 0 & 0 & 0 & 0 & 1 & \alpha^4 & 0 & 1 & \alpha^4 & 0 & 1 & \alpha^4 & \alpha & \alpha^6 & \alpha^7 & \alpha^6 & \alpha^7 & \alpha & \alpha^7 & \alpha & \alpha^6 \\
0 & 0 & 0 & 0 & 0 & 0 & 0 & 0 & 0 & 0 & 1 & 1 & 1 & \alpha^4 & \alpha^4 & \alpha^4 & \alpha^5 & \alpha^5 & \alpha^5 & \alpha^3 & \alpha^3 & \alpha^3 & \alpha^2 & \alpha^2 & \alpha^2 \\
0 & 0 & 0 & 0 & 0 & 0 & 0 & 0 & 0 & 0 & 0 & 0 & 0 & 0 & 1 & 1 & 1 & \alpha^7 & \alpha^7 & \alpha^7 & 1 & 1 & 1 & \alpha^5 & \alpha^5 & \alpha^5 \\
\end{bmatrix}.
$$

The swapping of rows also swaps the monomials but this does not affect the code. After Gaussian elimination we can see that the main diagonal of 1s does not continue at the twelfth column. Therefore, column 12 is swapped with column 13. Next, column 16 is swapped with column 13 so that the main diagonal is complete.

After applying Jordan elimination to **G**:

$$\mathbf{G} =$$

$$
\left[
\begin{array}{ccccccccccccc|ccccccccccccccc}
1 & 0 & 0 & 0 & 0 & 0 & 0 & 0 & 0 & 0 & 0 & 0 & 0 & \alpha^5 & \alpha^4 & \alpha & \alpha^6 & \alpha^5 & 0 & \alpha^4 & \alpha^7 & \alpha^5 & 1 & \alpha^7 & \alpha^7 & 0 & \alpha^7 \\
0 & 1 & 0 & 0 & 0 & 0 & 0 & 0 & 0 & 0 & 0 & 0 & 0 & \alpha^7 & \alpha^7 & \alpha & \alpha^5 & \alpha^6 & \alpha^3 & 1 & \alpha^5 & \alpha^2 & \alpha^5 & \alpha^2 & \alpha^3 & \alpha^6 & \alpha^2 \\
0 & 0 & 1 & 0 & 0 & 0 & 0 & 0 & 0 & 0 & 0 & 0 & 0 & \alpha^4 & \alpha^5 & \alpha & 1 & 1 & 1 & \alpha^5 & \alpha^3 & \alpha^5 & \alpha & \alpha^4 & \alpha^7 & \alpha^4 & \alpha^6 \\
0 & 0 & 0 & 1 & 0 & 0 & 0 & 0 & 0 & 0 & 0 & 0 & 0 & \alpha^4 & \alpha^7 & \alpha^4 & 1 & \alpha^2 & \alpha^3 & 0 & 0 & \alpha^7 & \alpha^7 & \alpha^2 & \alpha^4 & \alpha^3 & \alpha^7 \\
0 & 0 & 0 & 0 & 1 & 0 & 0 & 0 & 0 & 0 & 0 & 0 & 0 & \alpha^2 & \alpha^3 & \alpha^4 & \alpha^7 & \alpha^7 & \alpha^3 & \alpha^6 & \alpha^2 & 0 & \alpha^2 & \alpha^3 & \alpha^7 & 0 & \alpha^2 \\
0 & 0 & 0 & 0 & 0 & 1 & 0 & 0 & 0 & 0 & 0 & 0 & 0 & \alpha & 0 & \alpha^4 & \alpha^2 & 1 & 0 & \alpha^4 & \alpha^5 & \alpha^6 & \alpha^2 & \alpha & \alpha^2 & \alpha^6 & \alpha^4 \\
0 & 0 & 0 & 0 & 0 & 0 & 1 & 0 & 0 & 0 & 0 & 0 & 0 & \alpha^6 & \alpha^5 & \alpha^3 & \alpha^3 & \alpha^2 & \alpha^5 & 1 & \alpha^7 & 1 & \alpha & \alpha^7 & \alpha^7 & \alpha^2 & \alpha^7 \\
0 & 0 & 0 & 0 & 0 & 0 & 0 & 1 & 0 & 0 & 0 & 0 & 0 & \alpha^4 & 0 & \alpha^3 & 1 & \alpha^6 & \alpha^6 & \alpha^4 & \alpha^2 & \alpha^2 & \alpha^3 & \alpha^6 & 0 & \alpha^7 & \alpha^4 \\
0 & 0 & 0 & 0 & 0 & 0 & 0 & 0 & 1 & 0 & 0 & 0 & 0 & \alpha^3 & \alpha & \alpha^3 & \alpha & 0 & \alpha^4 & 1 & \alpha^4 & \alpha^3 & \alpha^6 & \alpha^2 & \alpha^6 & \alpha^2 & \alpha \\
0 & 0 & 0 & 0 & 0 & 0 & 0 & 0 & 0 & 1 & 0 & 0 & 0 & \alpha^4 & 1 & \alpha^4 & \alpha^4 & 1 & \alpha & \alpha^7 & \alpha^6 & \alpha^4 & 0 & 1 & \alpha^5 & \alpha^2 & \alpha^3 \\
0 & 0 & 0 & 0 & 0 & 0 & 0 & 0 & 0 & 0 & 1 & 0 & 0 & 1 & \alpha^4 & \alpha^4 & 1 & \alpha^4 & \alpha & \alpha^6 & \alpha^7 & \alpha^6 & \alpha^7 & \alpha & \alpha^7 & \alpha & \alpha^6 \\
0 & 0 & 0 & 0 & 0 & 0 & 0 & 0 & 0 & 0 & 0 & 1 & 0 & 1 & 1 & 0 & 0 & 0 & 1 & 1 & 1 & \alpha^5 & \alpha^5 & \alpha^5 & \alpha^7 & \alpha^7 & \alpha^7 \\
0 & 0 & 0 & 0 & 0 & 0 & 0 & 0 & 0 & 0 & 0 & 0 & 1 & 0 & 0 & 0 & 1 & 1 & \alpha^7 & \alpha^7 & \alpha^7 & 1 & 1 & 1 & \alpha^5 & \alpha^5 & \alpha^5
\end{array}
\right].
$$

The swapping of columns means that the points corresponding to those columns are also swapped. For example, when column 12 was swapped with column 13, $P_{12} = (\alpha, \alpha^5)$ was swapped with $P_{13} = (\alpha^7, \alpha^4)$. The reordered points are shown in Table 4.4.

Table 4.4　The reordered points from Table 4.3 after swapping columns.

$P_1 = (0,0)$	$P_2 = (0, \alpha^6)$	$P_3 = (0, \alpha^2)$	$P_4 = (1, 1)$	$P_5 = (1, \alpha)$	$P_6 = (1, \alpha^3)$
$P_7 = (\alpha^4, 1)$	$P_8 = (\alpha^4, \alpha)$	$P_9 = (\alpha^4, \alpha^3)$	$P_{10} = (\alpha, \alpha^4)$	$P_{11} = (\alpha, \alpha^7)$	$P_{12} = (\alpha^7, \alpha^4)$
$P_{13} = (\alpha^6, 1)$	$P_{14} = (\alpha^7, \alpha^7)$	$P_{15} = (\alpha^7, \alpha^5)$	$P_{16} = (\alpha, \alpha^5)$	$P_{17} = (\alpha^6, \alpha)$	$P_{18} = (\alpha^6, \alpha^3)$
$P_{19} = (\alpha^5, \alpha^4)$	$P_{20} = (\alpha^5, \alpha^7)$	$P_{21} = (\alpha^5, \alpha^5)$	$P_{22} = (\alpha^2, 1)$	$P_{23} = (\alpha^2, \alpha)$	$P_{24} = (\alpha^2, \alpha^3)$
$P_{25} = (\alpha^3, \alpha^4)$	$P_{26} = (\alpha^3, \alpha^7)$	$P_{27} = (\alpha^3, \alpha^5)$			

4.3 Decoding Algebraic–Geometric Codes

The decoding of algebraic–geometric codes can be achieved in two parts: first by determining the error locations, and then by determining the error magnitudes at these locations, which is similar to the decoding of Reed–Solomon codes in Chapter 3.

Sakata's algorithm was developed in 1988 [6] as a method of generating a set of polynomials whose coefficients formed recursive relationships among an array of finite field elements. This was an extension of the Berlekamp–Massey algorithm to two or more dimensions. In 1992, Justesen *et al.* [7] used Sakata's algorithm to generate a set of error-locating polynomials from a matrix containing syndrome values for algebraic–geometric codes in order to reduce decoding complexity. The syndromes

Table 4.5   Variables used in the Sakata algorithm.

Variable	Definition
(a, b)	A point in the syndrome array
$S_{a,b}$	A syndrome in the array at the point (a, b)
F	The set containing the error-locating polynomials
$f^{(i)}(x, y)$	The ith polynomial in the set F
G	The auxiliary set containing some of the polynomials that were in F earlier in the algorithm
$g^{(i)}(x, y)$	The ith polynomial in the set G
$h^{(i)}(x, y)$	The ith polynomial after modification
(a_g, b_g)	The point at which $g^{(i)}(x, y)$ was first placed in the set G
$d_f (d_g)$	The discrepancy in $f^{(i)}(x, y)$ $(g^{(i)}(x, y))$
$(t_1^{(i)}, t_2^{(i)})$	The powers of x and y respectively in the leading term of $f^{(i)}(x, y)$
$(u_1^{(i)}, u_2^{(i)})$	The powers of x and y respectively in the leading term of $g^{(i)}(x, y)$

in the two-dimensional syndrome array are defined as:

$$S_{ab} = \sum_{i=1}^{n} r_i x_i^a y_i^b = \sum_{i=1}^{n} (c_i + e_i) x_i^a y_i^b = \sum_{i=1}^{n} e_i x_i^a y_i^b, \tag{4.6}$$

where (a, b) is the location of the syndrome $S_{a,b}$ in the array, r_i is the ith received element, c_i is the ith coded symbol, e_i is the error magnitude in the ith position and (x_i, y_i) is the ith affine point.

The majority voting scheme [8] (explained later) was added to [7] by Sakata *et al.* [9] in order to increase the number of errors that could be corrected. The decoding algorithm in [9] has many variables, so, before it is explained, a list of definitions is provided in Table 4.5. Sakata's algorithm takes a two-dimensional array of syndromes calculated from the received word $\mathbf{r} = \{r_1, r_2, \ldots, r_n\}$, as given in (4.6). From this array a set F of error-locating polynomials is generated, of the form:

$$f^{(i)}(x, y) = \sum f_{k,l}^{(i)} x^k y^l, \tag{4.7}$$

where i is the ith polynomial in F and $f_{k,l}^{(i)}$ are the coefficients of the terms $x^k y^l$ in $f^{(i)}(x, y)$. Each syndrome from the array is read in and the polynomials in F are modified so that the coefficients in each polynomial $f^{(i)}(x, y)$ form recursive equations among known syndromes, up to the current syndrome.

The recursive relationship that must be satisfied is [9]:

$$\sum f_{k,l}^{(i)} S_{a-t_1^{(i)}+k, b-t_2^{(i)}+l} = 0, \tag{4.8}$$

where $t_1^{(i)}$ and $t_2^{(i)}$ are the powers of x and y, respectively, of the leading term of $f^{(i)}(x, y)$. A generalized syndrome array is shown in Figure 4.2. The syndromes are expressed in terms of the parameter j defined by (4.3). The known syndromes from

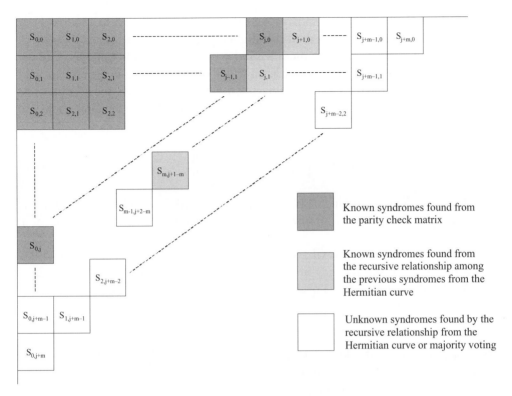

Figure 4.2   The generalized initial syndrome array.

$S_{0,0}$ to $S_{0,j}$ are calculated from (4.6). Further known syndromes $S_{j+1,0}$ to $S_{m,j-m+1}$ can be calculated by substituting the curve into (4.7) to obtain a recursive relationship among the known syndromes, in order to determine the value of all syndromes $S_{a,b}$ where $a \geq m$, the degree of the curve. The remaining unknown syndromes $S_{m-1,j-m+2}$ to $S_{0,j}$ can be found using (4.8) or the majority voting scheme (explained later).

For the Hermitian curve, $C(x, y) = x^{r+1} + y^r + y$ with $m = r + 1$, the recursion is:

$$\sum_{k,l} C_{k,l} S_{a-t_1^{(i)}+k,b-t_2^{(i)}+l} = 0$$
$$C_{0,1} S_{a-m+0,b-0+1} + C_{0,m-1} S_{a-m+0,b-0+m-1} + C_{m,0} S_{a-m+m,b-0+0} = 0,$$

where $C_{k,l}$ is a coefficient of $C(x, y)$, and k and l are the powers of the x and y term, respectively, for each term in $C(x, y)$. Since all the coefficients of $C(x, y)$ are equal to 1, this simplifies to:

$$S_{a,b} = S_{a-m,b+1} + S_{a-m,b+m-1}. \tag{4.9}$$

(4.9) is important for the decoding procedure and is used to find the known syndromes $S_{j+1,0}$ to $S_{m,j+1-m}$, as shown in Figure 4.2.

If all the polynomials in F satisfy (4.8) then there is no need to modify F. If a polynomial in F makes (4.7) nonzero, this polynomial is said to have a discrepancy, d_f, and may be used in the modification process. This is the same as Δ in the Berlekamp–Massey algorithm. The order in which each syndrome in the array is read in is determined by the total graduated degree order, denoted by $<_T$ [6]. Let (a, b) be a point in the syndrome array. The next point, (a', b'), according to the total graduated degree order, is [6]:

$$(a', b') = \begin{cases} (a - 1, b + 1), & \text{if } a > 0 \\ (b + 1, 0), & \text{if } a = 0 \end{cases}. \tag{4.10}$$

This gives a degree ordering of $\{(0, 0) <_T (1, 0) <_T (0, 1) <_T (2, 0) <_T (1, 1) \ldots\}$, meaning that the order the syndromes are read in from the array is $S_{0,0}, S_{1,0}, S_{0,1}, S_{2,0}, S_{1,1}$ and so on.

An important set is the auxiliary set G, which stores some of the polynomials that had a nonzero discrepancy at an earlier point in the algorithm (a_g, b_g). Each polynomial $g^{(i)}(x, y)$ in G has a quantity called the 'span', defined as:

$$\text{span}(g^{(i)}(x, y)) = \left(a_g - u_1^{(i)}, b_g - u_2^{(i)}\right), \tag{4.11}$$

where $u_1^{(i)}$ and $u_2^{(i)}$ are the degrees of x and y, respectively, of the leading term in $g^{(i)}(x, y)$. The span of a polynomial means that there is no polynomial with leading term $x^{a_g - u_1^{(i)}} y^{b_g - u_2^{(i)}}$ that can satisfy (4.8) at the point (a, b). The union of all sets less than or equal to each $\text{span}(g^{(i)}(x, y))$ in G defines the set $\text{span}(G)$. If there are φ polynomials in G then $\text{span}(G)$ is defined as:

$$\text{span}(G) = \sum_{i=1}^{\varphi} \left\{(k, l) | (k, l) \leq \text{span}\left(g^{(i)}(x, y)\right)\right\}, \tag{4.12}$$

where (k, l) are a pair of nonnegative integers. The span of G has interior and exterior corners. An interior corner is defined as a maximal point inside $\text{span}(G)$ with respect to the partial ordering, and an exterior corner is defined as a minimal point outside $\text{span}(G)$ with respect to the partial ordering denoted by $<$. For two integer pairs $a = (a_1, a_2)$ and $b = (b_1, b_2)$ [6]:

$$a < b \text{ if } a_1 \leq b_1 \wedge a_2 \leq b_2 \wedge a \neq b. \tag{4.13}$$

For example, $(3, 5) < (4, 5)$ because $3 \leq 5$ and $4 \leq 5$ and $(3, 5) \neq (4, 5)$. It is easier to see the corners by drawing $span(G)$. An example is shown in Figure 4.3 for $\text{span}(G) = \{(0, 0), (1, 0), (0, 1), (2, 0)\}$.

The interior corners, shown as the large black circles, are $(0, 1)$ and $(2, 0)$, since there are no points in $\text{span}(G)$ greater than either of these points. Similarly, the exterior corners, shown as the large white circles, are $(3, 0)$, $(1, 1)$ and $(0, 2)$, since there are no points outside $\text{span}(G)$ less than these points.

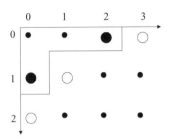

Figure 4.3   An example of span(G) showing interior (large black circles) and exterior (large white circles) corners.

The number of exterior corners gives the number of polynomials in F, and their values give the degrees of the polynomials. In this example there are three polynomials in F with leading terms x^3, xy and y^2. As the set F changes there is also a corresponding set, defined as [6]:

$$\Delta = \sum_{k=1}^{\lambda-1} \Delta_k, \tag{4.14}$$

where λ is the total number of polynomials in F and [6]:

$$\Delta_k = \left\{ (k, l) | (k, l) \le \left(t_1^{(k)} - 1, t_2^{(k+1)} - 1 \right) \right\}, \tag{4.15}$$

where (k, l) are a pair of nonnegative integers. It turns out that the set Δ is equal to span(G) and is used in this decoding algorithm for the majority-voting scheme described later.

4.3.1 Modification of the Sets F and G

In the Sakata algorithm the set G is first modified by calculating the spans of the polynomials currently in G, and also the polynomials in F that have a nonzero discrepancy. The polynomials in F with $d_f \neq 0$ are placed in a set F_N and are also added to G to create a new set G', that is $G' = G \cup F_N$. The span of each polynomial in G' is calculated using (4.11) and then used to calculate span(G') from (4.12). The interior corners are then found. A polynomial in G' with a span *equal* to one of the interior corners is kept in G' and the others are removed. If more than one polynomial has a span equal to the same interior corner then either one can be kept. The discrepancy d_g of the polynomials in G' and the point in the syndrome array (a_g, b_g) where their discrepancies were nonzero are also stored. The set G is now modified and will be used for the next point in the syndrome array.

The modification of F also uses *span(G')*. The exterior corners are found and these will give the number of polynomials and the leading terms of these polynomials in the modified set F'. For each exterior corner $(\varepsilon_1, \varepsilon_2)$ the polynomials in F are modified using one of three possible cases [10, 11]:

Case 1   If there is a polynomial $f^{(i)}(x, y)$ in the difference set F/F_N with $(t_1^{(i)}, t_2^{(i)}) = (\varepsilon_1, \varepsilon_2)$ then the new polynomial $h^{(i)}(x, y) \in F'$ is unchanged:

$$h^{(i)}(x, y) = f^{(i)}(x, y). \tag{4.16}$$

Case 2   Else if $\varepsilon_1 > a$ *or* $\varepsilon_2 > b$, find a polynomial $f^{(i)}(x, y) \in F_N$ with $(t_1^{(i)}, t_2^{(i)}) \le (\varepsilon_1, \varepsilon_2)$. The new minimal polynomial $h^{(i)}(x, y) \in F'$ is given by:

$$h^{(i)}(x, y) = x^{\varepsilon_1 - t_1^{(i)}} y^{\varepsilon_2 - t_2^{(i)}} f^{(i)}(x, y). \tag{4.17}$$

Case 3   Otherwise, find a polynomial $f^{(i)}(x, y) \in F_N$ with $(t_1^{(i)}, t_2^{(i)}) \le (\varepsilon_1, \varepsilon_2)$ and a polynomial $g^{(i)}(x, y) \in G$ with $\mathrm{span}(g^{(i)}(x, y)) \le (a - t_1^{(i)}, b - t_2^{(i)})$. Let $(p_1, p_2) = \mathrm{span}(g^{(i)}(x, y)) - (a - \varepsilon_1, b - \varepsilon_2)$. The new minimal polynomial $h^{(i)}(x, y) \in F'$ is then given by:

$$h^{(i)}(x, y) = x^{\varepsilon_1 - t_1^{(i)}} y^{\varepsilon_2 - t_2^{(i)}} f^{(i)}(x, y) - \frac{d_f}{d_g} x^{p_1} y^{p_2} g^{(i)}(x, y). \tag{4.18}$$

The full algorithm can now be described [8, 11, 12]:

1. Initialize $(a, b) = (0, 0)$, $F = \{1\}$, $G = \varnothing$, $D = \varnothing$.
2. $F_N = \varnothing$, $F' = \varnothing$.
   Calculate the set Δ from (4.14) and (4.15).
   If $|\Delta|$ exceeds the maximum number of errors the code can correct, terminate the algorithm.
   Else find the discrepancy d_f of each polynomial $f^{(i)}(x, y) \in F$ for the syndrome $S_{a,b}$ using (4.8). Every $f^{(i)}(x, y)$ with $d_f \ne 0$ is placed in F_N and $G' = G \cup F_N$.
   If all $f^{(i)}(x, y) \in F$ have no discrepancy then increment (a, b) with respect to the total order $<_T$ and go to step 2.
3. Find the span of each $g^{(i)}(x, y) \in G'$ using (4.10), then find the span of G' using (4.11).
4. Find the interior corners of $\mathrm{span}(G')$ and remove all $g^{(i)}(x, y) \in G'$ whose spans are not equal to any of the interior corners. The modification of G' is now complete. Also store the discrepancies d_g of each $g^{(i)}(x, y) \in G'$ and the point in the Sakata algorithm where each $g^{(i)}(x, y)$ had a nonzero discrepancy as (a_g, b_g).
5. Find the exterior corners of $\mathrm{span}(G')$ and apply Case 1, 2 or 3 to create the modified minimal polynomials $h^{(i)}(x, y)$ in the set F'.
6. Set $F = F'$, $G = G'$ and increment (a, b) with respect to the total order $<_T$.
7. If the last element in the syndrome array has been reached end the algorithm. The set F contains the error-locating polynomials. Else go to step 2.

In step 2, a stopping criterion was added by the authors [11] and the decoding algorithm is terminated when the number of elements in the set Δ exceeds the maximum number of errors that the code can correct. This is necessary because the majority voting scheme is unreliable if Δ is too large and it can choose an incorrect value for the unknown syndrome, affecting the remainder of the decoding algorithm.

All that remains is to substitute the points from the curve into one of the error-locating polynomials in F. The points that make the error-locating polynomial vanish are the error locations.

4.4 Majority Voting

The Berlekamp–Massey algorithm is able to correct up to the maximum number of errors, t, that the code is capable of correcting with the knowledge of $2t$ syndromes. However, the Sakata algorithm can only correct up to:

$$t \leq \left\lfloor \frac{d^* - \gamma - 1}{2} \right\rfloor, \tag{4.19}$$

with only the known syndromes. Unknown syndromes of the type $S_{a,b}$, $a \leq m$ can be calculated by substituting the curve into (4.8) to obtain a recursive relationship among the previous syndromes, such as for the Hermitian curve in (4.9), but this can only be used if the previous syndromes are actually known. For example, for a Hermitian code constructed from a Hermitian curve with degree $m = 5$, (4.9) can be used to calculate the syndrome $S_{7,0}$:

$$\begin{aligned} S_{7,0} &= S_{7-5,0} + S_{7-5,0+5-1} \\ &= S_{2,1} + S_{2,4} \end{aligned},$$

but only if the values of $S_{2,0}$ and $S_{2,4}$ are known. The unknown syndromes of the type $S_{a,b}$, $a < m$, are determined using the majority voting scheme, presented by Feng and Rao [8]. In this chapter, hard-decision decoding of AG codes is accomplished using the decoding algorithm by Sakata et al. [9], which uses majority voting in the following way: from [9] a candidate syndrome value v_i from the ith minimal polynomial $f^{(i)}(x, y) \in F$ can be calculated using (4.8):

$$\sum_{(k,l) \leq_T \left(t_1^{(i)}, t_2^{(i)}\right)} f_{k,l}^{(i)} S_{k+a-t_1^{(i)}, l+b-t_2^{(i)}} = -v_i. \tag{4.20}$$

However, it is not always possible to use (4.20) so a candidate syndrome value w_i can be calculated:

$$\sum_{(k,l) \leq_T \left(t_1^{(i)}, t_2^{(i)}\right)} f_{k,l}^{(i)} S_{k+a+m-t_1^{(i)}, l+b-m+1-t_2^{(i)}} - S_{a,b-m+2} = -w_i. \tag{4.21}$$

In some cases both criteria are satisfied and $f^{(i)}(x, y)$ can give two candidate values, v_i and w_i. If neither of these conditions is satisfied then the polynomial $f^{(i)}(x, y)$ is not used to find a candidate syndrome value. Which equation is used to find candidate values of the syndromes depends on two conditions:

1. If $a = t_1^{(i)}$ and $b = t_2^{(i)}$ then use (4.20).
2. If $a + m = t_1^{(i)}$ and $b - t_2^{(i)} = m - 1$ then use (4.21).

Now define a set $K = K_1 \cup K_2$ [9]:

$$
\begin{aligned}
K_1 &= \{(k, l) | 0 \le k \le a \wedge 0 \le l \le b\} \\
K_2 &= \{(k, l) | 0 \le k < m \wedge 0 \le l \le b - m + 1\}
\end{aligned}
\tag{4.22}
$$

where (k, l) are a pair of nonnegative integers. Also define two further sets:

$$
\begin{aligned}
A_i &= \left\{ (k, l) \in K \,|\, k + t_1^{(i)} \le a \wedge l + t_2^{(i)} \le b \right\} \\
B_i &= \left\{ (k, l) \in K \,|\, k + t_1^{(i)} \le a + m \wedge l + t_2^{(i)} \le b - m + 1 \right\}
\end{aligned}
\tag{4.23}
$$

Let $\delta_1, \delta_2, \delta_3, \ldots$ be the candidate syndrome values obtained from (4.20) or (4.21) with

$$
K_j = \left(\bigcup_{v_i = \delta_j} A_i \cup \bigcup_{w_i = \delta_j} B_i \right) \Big/ \Delta
\tag{4.24}
$$

associated with each candidate syndrome value. The set K_j containing the maximum number of elements implies that the corresponding candidate syndrome value δ_j is the correct value.

4.5 Calculating the Error Magnitudes

Previously, the error magnitudes were found by solving (4.6), but this can become too complex. Sakata *et al.* [9] used a two-dimensional inverse discrete Fourier transform (IDFT) that can calculate the error magnitudes with less operations but requires the knowledge of many unknown syndromes. The error magnitude at the ith point on the curve e_i is given as:

$$
e_i = \sum_{a=0}^{q-2} \sum_{b=0}^{q-2} S_{a,b} x_i^{-a} y_i^{-b},
\tag{4.25}
$$

where q is the size of the finite field. However, (4.25) can only be used if an error occurs at a point where both coordinates are nonzero. The Hermitian curves, for example, also have points with a zero coordinate, so another method must be used when errors occurs at these points. Liu [13] addressed this problem for Hermitian codes by using a one-dimensional IDFT, the properties of the Hermitian curve and knowledge of more unknown syndromes up to $S_{q-1,q-1}$. The points of the Hermitian curve are split into four types: all points with both terms nonzero are labelled $P_{(x,x)}$; all points with a zero x term and a nonzero y term are labelled $P_{(0,x)}$; all points with a nonzero x term and a zero y term are labelled $P_{(x,0)}$; the remaining point with both terms zero is labelled $P_{(0,0)}$. The following mapping is defined:

$$m \rightarrow \begin{cases} \alpha^m, & 0 \leq m \leq q - 2 \\ 0, & m = q - 1 \end{cases}. \tag{4.26}$$

For errors that occurred at points of type $P_{(0,x)}$ the one-dimensional IDFT is:

$$E_n = \sum_{i=0}^{q-2} S_{0,q-1-i}\alpha^{ni}, \tag{4.27}$$

where E_n is the sum of all the error values at the points with y-coordinate α^n, and α is a primitive element in the finite field. Fortunately, for Hermitian curves, if there is a point $(0, \alpha^n)$ then there will be no points of the form (α^m, α^n), so E_n is actually the error value that occurred at the point $(0, \alpha^n)$. Similarly, for errors that occurred at points with a zero y-coordinate, the one-dimensional IDFT is:

$$E_m = \sum_{i=0}^{q-2} S_{q-1-i,0}\alpha^{mi}, \tag{4.28}$$

where E_m is the sum of all the error values at the points with x-coordinate α^m, and α is a primitive element in the finite field. Again, for Hermitian curves, it turns out that there are no other points with x-coordinate α^m so E_m is the error value that occurred at the point $(\alpha^m, 0)$. Finally, if an error occurred at the point $(0, 0)$ we can simply use the properties of the Hermitian curve. We know that the syndrome $S_{0,0}$ is the sum of all the errors that occurred. To find the error value at the point $P_{(0,0)}$ we subtract all the error values that occurred at the points $P_{(x,x)}$, the error values that occurred at the points $P_{(0,x)}$ and all the error values that occurred at the points $P_{(x,0)}$. For the Hermitian curve we have the following relationships:

$$\sum_{P_i \in P_{(x,x)}} e_i = S_{q-1,q-1}$$

$$\sum_{P_i \in P_{(0,x)}} e_i = S_{0,q-1} - S_{q-1,q-1}. \tag{4.29}$$

$$\sum_{P_i \in P_{(x,0)}} e_i = S_{q-1,0} - S_{q-1,q-1}$$

Therefore, the error at the point $P_1 = (0, 0)$ is:

$$e_1 = \sum_i e_i - \sum_{P_i \in P_{(x,x)}} e_i - \sum_{P_i \in P_{(0,x)}} e_i - \sum_{P_i \in P_{(x,0)}} e_i \qquad (4.30)$$

$$= S_{0,0} - S_{q-1,q-1} - (S_{0,q-1} - S_{q-1,q-1}) - (S_{q-1,0} - S_{q-1,q-1})$$

For Hermitian codes defined over a finite field with a field characteristic of 2, this simplifies to:

$$e_1 = S_{0,0} + S_{0,q-1} + S_{q-1,0} + S_{q-1,q-1}. \qquad (4.31)$$

4.6 Complete Hard-Decision Decoding Algorithm for Hermitian Codes

The complete AG decoding algorithm used to obtain the simulation results presented later in this chapter can be found in [11]:

1. Find the known syndromes $S_{0,0}$ to $S_{0,j}$ using (4.6).
2. Use (4.9) to find more known syndromes $S_{j+1,0}$ to $S_{m,j+1-m}$.
3. Error location.

Apply the Sakata algorithm using known syndromes as the input:

For syndrome $= S_{0,0}$ to $S_{m,j+1-m}$
{
Run Sakata Algorithm
if syndrome is of the form $S_{a,b}, b \geq m - 1$ then find more unknown
    syndromes using (4.9)
}

Apply the Sakata Algorithm using unknown syndromes as the input:

For syndrome $= S_{m-1,j+2-m}$ to $S_{0,j+m}$
{
if syndrome is of the form $S_{a,b}, a \geq m$
    {
    Use (4.9) to find the value of the unknown syndrome
    }
else if syndrome is of the form $S_{a,b}, a < m$

```
            {
            Use the 'Majority Voting' scheme
            }
        Run Sakata Algorithm
        if syndrome is of the form $S_{a,b}, b \geq m - 1$ then find more unknown
            syndromes using (4.9)
        }
/* End of Sakata algorithm and majority voting */
```

Find the error locations by substituting each point of the Hermitian curve into one of the minimal (error-locating) polynomials in F. The roots of the minimal polynomial are the error locations.

4. Error magnitudes:

Find the remaining unknown syndromes:

```
For syndrome = $S_{j+m+1,0}$ to $S_{q-1,q-1}$
{
if syndrome is of the form $S_{a,b}, a \geq m$
        {
        Use (4.9) to find the value of the unknown syndrome
        }
else if syndrome is of the form $S_{a,b}, a < m$
        {
        Substitute the last minimal polynomial in the set $F$ into (4.8)
            to form a recursive relationship among the syndromes to the
            find the value of the unknown syndrome
        }
}
```

Use the inverse discrete Fourier transforms to find the error values:

- If the error location has both coordinates nonzero, use (4.25).
- Else if the error location has x-coordinate zero and y-coordinate nonzero, use (4.27).
- Else if the error location has y-coordinate zero and x-coordinate nonzero, use (4.28).
- Else if the error location occurred at the point $(0, 0)$, use (4.31).

5. The error values found from step 4 at the error locations found from step 3 are added to the received word to give the decoded code word. Since the code is systematic, the original message can be extracted by taking the first k symbols from the decoded code word.

Example 4.4: Using Sakata's algorithm with majority voting and an IDFT to decode a Hermitian code: In this example we use the $(64, 44, 15)$ Hermitian code defined over GF(16), which can correct up to seven symbol errors. The Hermitian curve used is:

$$C(x, y) = x^5 + y^4 + y, \tag{4.32}$$

which, from (2.12), has a genus $\gamma = 6$ and has 64 affine points that satisfy $C(x, y) = 0$, given in Table 4.6.

Table 4.6 The 64 affine points that satisfy C(x, y) = 0.

$P_1 = (0, 0)$	$P_2 = (0, 1)$	$P_3 = (0, \alpha^5)$	$P_4 = (0, \alpha^{10})$
$P_5 = (1, \alpha)$	$P_6 = (1, \alpha^4)$	$P_7 = (1, \alpha^2)$	$P_8 = (1, \alpha^8)$
$P_9 = (\alpha, \alpha^9)$	$P_{10} = (\alpha, \alpha^7)$	$P_{11} = (\alpha, \alpha^6)$	$P_{12} = (\alpha, \alpha^{13})$
$P_{13} = (\alpha^4, \alpha^9)$	$P_{14} = (\alpha^4, \alpha^7)$	$P_{15} = (\alpha^4, \alpha^6)$	$P_{16} = (\alpha^4, \alpha^{13})$
$P_{17} = (\alpha^2, \alpha^3)$	$P_{18} = (\alpha^2, \alpha^{14})$	$P_{19} = (\alpha^2, \alpha^{11})$	$P_{20} = (\alpha^2, \alpha^{12})$
$P_{21} = (\alpha^8, \alpha^3)$	$P_{22} = (\alpha^8, \alpha^{14})$	$P_{23} = (\alpha^8, \alpha^{11})$	$P_{24} = (\alpha^8, \alpha^{12})$
$P_{25} = (\alpha^5, \alpha^3)$	$P_{26} = (\alpha^5, \alpha^{14})$	$P_{27} = (\alpha^5, \alpha^{11})$	$P_{28} = (\alpha^5, \alpha^{12})$
$P_{29} = (\alpha^{10}, \alpha^9)$	$P_{30} = (\alpha^{10}, \alpha^7)$	$P_{31} = (\alpha^{10}, \alpha^6)$	$P_{32} = (\alpha^{10}, \alpha^{13})$
$P_{33} = (\alpha^3, \alpha)$	$P_{34} = (\alpha^3, \alpha^4)$	$P_{35} = (\alpha^3, \alpha^2)$	$P_{36} = (\alpha^3, \alpha^8)$
$P_{37} = (\alpha^{14}, \alpha^3)$	$P_{38} = (\alpha^{14}, \alpha^{14})$	$P_{39} = (\alpha^{14}, \alpha^{11})$	$P_{40} = (\alpha^{14}, \alpha^{12})$
$P_{41} = (\alpha^9, \alpha)$	$P_{42} = (\alpha^9, \alpha^4)$	$P_{43} = (\alpha^9, \alpha^2)$	$P_{44} = (\alpha^9, \alpha^8)$
$P_{45} = (\alpha^7, \alpha^9)$	$P_{46} = (\alpha^7, \alpha^7)$	$P_{47} = (\alpha^7, \alpha^6)$	$P_{48} = (\alpha^7, \alpha^{13})$
$P_{49} = (\alpha^6, \alpha)$	$P_{50} = (\alpha^6, \alpha^4)$	$P_{51} = (\alpha^6, \alpha^2)$	$P_{52} = (\alpha^6, \alpha^8)$
$P_{53} = (\alpha^{13}, \alpha^9)$	$P_{54} = (\alpha^{13}, \alpha^7)$	$P_{55} = (\alpha^{13}, \alpha^6)$	$P_{56} = (\alpha^{13}, \alpha)$
$P_{57} = (\alpha^{11}, \alpha^3)$	$P_{58} = (\alpha^{11}, \alpha^{14})$	$P_{59} = (\alpha^{11}, \alpha^{11})$	$P_{60} = (\alpha^{11}, \alpha^{12})$
$P_{61} = (\alpha^{12}, \alpha)$	$P_{62} = (\alpha^{12}, \alpha^4)$	$P_{63} = (\alpha^{12}, \alpha^2)$	$P_{64} = (\alpha^{12}, \alpha^8)$

From (4.3), the parameter j can have any value from $j = 3$ to $j = 12$. Choosing $j = 5$ and substituting this into (4.4) we get a message length of $k = 44$ and a designed minimum distance of $d^* = 15$. It is assumed that the received word has errors at the following affine points:

$e_6 = \alpha^{12}$ at $P_6 = (1, \alpha^4)$	$e_{13} = \alpha^{12}$ at $P_{13} = (\alpha^4, \alpha^9)$	$e_{27} = \alpha^{12}$ at $P_{27} = (\alpha^5, \alpha^{11})$
$e_{32} = \alpha^{12}$ at $P_{32} = (\alpha^{10}, \alpha^{13})$	$e_{33} = \alpha^{12}$ at $P_{33} = (\alpha^3, \alpha)$	$e_{45} = \alpha^{12}$ at $P_{45} = (\alpha^7, \alpha^9)$
$e_{51} = \alpha^{12}$ at $P_{51} = (\alpha^6, \alpha^2)$		

The known syndromes $S_{a,b}$, where $a + b \leq j = 5$, can be calculated using (4.6). The initial syndromes are shown in Figure 4.4. The syndromes $S_{6,0}$ and $S_{5,1}$ are found using the recursive relationship of (4.9):

$$S_{a,b} = S_{a-5,b+1} + S_{a-5,b+4}.$$

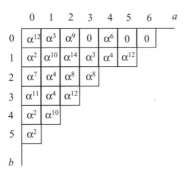

Figure 4.4   Initial syndrome array.

$$S_{a,b} = S_{a-5,b+1} + S_{a-5,b+4}$$

Therefore,

$$S_{6,0} = S_{1,1} + S_{1,4} = \alpha^{10} + \alpha^{10} = 0$$

$$S_{5,1} = S_{1,1} + S_{1,4} = \alpha^7 + \alpha^2 = (\alpha^3 + \alpha + 1) + \alpha^2 = \alpha^{12}.$$

The parameters in Sakata's algorithm are then initialized as follows: $(a, b) = (0, 0)$, $F = \{1\}$, $G = \emptyset$ and $\Delta = \emptyset$, $F_N = \emptyset$.

From (4.8), the discrepancy d_f of $f^{(1)} = 1$ when $S_{a,b} = \alpha^{12}$ is determined to be:

$$d_f = \sum_{(k,l) \leq_T (t_1^{(1)}, t_2^{(1)})} f_{k,l}^{(i)} S_{a-t_1^{(i)}+k, b-t_2^{(i)}+l} = f_{0,0}^{(1)} S_{0-0+0, 0-0+0} = 1.S_{0,0} = \alpha^{12}.$$

Hence, $F_N = \{1\}$ and $G' = G \cup F_N = \{1\}$, and we also store this discrepancy, $d_g = d_f$, and the point at which the discrepancy occurred, $(a_g, b_g) = (0, 0)$. Next, span(G') is found using (4.11) and (4.12). The x-degree and y-degree of $g(x, y) = 1$ are $u_1^{(1)} = 0$ and $u_2^{(1)} = 0$, respectively, and so span(1) = $(0 - 0, 0 - 0) = (0, 0)$.

Therefore,

$$\text{span}(G') = \sum_{i=1}^{\varphi} \left\{ (k, l) | (k, l) \leq \text{span}(g^{(i)}(x, y)) \right\}$$

$$= \sum_{i=1}^{1} \{ (k, l) | (k, l) \leq (0, 0) \} = \{(0, 0)\}.$$

If we represent span(G') graphically, as in Figure 4.5, we can see that it has only one interior corner, at $(0, 0)$, and two exterior corners, at $(1, 0)$ and $(0, 1)$.

Without performing any of the update procedures for sets F and G, we know that since there is only one interior corner, $(0, 0)$, G will contain a single polynomial

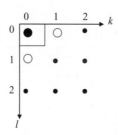

Figure 4.5   The set Δ after the first modification to the polynomials in F.

with a span equal to $(0, 0)$. Also, since there are two exterior corners, $(1, 0)$ and $(0, 1)$, F will contain two polynomials with leading monomials x and y, respectively. To update G, we keep only those polynomials in G' that have spans equal to all interior corners of span(G'). In this case, the polynomial $g = 1$ is the only polynomial in G and its span is equal to $(0, 0)$, so this is kept for G. This completes the updating of G.

To update F, we take each exterior corner of span(G') and apply Case 1, 2 or 3. For the exterior corner $(\varepsilon_1, \varepsilon_2) = (1, 0)$ we cannot apply Case 1 since the difference set F/F_N is empty. However, we can apply Case 2, since $\varepsilon_1 > a = 0$, so:

$$h^{(1)}(x, y) = x^{\varepsilon_1 - t_1^{(i)}} y^{\varepsilon_2 - t_2^{(i)}} f^{(1)}(x, y) = x^{1-0} y^{0-0} 1 = x.$$

Similarly, for the exterior corner $(\varepsilon_1, \varepsilon_2) = (0, 1)$, Case 2 can also be applied, since $\varepsilon_2 > b$, so:

$$h^{(2)}(x, y) = x^{\varepsilon_1 - t_1^{(i)}} y^{\varepsilon_2 - t_2^{(i)}} f^{(1)}(x, y) = x^{0-0} y^{1-0} 1 = y.$$

Therefore, for $(a, b) = (1, 0)$ we have $F = \{x, y), G = \{1\}$ and $\Delta = \{(0, 0)\}$. The outputs of the algorithm for all the syndromes up to the first unknown syndrome $S_{4,2}$ are given in Table 4.7.

Observations from Table 4.7:

- The set F always contains one more polynomial that the set G.
- For a t-error-correcting AG code, the number of elements in Δ never exceeds t if there are no more that t errors in the received word.
- Once $|\Delta| = t$, the number of polynomials in F and their leading monomials never change. The increasing size of Δ is shown in Figure 4.6.

The polynomials in F are now used to determine the value of the first unknown syndrome, S_{42}. The first polynomial in F is $f^{(1)}(x, y) = \alpha^{14} + \alpha x + \alpha^7 y + \alpha^4 x^2 + xy + \alpha^7 x^3 + \alpha x^2 y + x^4$, with leading degree $(t_1^{(1)}, t_2^{(2)}) = (4, 0)$. A candidate value

Table 4.7 Modification of the set of minimal polynomials using Sakata's algorithm for each syndrome in the array of Figure 4.4.

a, b	$S_{a,b}$	F	d_f	Δ	G	d_g	a_g, b_g
$0, 0$	α^{12}	1	α^{12}	æ	æ	—	—
$1, 0$	α^3	x	α^3	$\{(0,0)\}$	1	α^{12}	$0, 0$
		y	0				
$0, 1$	α^2	$\alpha^6 + x$	0	$\{(0,0)\}$	1	α^{12}	$0, 0$
		y	α^2				
$2, 0$	α^9	$\alpha^6 + x$	0	$\{(0,0)\}$	1	α^{12}	$0, 0$
		$\alpha^5 + y$	0				
$1, 1$	α^{10}	$\alpha^6 + x$	α	$\{(0,0)\}$	1	α^{12}	$0, 0$
		$\alpha^5 + y$	α				
$0, 2$	α^7	$\alpha^6 x + x^2$	0	$\{(0,0)(1,0),$	$\alpha^6 + x$	α	$1, 1$
		$\alpha^4 + \alpha^6 y + xy$	0	$(0,1)\}$	$\alpha^5 + y$	α	$1, 1$
		$\alpha^5 y + y^2$	0				
$3, 0$	0	$\alpha^6 x + x^2$	1	$\{(0,0)(1,0),$	$\alpha^6 + x$	α	$1, 1$
		$\alpha^4 + \alpha^6 y + xy$	0	$(0,1)\}$	$\alpha^5 + y$	α	$1, 1$
		$\alpha^5 y + y^2$	0				
$2, 1$	α^{14}	$\alpha^4 + \alpha^6 x + \alpha^{14} y + x^2$	α^7	$\{(0,0)(1,0),$	$\alpha^6 + x$	α	$1, 1$
		$\alpha^4 + \alpha^6 y + xy$	0	$(0,1)\}$	$\alpha^5 + y$	α	$1, 1$
		$\alpha^5 y + y^2$	0				
$1, 2$	α^4	$\alpha^6 + \alpha^{14} y + x^2$	0	$\{(0,0)(1,0),$	$\alpha^6 + x$	α	$1, 1$
		$\alpha^4 + \alpha^6 y + xy$	α	$(0,1)\}$	$\alpha^5 + y$	α	$1, 1$
		$\alpha^5 y + y^2$	α				
$0, 3$	α^{11}	$\alpha^6 + \alpha^{14} y + x^2$	0	$\{(0,0)(1,0),$	$\alpha^6 + x$	α	$1, 1$
		$\alpha^{12} + x + \alpha^6 y + xy$	0	$(0,1)\}$	$\alpha^5 + y$	α	$1, 1$
		$\alpha^5 + \alpha^{10} y + y^2$	1				
$4, 0$	α^6	$\alpha^6 + \alpha^{14} y + x^2$	0	$\{(0,0)(1,0),$	$\alpha^6 + x$	α	$1, 1$
		$\alpha^{12} + x + \alpha^6 y + xy$	0	$(0,1)\}$	$\alpha^5 + y$	α	$1, 1$
		$\alpha^{14} x + \alpha^{10} y + y^2$	0				
$3, 1$	α^3	$\alpha^6 + \alpha^{14} y + x^2$	α	$\{(0,0)(1,0),$	$\alpha^6 + x$	α	$1, 1$
		$\alpha^{12} + x + \alpha^6 y + xy$	α	$(0,1)\}$	$\alpha^5 + y$	α	$1, 1$
		$\alpha^{14} x + \alpha^{10} y + y^2$	0				
$2, 2$	α^8	$\alpha^6 + \alpha^{13} x + \alpha^{14} xy + x^3$	0	$\{(0,0), (1,0),$	$\alpha^6 + \alpha^{14} y + x^2$	α	$3, 1$
		$\alpha^5 + \alpha^{13} y + \alpha^{14} y^2 + x^2 y$	α^{12}	$(0,1), (2,0),$	$\alpha^{12} + x + \alpha^6 y$	α	$3, 1$
		$\alpha^{14} x + \alpha^{10} y + y^2$	α^{12}	$(1,1)\}$	$+ xy$		
$1, 3$	α^4	$\alpha^6 + \alpha^{13} x + \alpha^{14} xy + x^3$	0	$\{(0,0), (1,0),$	$\alpha^6 + \alpha^{14} y + x^2$	α	$3, 1$
		$\alpha^5 + \alpha^2 x + \alpha^{13} y + \alpha^{10} xy$	0	$(0,1), (2,0),$	$\alpha^{12} + x + \alpha^6 y$	α	$3, 1$
		$\quad + \alpha^{14} y^2 + \alpha^{11} x^3 + x^2 y$	α^{10}	$(1,1)\}$	$+ xy$		
		$\alpha^8 + \alpha^{10} x + \alpha^4 y + \alpha^{11} xy$					
		$\quad + y^2$					

(Continued)

Table 4.7 (*Continued*)

a, b	$S_{a,b}$	F	d_f	Δ	G	d_g	a_g, b_g
0, 4	α^{11}	$\alpha^6 + \alpha^{13}x + \alpha^{14}xy + x^3$	0	$\{(0,0), (1,0),$	$\alpha^6 + \alpha^{14}y + x^2$	α	3, 1
		$\alpha^5 + \alpha^2 x + \alpha^{13}y + \alpha^{10}xy$	0	$(0,1), (2,0),$	$\alpha^{12} + x + \alpha^6 y$	α	3, 1
		$+\, \alpha^{14}y^2 + \alpha^{11}x^3 + x^2 y$	0	$(1,1)\}$	$+\, xy$		
		$\alpha^2 + \alpha^{10}x + \alpha^5 y + \alpha^9 x^2$					
		$+\, \alpha^{11}xy + y^2$					
5, 0	0	$\alpha^6 + \alpha^{13}x + \alpha^{14}xy + x^3$	α^8	$\{(0,0), (1,0),$	$\alpha^6 + \alpha^{14}y + x^2$	α	3, 1
		$\alpha^5 + \alpha^2 x + \alpha^{13}y + \alpha^{10}xy$	0	$(0,1), (2,0),$	$\alpha^{12} + x + \alpha^6 y$	α	3, 1
		$+\, \alpha^{14}y^2 + \alpha^{11}x^3 + x^2 y$	0	$(1,1)\}$	$+\, xy$		
		$\alpha^2 + \alpha^{10}x + \alpha^5 y + \alpha^9 x^2$					
		$+\, \alpha^{11}xy + y^2$					
4, 1	α^4	$\alpha^{12} + \alpha^5 x + \alpha^{13}y + \alpha xy + x^3$	α^8	$\{(0,0), (1,0),$	$\alpha^6 + \alpha^{14}y + x^2$	α	3, 1
		$\alpha^5 + \alpha^2 x + \alpha^{13}y + \alpha^{10}xy$	α^4	$(0,1), (2,0),$	$\alpha^{12} + x + \alpha^6 y$	α	3, 1
		$+\, \alpha^{14}y^2 + \alpha^{11}x^3 + x^2 y$	0	$(1,1)\}$	$+\, xy$		
		$\alpha^2 + \alpha^{10}x + \alpha^5 y + \alpha^9 x^2$					
		$+\, \alpha^{11}xy + y^2$					
3, 2	α^8	$\alpha + \alpha^5 x + y + \alpha^7 x^2 + \alpha xy$	α^4	$\{(0,0), (1,0),$	$\alpha^6 + \alpha^{14}y + x^2$	α	3, 1
		$+\, x^3$	α^4	$(0,1), 2,0),$	$\alpha^{12} + x + \alpha^6 y$	α	3, 1
		$\alpha^{10} + \alpha^6 x + \alpha^{10}y + \alpha^{12}xy$	α^4	$(1,1)\}$	$+\, xy$		
		$+\, \alpha^{14}y^2 + \alpha^{11}x^3 + x^2 y$					
		$\alpha^2 + \alpha^{10}x + \alpha^5 y + \alpha^9 x^2$					
		$+\, \alpha^{11}xy + y^2$					
2, 3	α^{12}	$\alpha x + \alpha^5 x^2 + xy + \alpha^7 x^3$	0	$\{(0,0), (1,0),$	$\alpha + \alpha^5 x + y$	α^4	3, 2
		$+\, \alpha x^2 y + x^4$	α	$(0,1), (2,0),$	$+\, \alpha^7 x^2 + \alpha xy$	α	3, 1
		$\alpha^{13} + \alpha^6 x + \alpha^4 y + \alpha^3 x^2$	α^4	$(1,1), (0,2),$	$+\, x^3$	α^4	3, 2
		$+\, \alpha^{12}xy + \alpha^{14}y^2 + \alpha^{11}x^3$	α^4	$(3,0)\}$	$\alpha^6 + \alpha^{14}y + x^2$		
		$+\, x^2 y$			$\alpha^2 + \alpha^{10}x$		
		$1 + \alpha^6 x + \alpha^9 y + \alpha^{10}x^2$			$+\, \alpha^5 y + \alpha^9 x^2$		
		$+\, \alpha^{11}xy + \alpha^9 x^3 + \alpha^{11}x^2 y$			$+\, \alpha^{11}xy + y^2$		
		$+\, xy^2$					
		$\alpha^2 y + \alpha^{10}xy + \alpha^5 y^2 + \alpha^9 x^2 y$					
		$+\, \alpha^{11}xy^2 + y^3$					
1, 4	α^{10}	$\alpha x + \alpha^5 x^2 + xy + \alpha^7 x^3$	0	$\{(0,0), (1,0),$	$\alpha + \alpha^5 x + y$	α^4	3, 2
		$+\, \alpha x^2 y + x^4$	0	$(0,1), (2,0),$	$+\, \alpha^7 x^2 + \alpha xy$	α	3, 1
		$\alpha^3 x + \alpha^6 y + \alpha^7 x^2 + \alpha xy$	α^2	$(1,1), (0,2),$	$+\, x^3$	α^4	3, 2
		$+\, \alpha^{14}y^2 + x^3 + x^2 y$	α^7	$(3,0)\}$	$\alpha^6 + \alpha^{14}y + x^2$		
		$\alpha^7 + \alpha^6 x + \alpha^{11}y + \alpha^{12}x^2$			$\alpha^2 + \alpha^{10}x$		
		$+\, \alpha^{11}xy + \alpha^9 x^3 + \alpha^{11}x^2 y$			$+\, \alpha^5 y + \alpha^9 x^2$		
		$+\, xy^2$			$+\, \alpha^{11}xy + y^2$		
		$\alpha^2 x + \alpha^2 y + \alpha^{10}x^2 + xy$					
		$+\, \alpha^5 y^2 + \alpha^9 x^3 + \alpha^2 x^2 y$					
		$+\, \alpha^{12}xy^2 + y^3$					

(*Continued*)

Table 4.7　(*Continued*)

a, b	$S_{a,b}$	F	d_f	Δ	G	d_g	a_g, b_g
0, 5	α^2	$\alpha x + \alpha^5 x^2 + xy + \alpha^7 x^3$	0	$\{(0,0),(1,0),$	$\alpha + \alpha^5 x + y$	α^4	3, 2
		$\quad + \alpha x^2 y + x^4$	0	$(0,1),(2,0),$	$\quad + \alpha^7 x^2 + \alpha xy$	α	3, 1
		$\alpha^3 x + \alpha^6 y + \alpha^7 x^2 + \alpha xy$	0	$(1,1),(0,2),$	$\quad + x^3$	α^4	3, 2
		$\quad + \alpha^{14} y^2 + x^3 + x^2 y$	α^7	$(3,0)\}$	$\alpha^6 + \alpha^{14} y + x^2$		
		$\alpha + \alpha^2 x + \alpha^4 y + \alpha^{14} x^2$			$\alpha^2 + \alpha^{10} x$		
		$\quad + \alpha^{10} xy + \alpha^{10} x^3 + \alpha^{11} x^2 y$			$\quad + \alpha^5 y + \alpha^9 x^2$		
		$\quad + xy^2$			$\quad + \alpha^{11} xy + y^2$		
		$\alpha^{12} + \alpha^2 x + \alpha y + \alpha^7 x^2 + xy$					
		$\quad + \alpha^5 y^2 + \alpha^9 x^3 + \alpha^2 x^2 y$					
		$\quad + \alpha^{12} xy^2 + y^3$					
6, 0	0	$\alpha x + \alpha^5 x^2 + xy + \alpha^7 x^3$	0	$\{(0,0),(1,0),$	$\alpha + \alpha^5 x + y$	α^4	3, 2
		$\quad + \alpha x^2 y + x^4$	0	$(0,1),(2,0),$	$\quad + \alpha^7 x^2 + \alpha xy$	α	3, 1
		$\alpha^3 x + \alpha^6 y + \alpha^7 x^2 + \alpha xy$	0	$(1,1),(0,2),$	$\quad + x^3$	α^4	3, 2
		$\quad + \alpha^{14} y^2 + x^3 + x^2 y$	0	$(3,0)\}$	$\alpha^6 + \alpha^{14} y + x^2$		
		$\alpha + \alpha^2 x + \alpha^4 y + \alpha^{14} x^2$			$\alpha^2 + \alpha^{10} x$		
		$\quad + \alpha^{10} xy + \alpha^{10} x^3 + \alpha^{11} x^2 y$			$\quad + \alpha^5 y + \alpha^9 x^2$		
		$\quad + xy^2$			$\quad + \alpha^{11} xy + y^2$		
		$\alpha^6 + x + \alpha^9 y + \alpha^6 x^2 + \alpha xy$					
		$\quad + \alpha^5 y^2 + \alpha x^3 + \alpha^2 x^2 y$					
		$\quad + \alpha^{12} xy^2 + y^3$					
5, 1	α^{12}	$\alpha x + \alpha^5 x^2 + xy + \alpha^7 x^3$	α^9	$\{(0,0),(1,0),$	$\alpha + \alpha^5 x + y$	α^4	3, 2
		$\quad + \alpha x^2 y + x^4$	α	$(0,1),(2,0),$	$\quad + \alpha^7 x^2 + \alpha xy$	α	3, 1
		$\alpha^3 x + \alpha^6 y + \alpha^7 x^2 + \alpha xy$	0	$(1,1),(0,2),$	$\quad + x^3$	α^4	3, 2
		$\quad + \alpha^{14} y^2 + x^3 + x^2 y$	0	$(3,0)\}$	$\alpha^6 + \alpha^{14} y + x^2$		
		$\alpha + \alpha^2 x + \alpha^4 y + \alpha^{14} x^2$			$\alpha^2 + \alpha^{10} x$		
		$\quad + \alpha^{10} xy + \alpha^{10} x^3 + \alpha^{11} x^2 y$			$\quad + \alpha^5 y + \alpha^9 x^2$		
		$\quad + xy^2$			$\quad + \alpha^{11} xy + y^2$		
		$\alpha^6 + x + \alpha^9 y + \alpha^6 x^2 + \alpha xy$					
		$\quad + \alpha^5 y^2 + \alpha x^3 + \alpha^2 x^2 y$					
		$\quad + \alpha^{12} xy^2 + y^3$					
4, 2	?	$\alpha^{14} + \alpha x + \alpha^7 y + \alpha^4 x^2 + xy$	?	$\{(0,0),(1,0),$	$\alpha + \alpha^5 x + y$	α^4	3, 2
		$\quad + \alpha^7 x^3 + \alpha x^2 y + x^4$	?	$(0,1),(2,0),$	$\quad + \alpha^7 x^2 + \alpha xy$	α	3, 1
		$\alpha^{14} + \alpha^4 x + \alpha^3 y + \alpha^{10} x^2$	?	$(1,1),(0,2),$	$\quad + x^3$	α^4	3, 2
		$\quad + \alpha^{10} xy + \alpha^5 y^2 + x^3 + x^2 y$	?	$(3,0)\}$	$\alpha^6 + \alpha^{14} y + x^2$		
		$\alpha + \alpha^2 x + \alpha^4 y + \alpha^{14} x^2$			$\alpha^2 + \alpha^{10} x$		
		$\quad + \alpha^{10} xy + \alpha^{10} x^3 + \alpha^{11} x^2 y$			$\quad + \alpha^5 y + \alpha^9 x^2$		
		$\quad + xy^2$			$\quad + \alpha^{11} xy + y^2$		
		$\alpha^6 + x + \alpha^9 y + \alpha^6 x^2 + \alpha xy$					
		$\quad + \alpha^5 y^2 + \alpha x^3 + \alpha^2 x^2 y$					
		$\quad + \alpha^{12} xy^2 + y^3$					

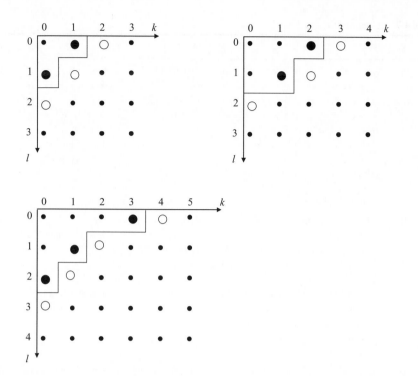

Figure 4.6 Graphical representation of the set Δ at different stages of Sakata's algorithm from Table 4.7.

for S_{42} can be obtained by using (4.20):

$$\sum_{(k,l) \leq_T (4,0)} f_{k,l}^{(1)} S_{k+4-4,l+2-0} = -v_1$$

$$= \alpha^{14} S_{0,2} + \alpha S_{1,2} + \alpha^7 S_{0,3} + \alpha^4 S_{2,2} + 1.S_{1,3} + \alpha^7 S_{3,2} + \alpha S_{2,3}$$

$$= \alpha^{14} \alpha^7 + \alpha \alpha^4 + \alpha^7 \alpha^{11} + \alpha^4 \alpha^8 + \alpha^4 + \alpha^7 \alpha^8 + \alpha \alpha^{12}$$

$$= \alpha^6 + \alpha^5 + \alpha^3 + \alpha^{12} + \alpha^4 + 1 + \alpha^{13}$$

$$= (\alpha^3 + \alpha^2) + (\alpha^2 + \alpha) + \alpha^3 + (\alpha^3 + \alpha^2 + \alpha + 1) + (\alpha + 1) + 1$$
$$+ (\alpha^3 + \alpha^2 + 1)$$

$$v_1 = \alpha.$$

(4.20) can also be used to obtain candidate values for the second and third polynomials in F; these are $v_2 = \alpha$ and $v_3 = \alpha$.

However, the fourth polynomial in F, $\alpha^6 + x + \alpha^9 y + \alpha^6 x^2 + \alpha xy + \alpha^5 y^2 + \alpha x^3 + \alpha^2 x^2 y + \alpha^{12} xy^2 + y^3$, which has leading degree $(t_1^{(1)}, t_2^{(2)}) = (0, 3)$, cannot be used to obtain a candidate value using either (4.20) or (4.21). Since the contributing polynomials all give the same candidate value, the syndrome $S_{42} = \alpha$. For S_{33}, all contributing polynomials also give the same candidate value of α^6.

The first unknown syndrome value where all contributing polynomials give different candidate values is S_{24}. Since the first polynomial in F has leading degree

$(t_1^{(1)}, t_2^{(2)}) = (4, 0)$, (4.20) can be used to obtain a candidate value:

$$\sum_{(k,l) \leq_T (4,0)} f_{k,l}^{(1)} S_{k+2+5-4, l+4-5+1-0} - S_{2,4-5+2} = -w_1$$

$$= \alpha^{14} S_{3,0} + \alpha S_{4,0} + \alpha^7 S_{3,1} + \alpha^4 S_{5,0} + 1 \cdot S_{4,1} + \alpha^7 S_{6,0} + \alpha S_{5,1} + S_{2,1}$$

$$= \alpha^{14} \cdot 0 + \alpha \cdot \alpha^6 + \alpha^7 \cdot \alpha^3 + \alpha^4 \cdot 0 + 1 \cdot \alpha^4 + \alpha^7 \cdot 0 + \alpha \cdot \alpha^{12} + \alpha^{14}$$

$$= \alpha^7 + \alpha^{10} + \alpha^4 + \alpha^{13} + \alpha^{14}$$

$$= (\alpha^3 + \alpha + 1) + (\alpha^2 + \alpha + 1) + (\alpha + 1) + (\alpha^3 + \alpha^2 + 1) + (\alpha^3 + 1)$$

$$= \alpha^3 + \alpha + 1$$

$$w_1 = \alpha^7.$$

For the other three contributing polynomials in F, (4.20) can be used to obtain the candidate value of α^{11}. Now, from (4.23):

$$K_1 = \{(k, l) \mid 0 \leq k \leq 2 \wedge 0 \leq l \leq 4\}$$
$$= \{(0, 0), (0, 1), (0, 2), (0, 3), (0, 4), (1, 1), (1, 2), (1, 3), (1, 4),$$
$$\times (2, 0), (2, 1), (2, 2), (2, 3), (2, 4)\}$$
$$K_2 = \{(k, l) \mid 0 \leq k < 5 \wedge 0 \leq l \leq 4 - 5 + 1\}$$
$$= \{(0, 0), (1, 0), (2, 0), (3, 0), (4, 0)\}$$

and the set $K = K_1 \cup K_2$ is:

$$K = \{(0, 0), (0, 1), (0, 2), (0, 3), (0, 4), (1, 0), (1, 1), (1, 2), (1, 3), (1, 4),$$
$$\times (2, 0), (2, 1), (2, 2), (2, 3), (2, 4), (3, 0), (4, 0)\}.$$

From (4.22) the sets A and B are:

$$A_1 = \{(k, l) \in K \mid k + 4 \leq 2 \wedge l + 0 \leq 4\} = \emptyset$$
$$A_2 = \{(k, l) \in K \mid k + 2 \leq 2 \wedge l + 1 \leq 4\} = \{(0, 0), (0, 1), (0, 2), (0, 3)\}$$
$$A_3 = \{(k, l) \in K \mid k + 1 \leq 2 \wedge l + 2 \leq 4\} = \{(0, 0), (0, 1), (0, 2), (1, 0), (1, 1), (1, 2)\}$$
$$A_4 = \{(k, l) \in K \mid k + 0 \leq 2 \wedge l + 3 \leq 4\} = \{(0, 0), (0, 1), (1, 0), (1, 1), (2, 0), (2, 1)\}$$
$$B_1 = \{(k, l) \in K \mid k + 4 \leq 7 \wedge l + 0 \leq 0\} = \{(0, 0), (1, 0), (2, 0), (3, 0)\}$$
$$B_2 = \{(k, l) \in K \mid k + 2 \leq 7 \wedge l + 1 \leq 0\} = \emptyset$$
$$B_3 = \{(k, l) \in K \mid k + 1 \leq 7 \wedge l + 2 \leq 0\} = \emptyset$$
$$B_4 = \{(k, l) \in K \mid k + 0 \leq 7 \wedge l + 3 \leq 0\} = \emptyset$$

The candidate syndromes are $\delta_1 = \alpha^7$ and $\delta_2 = \alpha^{11}$. Finally, from (4.24):

$$K_1 = \left(\bigcup_{v_i = \alpha^7} A_i \cup \bigcup_{w_i = \alpha^7} B_i \right) / \{(0, 0), (1, 0), (0, 1), (2, 0), (1, 1), (0, 2), (3, 0)\}$$

$$= \{(0, 0), (1, 0), (2, 0), (3, 0)\} / \{(0, 0), (1, 0), (0, 1), (2, 0), (1, 1), (0, 2), (3, 0)\}$$

$$= \emptyset$$

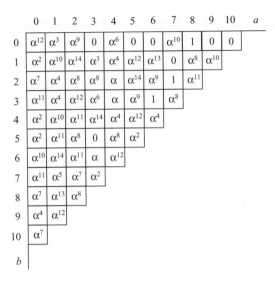

	0	1	2	3	4	5	6	7	8	9	10	a
0	α^{12}	α^3	α^9	0	α^6	0	0	α^{10}	1	0	0	
1	α^2	α^{10}	α^{14}	α^3	α^4	α^{12}	α^{13}	0	α^8	α^{10}		
2	α^7	α^4	α^8	α^8	α	α^{14}	α^9	1	α^{11}			
3	α^{11}	α^4	α^{12}	α^6	α	α^9	1	α^8				
4	α^2	α^{10}	α^{11}	α^{14}	α^4	α^{12}	α^4					
5	α^2	α^{11}	α^8	0	α^8	α^2						
6	α^{10}	α^{14}	α^{11}	α	α^{12}							
7	α^{11}	α^5	α^7	α^2								
8	α^7	α^{13}	α^8									
9	α^4	α^{12}										
10	α^7											
b												

Figure 4.7 The syndrome array after majority voting has been completed.

$$K_2 = \left(\bigcup_{v_i = \alpha^{11}} A_i \cup \bigcup_{w_i = \alpha^{11}} B_i \right) \backslash \{(0, 0), (1, 0), (0, 1), (2, 0), (1, 1), (0, 2), (3, 0)\}$$
$$= A_2 \cup A_3 \cup A_4 \backslash \{(0, 0), (1, 0), (0, 1), (2, 0), (1, 1), (0, 2), (3, 0)\}$$
$$= \{(0, 0), (0, 1), (0, 2), (0, 3), (1, 0), (1, 1), (1, 2), (2, 0), (2, 1)\}/$$
$$\{(0, 0), (1, 0), (0, 1), (2, 0), (1, 1), (0, 2), (3, 0)\}$$
$$= \{(0, 3), (1, 2), (2, 1)\}$$

Therefore, there are $|K_1| = 0$ votes for the candidate value α^7 and $|K_2| = 3$ votes for the candidate value α^{11}, and so $S_{24} = \alpha^{11}$. By employing majority voting up to the last unknown syndrome, $S_{0,10}$, we obtain the syndrome array shown in Figure 4.7. The final set of minimal polynomials is:

$$F = \begin{cases} f^{(1)}(x, y) = \alpha^{14} + \alpha x + \alpha^7 y + \alpha^4 x^2 + xy + \alpha^7 x^3 + \alpha x^2 y + x^4 \\ f^{(2)}(x, y) = \alpha^{14} + \alpha^4 x + \alpha^3 y + \alpha^{10} x^2 + \alpha^{10} xy + \alpha^5 y^2 + x^3 + x^2 y \\ f^{(3)}(x, y) = \alpha + \alpha^2 x + \alpha^4 y + \alpha^{14} x^2 + \alpha^{10} xy + \alpha^{10} x^3 + \alpha^{11} x^2 y + xy^2 \\ f^{(4)}(x, y) = \alpha^6 + x + \alpha^9 y + \alpha^6 x^2 + \alpha xy + \alpha^5 y^2 + \alpha x^3 + \alpha^2 x^2 y \\ \qquad\qquad + \alpha^{12} xy^2 + y^3 \end{cases}$$

All four polynomials in F are error-locating polynomials, which means that those affine points that cause any of these polynomials to vanish are possible error locations. If we take the polynomial $f^{(4)}(x, y)$, there are 12 points that cause it to vanish:

$P_6 = (1, \alpha^4)$	$P_{11} = (\alpha, \alpha^6)$	$P_{13} = (\alpha^4, \alpha^9)$	$P_{27} = (\alpha^5, \alpha^{11})$	$P_{31} = (\alpha^{10}, \alpha^6)$
$P_{32} = (\alpha^{10}, \alpha^{13})$	$P_{33} = (\alpha^3, \alpha)$	$P_{45} = (\alpha^7, \alpha^9)$	$P_{51} = (\alpha^6, \alpha^2)$	$P_{52} = (\alpha^6, \alpha^8)$
$P_{57} = (\alpha^{11}, \alpha^3)$	$P_{63} = (\alpha^{12}, \alpha^2)$			

The remaining syndromes from $S_{11,0}$ to $S_{15,15}$ are found either by using (4.9) for $S_{a,b}$, $a \geq 5$, or by substituting the minimal polynomial $f^{(4)}(x, y)$ in F into (4.8) to obtain the recursive equation:

$$\sum_{(k,l) \leq_T (0,3)} f^{(4)}_{k,l} S_{a-0+k,b-3+l} = 0$$

$$f_{0,0} S_{a,b-3} + f_{1,0} S_{a+1,b-3} + f_{0,1} S_{a,b-2} + f_{2,0} S_{a+2,b-3} + f_{1,1} S_{a+1,b-2}$$
$$+ f_{0,2} S_{a,b-1} + f_{3,0} S_{a+3,b-3} + f_{2,1} S_{a+2,b-2} + f_{1,2} S_{a+1,b-1} + f_{0,3} S_{a,b}.$$

Making $S_{a,b}$ the subject of the equation gives:

$$S_{a,b} = \alpha^6 S_{a+1,b-3} + S_{a+1,b-3} + \alpha^9 S_{a,b-2} + \alpha^6 S_{a+2,b-3} + \alpha S_{a+1,b-2}$$
$$+ \alpha^5 S_{a,b-1} + \alpha S_{a+3,b-3} + \alpha^2 S_{a+2,b-2} + \alpha^{12} S_{a+1,b-1}. \qquad (4.33)$$

So the syndromes $S_{11,0}$, $S_{10,1}$, $S_{9,2}$, $S_{8,3}$, $S_{7,4}$, $S_{6,5}$ and $S_{5,6}$ are calculated using (4.9) and have the values α^{11}, α^{13}, α^{13}, α^{13}, 0 and α^5, respectively. To calculate $S_{4,7}$ we would use (4.33):

$$S_{4,7} = \alpha^6 S_{4,4} + S_{5,4} + \alpha^9 S_{4,5} + \alpha^6 S_{6,4} + \alpha S_{5,5} + \alpha^5 S_{4,6} + \alpha S_{7,4}$$
$$+ \alpha^2 S_{6,5} + \alpha^{12} S_{5,6}$$
$$= \alpha^6 \alpha^4 + \alpha^{12} + \alpha^9 \alpha^8 + \alpha^6 \alpha^4 + \alpha \alpha^2 + \alpha^5 \alpha^{12} + \alpha \cdot 0 + \alpha^2 \alpha^5 + \alpha^{12} \alpha^8$$
$$= \alpha^{10} + \alpha^{12} + \alpha^2 + \alpha^{10} + \alpha^3 + \alpha^2 + \alpha^7 + \alpha^5$$
$$= \alpha^{12} + \alpha^3 + \alpha^7 + \alpha^5$$
$$= (\alpha^3 + \alpha^2 + \alpha + 1) + \alpha^3 + (\alpha^3 + \alpha + 1) + (\alpha^2 + \alpha)$$
$$= \alpha^3 + \alpha$$
$$= \alpha^9$$

The complete syndrome array is given in Figure 4.8.

All that remains is to find the error magnitudes at the possible error locations. Taking the point $P_6 = (1, \alpha^4)$, we use (4.24) to determine the magnitude:

$$e_6 = \sum_{a=0}^{14} \sum_{b=0}^{14} S_{a,b} x_6^{-a} y_6^{-b}$$

$$= \sum_{a=0}^{q-2} \sum_{b=0}^{q-2} S_{a,b} 1^{-a} \cdot \alpha^{-4b}$$

$$= S_{0,0} 1^0 \alpha^0 + S_{1,0} 1^0 \alpha^{-4} + S_{2,0} 1^0 \alpha^{-8} + \cdots S_{14,12} 1^{-14} \alpha^{-3}$$

$$+ S_{14,13} 1^{-14} \alpha^{-7} + S_{14,14} 1^{-14} \alpha^{-11}$$

This gives a value of $e_6 = \alpha^{12}$. Similar calculations for the remaining points give:

$e_6 = \alpha^{12}$ at $P_6 = (1, \alpha^4)$	$e_{11} = 0$ at $P_{11} = (\alpha, \alpha^6)$	$e_{13} = \alpha^{12}$ at $P_{13} = (\alpha^4, \alpha^9)$
$e_{27} = \alpha^{12}$ at $P_{27} = (\alpha^5, \alpha^{11})$	$e_{31} = 0$ at $P_{31} = (\alpha^{10}, \alpha^6)$	$e_{32} = \alpha^{12}$ at $P_{32} = (\alpha^{10}, \alpha^{13})$
$e_{33} = \alpha^{12}$ at $P_{33} = (\alpha^3, \alpha)$	$e_{45} = \alpha^{12}$ at $P_{45} = (\alpha^7, \alpha^9)$	$e_{51} = \alpha^{12}$ at $P_{51} = (\alpha^6, \alpha^2)$
$e_{52} = 0$ at $P_{52} = (\alpha^6, \alpha^8)$	$e_{57} = 0$ at $P_{57} = (\alpha^{11}, \alpha^3)$	$e_{63} = 0$ at $P_{63} = (\alpha^{12}, \alpha^2)$

	0	1	2	3	4	5	6	7	8	9	10	11	12	13	14	15	a
0	α^{12}	α^3	α^9	0	α^6	0	0	α^{10}	1	0	0	α^{11}	0	α^4	α^8	α^{12}	
1	α^2	α^{10}	α^{14}	α^3	α^4	α^{12}	α^{13}	0	α^8	α^{10}	α^{13}	α^6	α^5	0	α^2	α^2	
2	α^7	α^4	α^8	α^8	α	α^{14}	α^9	1	α^{11}	α^{13}	α^{12}	0	α^{14}	α^7	0	α^7	
3	α^{11}	α^4	α^{12}	α^6	α	α^9	1	α^8	α^{13}	α^{14}	α^{10}	α^9	α^{13}	α^{13}	1	α^{11}	
4	α^2	α^{10}	α^{11}	α^{14}	α^4	α^{12}	α^4	0	α^5	α	α	α^2	α^6	0	α^3	α^2	
5	α^2	α^{11}	α^8	0	α^8	α^2	α^5	α^{10}	α^{11}	α^{14}	α^7	α^{10}	0	α^4	α^4	α^2	
6	α^{10}	α^{14}	α^{11}	α	α^{12}	α^8	1	α^6	α^5	α^{14}	α^6	α^{14}	α^3	α^3	α^4	α^{10}	
7	α^{11}	α^5	α^7	α^2	α^9	α^3	α^{14}	α^{13}	α^7	α^4	α^6	α^{10}	α^4	α^6	α^9	α^{11}	
8	α^7	α^{13}	α^8	α^5	α^{10}	α^5	α	α^7	α^{11}	1	α^{12}	0	α^{11}	α^9	0	α^7	
9	α^4	α^{12}	α^{14}	α^6	α^5	α^{11}	α^5	α^6	α^8	α^9	α^3	α^{10}	α^{13}	α^3	α^5	α^4	
10	α^7	α^{10}	α^{10}	α	α^4	α^2	0	α^8	α^4	0	α^7	α^{11}	α^{13}	α^{11}	α^8	α^7	
11	α^4	α^2	α^3	α^{13}	α^2	α^9	α^8	α^3	α	α^7	1	α^5	α^8	α^{11}	α^9	α^4	
12	α^8	α^{13}	α	α	α^{10}	1	α^5	α	α^{12}	α^9	α^4	α^9	α^3	α^{11}	1	α^8	
13	α^8	1	α^7	α^{10}	α^{14}	α^6	α^{10}	α^3	α^7	α^5	α^3	1	α^2	α	α^4	α^8	
14	α^{10}	α^2	α^{13}	α^{11}	α^2	1	α^7	α^8	α^6	α^{11}	α^9	1	α	α^6	α^{14}	α^{10}	
15	α^{12}	α^3	α^9	0	α^6	0	0	α^{10}	1	0	0	α^{11}	0	α^4	α^8	α^{12}	

b

Figure 4.8 The complete syndrome array.

4.7 Simulation Results

In this section, we present some simulation results evaluating the performance of Hermitian codes in comparison with Reed–Solomon codes. It is obvious that if we compare Hermitian codes with Reed–Solomon codes defined over the same finite field then the Hermitian codes will always perform better due to their increased length and larger Hamming distance. However, it is interesting to compare Hermitian codes defined over smaller finite fields and Reed–Solomon codes defined over larger finite fields. A Hermitian code defined over GF(64) has much better parameters than a Reed–Solomon code defined over GF(64), but it also has better parameters than a Reed–Solomon code defined over GF(256). For example, the (512, 314, 171) Hermitian code defined over GF(64) of rate 0.6 has a much larger minimum Hamming distance than the (63, 39, 25) Reed–Solomon code defined over GF(64), and also a larger Hamming distance than the (255, 153, 103) Reed–Solomon code defined over GF(256). This illustrates another advantage of Hermitian codes in that improved performance can be achieved with smaller alphabets, consequently reducing the complexity of the finite field arithmetic for hardware implementation.

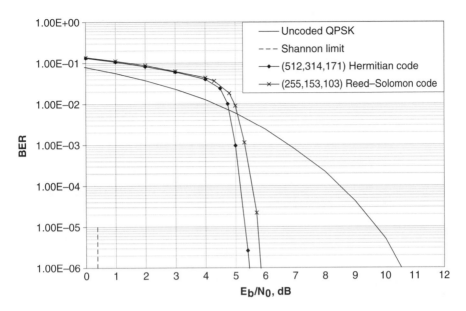

Figure 4.9   Performance comparison of the (512, 314, 171) Hermitian code defined over GF(64) with the (255, 153, 103) Reed–Solomon code defined over GF(256) on the AWGN channel.

A comparison of the (512, 314, 171) Hermitian code and (255, 153, 103) Reed–Solomon code is now presented in terms of simulation results on the AWGN channel and a Rayleigh fading channel. In Figure 4.9, the performance of the two codes is shown on the AWGN channel, with the Hermitian code achieving only a 0.3 dB coding gain over the Reed–Solomon code.

However, under slow fading conditions where bursts of errors are more likely to occur it can be shown that the Hermitian code can significantly outperform the Reed–Solomon code. In Figure 4.10 the performance of the two codes is shown on a Rayleigh fading channel that is frequency nonselective, with a carrier frequency of 1.9 GHz, a velocity of 20 m/s and perfect channel estimation for a data rate of 30 kbps. It can be seen that the Hermitian code achieves a coding gain of 4.8 dB over the Reed–Solomon code.

In Figure 4.11 the performance of the two codes is shown on the same Rayleigh fading channel, but with an increased data rate of 100 kbps. Under more harsh conditions the Hermitian code still achieves a coding gain of 3.1 dB over the Reed–Solomon code, at a BER of 10^{-6}.

Finally, we also present the performance of the same Hermitian codes on magnetic recording channels [12], as described in Chapter 1. In Figure 4.12, the performances of the (64, 39, 20) Hermitian code and the (512, 314, 171) Hermitian code are compared with the (15, 9, 7), (63, 39, 25) and (255, 153, 103) Reed–Solomon codes on a magnetic recording channel with a recording linear density of $D_s = 2$, which is a measure of the intersymbol interference (ISI). The signal-to-noise ratio E_t/N_0 is the ratio of the energy

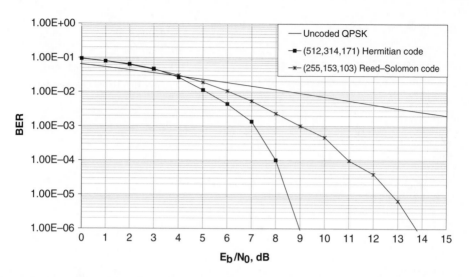

Figure 4.10 Performance comparison of the $(512, 314, 171)$ Hermitian code defined over GF(64) with the $(255, 153, 103)$ Reed–Solomon code defined over GF(256) on a Rayleigh fading channel.

in the Lorentzian Pulse E_t to the noise power spectral density N_0 of the electronics noise. As stated before, the Hermitian codes outperform the Reed–Solomon codes defined over the same finite fields due to their increased lengths. Interestingly, the $(512, 314, 171)$ Hermitian code defined over GF(64) still outperforms the $(255, 153, 103)$ Reed–Solomon code defined over GF(256) on magnetic recording channels.

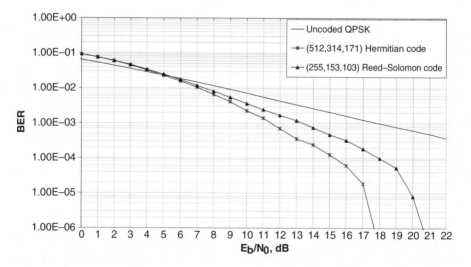

Figure 4.11 Performance comparison of the $(512, 314, 171)$ Hermitian code defined over GF(64) with the $(255, 153, 103)$ Reed–Solomon code defined over GF(256) on a Rayleigh fading channel.

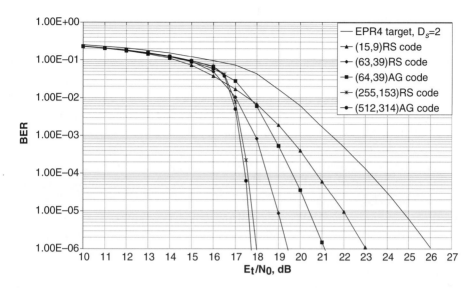

Figure 4.12 Performance comparison of Hermitian codes and Reed–Solomon codes on a magnetic recording channel with a recording linear density of $D_s = 2$.

Figure 4.13 shows the performance of the same codes on a magnetic recording channel with a recording linear density of $D_s = 3$, representing an increase in ISI. Overall, the performance of all the codes is worse but the Hermitian codes still outperform the Reed–Solomon codes.

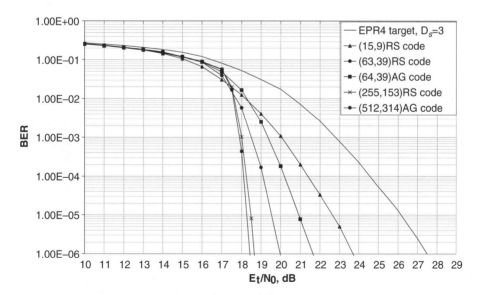

Figure 4.13 Performance comparison of Hermitian codes and Reed–Solomon codes on a magnetic recording channel with a recording linear density of $D_s = 3$.

4.8 Conclusions

A class of non-binary block code called the algebraic–geometric code has been introduced. Such codes are constructed from the affine points of a projective curve and a set of rational functions defined on that curve. Depending on the choice of curve, very long AG codes can be constructed with very large minimum Hamming distances. The well-known Reed–Solomon codes are the simplest class of AG code, constructed from the affine points of the projective line. Consequently, they have the shortest block lengths of all AG codes, and there are not many Reed–Solomon codes that can be constructed. However, Reed–Solomon codes are maximum distance separable (MDS) unlike other AG codes, where the genus of the curve reduces the actual minimum Hamming distance.

Despite this genus penalty, AG codes still have much larger minimum Hamming distances than Reed–Solomon codes defined over the same finite field and consequently AG codes can correct much longer bursts of errors, which are common in data storage channels and slow fading channels. A disadvantage of AG codes is their higher decoding complexity. Sakata's algorithm is more complex than the Berlekamp–Massey algorithm, but Kotter recently showed how several Berlekamp–Massey decoders could be implemented in parallel, replacing Sakata's algorithm and having a running time equal to a single Berlekamp–Massey decoder.

In this chapter, we have covered the construction of AG codes from Hermitian curves. A detailed explanation of the decoding of AG codes has also been presented, in particular the use of Sakata's algorithm to locate errors, and the use of inverse discrete Fourier transforms to determine the error magnitudes at these locations. Finally, some simulation results have been presented on the AWGN and Rayleigh fading channels, showing how a Hermitian curve defined over GF(64) can outperform a Reed–Solomon code defined over the much larger finite field GF(256).

Reed–Solomon codes are still the most commonly used coding scheme today, but they are rapidly being replaced in communication systems with more powerful coding schemes such as turbo and LDPC codes. However, in optical storage, the latest devices, such as Blu-Ray and HD-DVD, still use error-correction schemes involving Reed–Solomon codes due to their good burst-error-correcting performance and efficient decoders. Eventually, as storage density increases and consequently the effects of ISI become more severe, Reed–Solomon codes will not be good enough and they will need to be replaced. AG codes could be a possible candidate for the error-correcting schemes in future data storage devices.

References

[1] Goppa, V.D. (1981) Codes on algebraic curves. *Soviet Math. Dokl.*, **24**, 75–91.
[2] Justesen, J., Larsen, K.J., Jensen, H. E. *et al.* (1989) Construction and decoding of a class of algebraic geometry codes. *IEEE Trans. Inform. Theory*, **IT-35**, 811–21.
[3] Blake, I.F., Heegard, C., Hoholdt, T. and Wei, V. (1998) Algebraic geometry codes. *IEEE Trans. Inform. Theory*, **44** (6), 2596–618.

[4] Atkinson, K.A. (1989) *An Introduction to Numerical Analysis*, 2nd edn, John Wiley & Sons, Inc., New York, ISBN 0-471-50023-2.

[5] Heegard, C., Little, J. and Saints, K. (1995) Systematic encoding via Grobner bases for a class of algebraic–geometric Goppa codes. *IEEE Trans. Inform. Theory*, **41** (6), 1752–61.

[6] Sakata, S. (1988) Finding a minimal set of linear recurring relations capable of generating a given finite two-dimensional array. *J. Symbolic Computation*, **5**, 321–37.

[7] Justesen, J., Larsen, K.J., Jensen, H.E. and Hoholdt, T. (1992) Fast decoding of codes from algebraic plane curves. *IEEE Trans. Inform. Theory*, **IT-38** (6), 1663–76.

[8] Feng, G.L. and Rao, T.R.N. (1993) Decoding algebraic geometric codes up to the designed minimum distance. *IEEE Trans. Inform. Theory*, **IT-39**, 37–46.

[9] Sakata, S., Justesen, J., Madelung, Y. *et al.* (1995) Fast decoding of algebraic–geometric codes up to the designed minimum distance. *IEEE Trans. Inform. Theory*, **IT-41** (5), 1672–7.

[10] Saints, K. and Heegard, C. (1995) Algebraic–geometric codes and multidimensional cyclic codes: a unified theory and algorithms for decoding using Grobner bases. *IEEE Transactions on Information Theory*, **41** (6), 1733–51.

[11] Johnston, M. and Carrasco, R.A. (2005) Construction and performance of algebraic–geometric codes over the AWGN and fading channels. *IEE Proc. Commun.*, **152** (5), 713–22.

[12] Carrasco, R.A. and Johnston, M. (2007) Hermitian codes on magnetic recording channels. 9th International Symposium on Communication Theory and Applications, Ambleside, Lake District, UK.

[13] Liu, C.-W. (1999) Determination of error values for decoding Hermitian codes with the inverse affine fourier transform. *IEICE Trans. Fundamentals*, **E82-A** (10), 2302–5.

5

List Decoding

5.1 Introduction

In this chapter, we discuss an alternative decoding algorithm which produces a list of candidate messages instead of just one. This class of algorithm is known as a *List Decoding* algorithm and was developed by Elias [1, 2] and Wozencraft [3]. A list decoder has the advantage of being able to correct more errors than a conventional decoding algorithm that only returns a single decoded message, from now on known as a unique decoding algorithm. This is unfortunately at the expense of increased decoding complexity.

The chapter begins by introducing the Guruswami–Sudan algorithm [4] for the list decoding of Reed–Solomon codes. This algorithm has two processes: interpolation and factorization. It is shown that most of the decoding complexity of the list decoding algorithm is due to the interpolation process, but we present a method to reduce this complexity. This is followed by an explanation of the soft-decision list decoding of Reed–Solomon codes using the Kotter–Vardy algorithm [5].

We also describe how to modify the Guruswami–Sudan algorithm to list decode AG codes and again show we can still reduce the complexity of the interpolation process. The Kotter–Vardy algorithm is then extended for the soft-decision list decoding of AG codes. The performance of hard- and soft-decision list decoding of Reed–Solomon codes and AG codes is evaluated, and simulation result are presented showing the performance of both coding schemes on the AWGN and fading channels. The idea of list decoding is as follows: given a received word R, reconstruct a list of all code words with a distance τ to the received word R, in which τ can be greater than $\tau_{unique} = \lfloor \frac{d-1}{2} \rfloor$.

This idea is illustrated in Figure 5.1, where c_1, c_2 and c_3 are three independent code words at a distance d from each other. A received word r_1, which has distance less than τ_{unique} to code word c_1, can be decoded by the unique decoding algorithm, which results in c_1. However, for received word r_2, which has distance greater than τ_{unique} to any of the code words, the unique decoding algorithm will fail to decode it. But

Non-Binary Error Control Coding for Wireless Communication and Data Storage   Rolando Antonio Carrasco and Martin Johnston
© 2008 John Wiley & Sons, Ltd

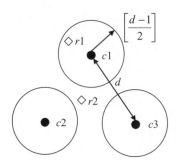

Figure 5.1   The idea of list decoding.

using the list decoding algorithm, a list of possible transmitted code words will be produced. For example, a decoded output list $\{c_1, c_2, c_3\}$ is produced by the decoder, then the code word that has the minimal distance to r_2 is chosen from the list and decoding is completed.

5.2 List Decoding of Reed–Solomon Codes Using the Guruswami–Sudan Algorithm

5.2.1 Weighted Degrees and Lexicographic Orders

To apply the list decoding algorithm we must first impose an ordering on the powers of a set of monomials. If we define the weighted (u, v)-degree of a monomial $x^a y^b$ by:

$$\deg_{u,v}(x^a y^b) = ua + vb \tag{5.1}$$

then we can order a sequence of monomials by their weighted degrees. The *lexicographic order* of the monomials is then defined as:

$$x^{a_1} y^{b_1} < x^{a_2} y^{b_2} \tag{5.2}$$

if $\deg_{u,v}(x^{a_1} y^{b_1}) < \deg_{u,v}(x^{a_2} y^{b_2})$ or $\deg_{u,v}(x^{a_1} y^{b_1}) = \deg_{u,v}(x^{a_2} y^{b_2})$ and $a_1 > a_2$. This has already been used in Chapter 4 to order the monomial basis used to construct algebraic–geometric codes. The lexicographic order of the weighted $(1, 1)$-degree of the sequence of monomials $x^a y^b$ is actually the Total Graduated Degree Order $<_T$ defined in (4.9).

To list decode a (n, k) Reed–Solomon code we use the $(1, k - 1)$-weighted degree. For example, in order to decode a $(7, 5)$ RS code defined in GF(8), $(1, 4)$-lexicographic order is used. The generation of this order is shown in Table 5.1. The entries in Table 5.1a and 5.1b represent the $(1, 4)$-weighted degree and $(1, 4)$-lexicographic order of monomials M with x degree a and y degree b, respectively. Applying (5.1) with $u = 1$ and $v = 4$, we can generate the $(1, 4)$-weighted degree of monomials M shown by Table 5.1a. For example, for monomial $x^2 y$, its $(1, 4)$-weighted degree is

Table 5.1 (a) $(1, 4)$-weighted degree of monomial $x^a y^b$; (b) $(1, 4)$-lexicographic order of monomial $x^a y^b$.

a/b	0	1	2	3	4	5	6	7	8	9	10	11	12	...
							(a)							
0	0	1	2	3	4	5	6	7	8	9	10	11	12	...
1	4	5	6	7	8	9	10	11	12					
2	8	9	10	11	12					...				
3	12						...							
⋮	⋮													

a/b	0	1	2	3	4	5	6	7	8	9	10	11	12	...
							(b)							
0	0	1	2	3	4	6	8	10	12	15	18	21	24	...
1	5	7	9	11	13	16	19	22	25					
2	14	17	20	23	26					...				
3	27						...							
⋮	⋮													

$\deg_{1,4}(x^2 y) = 2 \cdot 1 + 4 = 6$, which is shown in Table 5.1a. Based on Table 5.1a, and applying the above lexicographic order rule, we can generate the $(1, 4)$-lexicographic order of monomials M shown in Table 5.1b and denoted as $\mathrm{ord}(M)$. From Table 5.1b, it can be observed that $x^4 < x^2 y < y^2$, since $\mathrm{ord}(x^4) = 4$, $\mathrm{ord}(x^2 y) = 9$ and $\mathrm{ord}(y^2) = 14$.

With the definition of the weighted degree and order of a monomial, we can define the weighted degree of a nonzero bivariate polynomial in $F_q[x, y]$ as the weighted degree of its leading monomial, M_L. Any nonzero bivariate polynomial $Q(x, y)$ can be written as:

$$Q(x, y) = Q_0 M_0 + Q_1 M_1 + \cdots + Q_L M_L, \qquad (5.3)$$

with bivariate monomial $M = x^a y^b$ ordered as $M_0 < M_1 < \cdots < M_L$, Q_0, $Q_1, \ldots, Q_L \in \mathrm{GF}(q)$ and $Q_L \neq 0$. The $(1, k - 1)$-weighted degree of $Q(x, y)$ can be defined as:

$$\deg_{1,k-1}(Q(x, y)) = \deg_{1,k-1}(M_L) \qquad (5.4)$$

L is called the leading order (lod) of polynomial $Q(x, y)$, defined as:

$$\mathrm{lod}(Q(x, y)) = \mathrm{ord}(M_L) = L. \qquad (5.5)$$

For example, given a polynomial $Q(x, y) = 1 + x^2 + x^2 y + y^2$, applying the above $(1, 4)$-lexicographic order, it has leading monomial $M_L = y^2$. Therefore, $\deg_{1,4}(Q(x, y)) = \deg_{1,4}(y^2) = 8$ and $\mathrm{lod}(Q(x, y)) = \mathrm{ord}(y^2) = 14$. Consequently, any two nonzero polynomials Q and H ($Q, H \in F_q[x, y]$) can be compared with respect to their leading order:

$$Q \leq H, \text{ if } \mathrm{lod}(Q) \leq \mathrm{lod}(H) \tag{5.6}$$

$S_x(T)$ and $S_y(T)$ are denoted as the highest degree of x and y respectively under the $(1, k - 1)$-lexicographic order such that:

$$S_x(T) = \max\{a : \mathrm{ord}(x^a y^0) \leq T\} \tag{5.7}$$
$$S_y(T) = \max\{b : \mathrm{ord}(x^0 y^b) \leq T\}, \tag{5.8}$$

where T is any nonnegative integer. The error-correction capability, τ_m, and the maximum number of candidate messages, l_m, in the output list with respect to a certain multiplicity, m, of the GS algorithm can be stated as [4]:

$$\tau_m = n - 1 - \left\lfloor \frac{S_x(C)}{m} \right\rfloor \tag{5.9}$$

$$l_m = S_y(C), \tag{5.10}$$

in which T in (5.7) and (5.8) is replaced by the nonnegative integer C, defined as:

$$C = n \binom{m + 1}{2} = n \frac{(m + 1)!}{(m + 1 - 2)!2!} = \frac{n(m + 1)m}{2} \tag{5.11}$$

C represents the number of iterations in the interpolation process, or alternatively the maximum number of Hasse derivative evaluations of an individual interpolated polynomial. These parameters will be proven in Section 5.2.5, when the factorization theorem is presented. τ_m and l_m grow monotonically with multiplicity m [4]:

$$\tau_{m_1} \leq \tau_{m_2} \tag{5.12}$$
$$l_{m_1} < l_{m_2}, \tag{5.13}$$

if $m_1 < m_2$.

5.2.2 Interpolation Theorem

From an algebraic geometric point of view, the interpolation process in the GS algorithm requires a hard decision to be made on the received word $(r_0, r_1, \ldots, r_{n-1})$ and the set of affine points corresponding to the location of each symbol in the received word. For Reed–Solomon codes, the affine points are simply the elements of the finite field over which it is defined. Each affine point and received symbol is paired up to form a set of interpolation points (x_i, r_i), where $i = 0, 1, \ldots, n - 1$. The idea of interpolation is to generate a polynomial $Q(x, y)$ that intersects all these points a given number of times. The number of times $Q(x, y)$ intersects each point is called its *multiplicity*, denoted as m, and this parameter determines the error-correcting capability of the GS algorithm. The interpolation polynomial $Q(x, y)$ can be expressed as [4, 6]:

$$Q(x, y) = \sum_{a,b} Q_{ab} x^a y^b, \tag{5.14}$$

where $Q_{ab} \in \mathrm{GF}(q)$.

For $Q(x, y)$ to be satisfied at the point (x_i, r_i), $Q(x_i, r_i) = 0$. We can write x^a and y^b as:

$$x^a = (x - x_i + x_i)^a = \sum_{u \le a} \binom{a}{u} x_i^{a-u} (x - x_i)^u \quad \text{and} \quad y^b = (y - r_i + r_i)^b$$

$$= \sum_{v \le b} \binom{b}{v} r_i^{b-v} (y - r_i)^v$$

and therefore:

$$Q(x, y) = \sum_{a,b} Q_{ab} \sum_{u \le a, v \le b} \binom{a}{u} x_i^{a-u} (x - x_i)^u \binom{b}{v} r_i^v (y - r_i)^v$$

$$= \sum_{u,v} Q_{uv} (x - x_i)^u (y - r_i)^v \tag{5.15}$$

where:

$$Q_{uv}^{(x_i, r_i)} = \sum_{a \ge u, b \ge v} \binom{a}{u} \binom{b}{v} x_i^{a-u} r_i^{b-v}. \tag{5.16}$$

(5.16) is the (u, v)-*Hasse Derivative* evaluation at the point (x_i, r_i) of $Q(x, y)$ and defines the constraints on the coefficients of $Q(x, y)$ in order to have a zero of multiplicity m at that point.

Example 5.1: Given the polynomial $Q(x, y) = \alpha^5 + \alpha^5 x + y + xy$ defined over GF(8), we can use (5.16) to show that $Q(x, y)$ has a zero of multiplicity of at least $m = 2$ at the point $(1, \alpha^5)$. In order to have a zero of at least multiplicity $m = 2$, the (u, v)-Hasse Derivative at this point must be zero for all $u + v < m$, that is (u, v) can be $(0, 0)$, $(1, 0)$ and $(0, 1)$.

$$Q_{00}^{(1,\alpha^5)} = Q_{00} \binom{0}{0} \binom{0}{0} 1^{0-0} (\alpha^5)^{0-0} + Q_{10} \binom{1}{0} \binom{0}{0} 1^{1-0} (\alpha^5)^{0-0}$$

$$+ Q_{01} \binom{0}{0} \binom{1}{0} 1^{0-0} (\alpha^5)^{1-0} + Q_{11} \binom{1}{0} \binom{1}{0} 1^{1-0} (\alpha^5)^{1-0}$$

$$= 1 + 1 + \alpha^5 + \alpha^5 = 0$$

$$Q_{10}^{(1,\alpha^5)} = Q_{10} \binom{1}{1} \binom{0}{0} 1^{1-1} (\alpha^5)^{0-0} + Q_{11} \binom{1}{1} \binom{1}{0} 1^{1-1} (\alpha^5)^{1-0} = \alpha^5 + \alpha^5 = 0$$

$$Q_{01}^{(1,\alpha^5)} = Q_{01} \binom{0}{0} \binom{1}{1} 1^{0-0} (\alpha^5)^{1-1} + Q_{11} \binom{1}{0} \binom{1}{1} 1^{1-0} (\alpha^5)^{1-1} = 1 + 1 = 0.$$

Therefore, since $Q_{00}^{(1,\alpha^5)} = Q_{10}^{(1,\alpha^5)} = Q_{01}^{(1,\alpha^5)} = 0, Q(x, y)$ must have a multiplicity of at least $m = 2$.

We denote the Hasse derivative operator by D_{uv} where [7]:

$$D_{uv}Q(x_i, r_i) = \sum_{a \geq u, b \geq v} Q_{ab} \binom{a}{u} \binom{b}{v} x_i^{a-u} r_i^{b-v}. \qquad (5.17)$$

Therefore, the interpolation of the GS algorithm can be generalized as: Find a minimal $(1, k-1)$-weighted degree polynomial $Q(x, y)$ that satisfies:

$$Q(x, y) = \min_{\text{lod}(Q)} \{Q(x, y) | D_{uv}Q(x_i, r_i) = 0 \quad \text{for} \quad i = 0, 1, \dots, n \quad \text{and}$$

$$u + v < m\}. \qquad (5.18)$$

5.2.3 Iterative Polynomial Construction

To find the interpolated polynomial of (5.18), an iterative polynomial construction algorithm [4, 8–12] is employed. In this algorithm a group of polynomials is initialized, tested by applying the Hasse derivative (5.16) and modified interactively. The interactive modification between two polynomials is based on the following two properties of the Hasse derivative [7, 10]:

Property 1: Linear functional of Hasse derivative   If $H, Q \in F_q[x, y]$, δ_1 and $\delta_2 \in$ GF(q), then:

$$D(\delta_1 H + \delta_2 Q) = \delta_1 D(H) + \delta_2 D(Q). \qquad (5.19)$$

Property 2: Bilinear Hasse derivative   If $H, Q \in F_q[x, y]$, then:

$$[H, Q]_D = HD(Q) - QD(H). \qquad (5.20)$$

If the Hasse derivative evaluation of $D(Q) = \delta_1$ and of $D(H) = \delta_2$ $(d_1, d_2 \neq 0)$, based on Property 1, it is straightforward to prove that the Hasse derivative evaluation of (5.20) is zero, as follows:

$$D([H, Q]_D) = D(HD(Q) - QD(H)) = D(\delta_1 H - \delta_2 Q).$$

Using Property 1:

$$D(\delta_1 H - \delta_2 Q) = \delta_1 D(H) - \delta_2 D(Q) = \delta_1 \delta_2 - \delta_2 \delta_1 = 0.$$

Therefore:

$$D\left([H, Q]_D\right) = 0. \tag{5.21}$$

If $\text{lod}(H) > \text{lod}(Q)$, the new constructed polynomial from (5.20) has leading order $\text{lod}(H)$. Therefore, by performing the bilinear Hasse derivative over two polynomials with nonzero Hasse derivatives, we can reconstruct a polynomial which has a Hasse derivative of zero. Based on this principle, the implementation of an algorithm for interpolation will iteratively modify a set of polynomials through all n points and with every possible (u, v) pair under each point.

With multiplicity m, there are $\binom{m+1}{2}$ pairs of (u, v), which are arranged as:
$(u, v) = (0, 0), (0, 1), \ldots, (0, m-1), (1, 0), (1, 1), \ldots, (1, m-2), \ldots, (m-1, 0)$.
Therefore, when decoding a (n, k) Reed–Solomon code with multiplicity m, there are $C = n\binom{m+1}{2}$ iterations required to construct a polynomial defined by (5.18).

At the start of the algorithm, a group of polynomials is initialized as:

$$G_0 = \{Q_{0,j} = y^j j = 0, 1, \ldots, l_m\}, \tag{5.22}$$

where l_m is the maximum number of messages in the output list defined by (5.10). If M_L denotes the leading monomial of polynomial Q, it is important to point out that:

$$Q_{0,j} = \min\{Q(x, y) \in F_q[x, y] | \deg y(M_L) = j\}. \tag{5.23}$$

Let i_k denotes the iteration index of the algorithm, where $i_k = i\binom{m+1}{2} + r, i = 0, 1, \ldots, n-1$ and $r = 0, 1, \ldots, \binom{m+1}{2} - 1$. For iteration i_k of the algorithm, each polynomial $Q_{i_k,j}$ in group G_{i_k} is tested by (5.17), and the value of each after Hasse derivative evaluation is denoted by Δ_j as:

$$\Delta_j = D_{i_k}(Q_{i_k,j}). \tag{5.24}$$

Those polynomials with $\Delta_j = 0$ do not need to be modified. However, those polynomials with $\Delta_j \neq 0$ need to be modified based on (5.20). In order to construct a group of polynomials which satisfy:

$$Q_{i_k+1,j} = \min\{Q \in F_q[x, y] | D_{i_k}(Q_{i_k+1,j}) = 0, D_{i_k-1}(Q_{i_k+1,j}) = 0, \ldots,$$
$$D_0(Q_{i_k+1,j}) = 0, \text{ and } \deg y(M_L) = j\}, \tag{5.25}$$

the minimal polynomial among those polynomials with $\Delta_j \neq 0$ is chosen. Denote the index of the minimal polynomial as j' and record it as Q':

$$j' = \text{index}(\min\{Q_{i_k,j}|\Delta_j = 0\}) \qquad (5.26)$$

$$Q' = Q_{i_k,j'}. \qquad (5.27)$$

For the remaining polynomials with $\Delta_j \neq 0$ but $j \neq j'$, (5.20) is used to modify them without the leading order being increased:

$$Q_{i_k+1,j'} = [Q_{i_k,}, Q']_{D_{i_k}} = \Delta_j Q_{i_k,j} - Q'. \qquad (5.28)$$

Based on (5.21), we know that $D_{i_k}(Q_{i_k+1,j}) = 0$. As $\text{lod}(Q_{i_k,j}) > \text{lod}(Q')$, it follows that $\text{lod}(Q_{i_k+1,j}) = \text{lod}(Q_{i_k,j})$.

Q' is modified by (5.20) with the leading order increasing:

$$Q_{i_k+1,j'} = [xQ', Q']_{D_{i_k}} = (x - x_i)Q', \qquad (5.29)$$

where x_i is the x-coordinate of current interpolating point (x_i, r_i). $\Delta_{j*} = D_{i_k}(Q') \neq 0$ and so, as $D_{i_k}(xQ') \neq 0$, $D_{i_k}(Q_{i_k+1,j'}) = 0$. As $\text{lod}(xQ') > \text{lod}(Q')$, $\text{lod}(Q_{i_k+1,j'}) = \text{lod}(xQ') > \text{lod}(Q_{i_k,j'})$. Therefore whenever (5.29) is performed, we have: $\text{lod}(Q_{i_k+1,j}) > \text{lod}(Q_{i_k,j})$.

After C iterative modifications, the minimal polynomial in group G_c is the interpolated polynomial that satisfies (5.18), and it is chosen to be factorized in the next step:

$$Q(x, y) = \min\{Q_{C,j}|Q_{C,j} \in G_C\}. \qquad (5.30)$$

5.2.4 Complexity Reduced Modification

Based on the above analysis, it can be observed that when decoding a (n, k) Reed–Solomon code with multiplicity m, $l_m + 1$ bivariate polynomials are being interactively modified over C iterative steps in which Hasse derivative evaluation (5.17) and bilinear Hasse derivative modification (5.20) are being performed. This process has a complexity of approximately $O(n^2 m^4)$ [4] and is responsible for the GS algorithm's high decoding complexity. Therefore, reducing the complexity of interpolation is essential to improving the algorithm's efficiency.

The leading order of the polynomial group G_{i_k} is defined as the minimal leading order (lod) among the group's polynomials [13]:

$$\text{lod}(G_{i_k}) = \min\{\text{lod}(Q_{i_k,j})|Q_{i_k,j} \in G_{i_k}\}. \qquad (5.31)$$

Based on the initialization defined in (5.22), the leading order of polynomial group G_0 is $\text{lod}(G_0) = \text{lod}(Q_{0,0}) = 0$. In the i_k modification, if no polynomial needs to be modified then the polynomial group is unchanged; $\text{lod}(G_{i_k+1}) = \text{lod}(G_{i_k})$. When a polynomial needs to be modified, (5.29) must be used. If M_L is the leading monomial of Q^*, we have [13]:

$$\text{lod}(xQ^*) = \text{lod}(Q^*) + \left\lfloor \frac{\deg_x Q^*}{k-1} \right\rfloor + \deg_y(M_L) + 1 \qquad (5.32)$$

when k is the dimension of the code. Based on (5.32), it can be seen that $\text{lod}(G_{i_k})$ will be increased if Q^* is the minimal polynomial in the group G_{i_k}. The leading order increase guarantees that in the i_k iterative step, the leading order of the polynomials group G_{i_k} is always less than or equal to i_k:

$$\text{lod}(G_{i_k}) \leq i_k. \qquad (5.33)$$

Based on (5.33), after C iterative steps we have:

$$\text{lod}(G_C) \leq C. \qquad (5.34)$$

From (5.30) we know that only the minimal polynomial is chosen from the polynomial group G_C as $Q(x, y) = \{Q_{c,j} | Q_{c,j} \in G_c \text{ and } \text{lod}(Q_{c,j}) = \text{lod}(G_c)\}$, therefore:

$$\text{lod}(Q(x, y)) \leq C, \qquad (5.35)$$

which means the interpolated polynomial $Q(x, y)$ has leading order less than or equal to C. Those polynomials with leading order over C will not be candidates for $Q(x, y)$. Therefore, during the iterative process, we can modify the group of polynomials by eliminating those with leading order greater than C, as [13]:

$$G_{i_k} = \{Q_{i_k,j} | \text{lod}(Q_{i_k,j}) = C\}. \qquad (5.36)$$

We now prove this modification will not affect the final result. In iteration i_k, if there is a polynomial $Q_{i_k,j}$ with $\text{lod}(Q_{i_k,j}) > C$, it may be modified by either (5.28) or (5.29), which will result in its leading order being unchanged or increased. Therefore, at the end $\text{lod}(Q_{c,j}) > C$, and based on (5.35) it cannot be $Q(x, y)$. However, if $Q_{i_k,j}$ is the minimal polynomial defined by (5.27), this implies that those polynomials with leading order less than C do not need to be modified. If $Q_{i_k,j}$ is not the minimal polynomial defined by (5.27), $Q_{i_k,j}$ will not be chosen to perform bilinear Hasse derivative (5.28) with other polynomials. Therefore, $Q(x, y)$ has no information introduced from $Q_{i_k,j}$, since $\text{lod}(Q_{i_k,j}) > C$. As a result, eliminating the polynomials with leading order greater than C will not affect the final outcome.

This complexity modification scheme can be generally applied to the iterative interpolation process, for example to soft-decision list decoding of Reed–Solomon codes and hard/soft-decision list decoding of Hermitian codes. Based on the total number of iterations C for interpolation, the interpolated polynomial's leading order always satisfies $\text{lod}(Q(x, y)) \leq C$. It implies that those polynomials in the group G can be eliminated once their leading order is greater than C.

This modification can reduce some unnecessary computation in terms of avoiding Hasse derivative evaluation (5.24) and bilinear Hasse derivative modification (5.28) and (5.29) of polynomials with leading order over C. Based on the above analysis, the modified interpolation process can be summarized as:

Algorithm 5.1: Interpolation for list decoding a (n, k) Reed–Solomon code [13, 14]

1. Initialize a group of polynomials by (5.22), and set the index of the interpolated point $i = 0$.
2. Set interpolation point to (x_i, r_i).
3. For each (u, v) where $u + v < m$.
   {
4. Modify the polynomial group by (5.36).
5. Perform Hasse derivative evaluation (5.24) for each polynomial in the group.
6. If all the polynomials' Hasse derivative evaluations are zero, choose another pair (u, v) and go to step 3.
7. Find the minimal polynomial defined by (5.26) and (5.27).
8. For the minimal polynomial, modify it by (5.29). For the other polynomials with nonzero Hasse derivative evaluation, modify them by (5.28).
   }
9. $i = i + 1$.
10. If $i = n$, stop the process and choose $Q(x, y)$ defined by (5.30). Else go to step 2.

Example 5.2 shows the modified interpolation process.

Example 5.2: Decoding the (7, 2) Reed–Solomon code defined over GF(8) with multiplicity m $= 2$ As $C = 7\binom{3}{1} = 21$, based on (5.9) and (5.10) we have $\tau_2 = 3$ and $l_2 = 5$. The transmitted code word is generated by evaluating the message polynomial $f(x) = \alpha + \alpha^6 x$ over the set of points $x = (1, \alpha, \alpha^3, \alpha^2, \alpha^6, \alpha^4, \alpha^5)$, and the corresponding received word is $R = (\alpha^5, \alpha^3, \alpha^4, 0, \alpha^6, \alpha^2, \alpha^2)$, where α is a primitive element in GF(8) satisfying $\alpha^3 + \alpha + 1 = 0$. Construct a bivariate polynomial that has a zero of multiplicity $m = 2$ over the n points $(x_i, r_i)|_{i=0}^{n-1}$.

At the beginning, six polynomials are initialized as:

$Q_{0,0} = 1, Q_{0,1} = y, Q_{0,2} = y^2, Q_{0,3} = y^3, Q_{0,4} = y^4$ and $Q_{0,5} = y^5$. Their leading orders are $\text{lod}(Q_{0,0}) = 0, \text{lod}(Q_{0,1}) = 2, \text{lod}(Q_{0,2}) = 5, \text{lod}(Q_{0,3}) = 9, \text{lod}(Q_{0,4}) = 14$ and $\text{lod}(Q_{0,5}) = 20$ respectively. $\text{lod}(G_0) = \text{lod}(Q_{0,0}) = 0$.

When $i = 0$ and $(u, v) = (0, 0)$, $i_k = 0$. No polynomial is eliminated from the group G_0.

Perform Hasse derivative evaluation for each of the polynomials in G_0 as:

$$\Delta_0 = D_{(0,0)}^{(1,\alpha^5)} Q_{0,0} = 1, \Delta_1 = D_{(0,0)}^{(1,\alpha^5)} Q_{0,1} = \alpha^5$$

$$\Delta_2 = D_{(0,0)}^{(1,\alpha^5)} Q_{0,2} = \alpha^3, \Delta_3 = D_{(0,0)}^{(1,\alpha^5)} Q_{0,3} = \alpha$$

$$\Delta_4 = D_{(0,0)}^{(1,\alpha^5)} Q_{0,4} = \alpha^7, \Delta_5 = D_{(0,0)}^{(1,\alpha^5)} Q_{0,5} = \alpha.$$

Find the minimal polynomial with $\Delta_j \neq 0$ as:

$$j' = 0 \quad \text{and} \quad Q' = Q_{0,0}.$$

Modify polynomials in G_0 with $\Delta_j \neq 0$ as:

$$Q_{1,0} = \Delta_0(x - x_0)Q' = 1 + x, \text{ and } \text{lod}(Q_{1,0}) = 1$$
$$Q_{1,1} = \Delta_0 Q_{0,1} - \Delta_1 Q' = \alpha^5 + y \text{ and } \text{lod}(Q_{1,1}) = 2$$
$$Q_{1,2} = \Delta_0 Q_{0,2} - \Delta_2 Q' = \alpha^3 + y^2 \text{ and } \text{lod}(Q_{1,2}) = 5$$
$$Q_{1,3} = \Delta_0 Q_{0,3} - \Delta_3 Q' = \alpha + y^3 \text{ and } \text{lod}(Q_{1,3}) = 9$$
$$Q_{1,4} = \Delta_0 Q_{0,4} - \Delta_4 Q' = \alpha^6 + y^4 \text{ and } \text{lod}(Q_{1,4}) = 14$$
$$Q_{1,5} = \Delta_0 Q_{0,5} - \Delta_5 Q' = \alpha^4 + y^5 \text{ and } \text{lod}(Q_{1,5}) = 20$$
$$\text{lod}(G_1) = \text{lod}(Q_{1,0}) = 1.$$

When $i = 0$ and $(u, v) = (0, 1)$, $i_k = 1$. No polynomial is eliminated from the group G_1.

Perform Hasse derivative evaluation for each of the polynomial in G_1 as:

$$\Delta_0 = D_{(0,1)}^{(1,\alpha^5)}(Q_{1,0}) = 0. \ \Delta_1 = D_{(0,1)}^{(1,\alpha^5)}(Q_{1,1}) = 1$$

$$\Delta_2 = D_{(0,1)}^{(1,\alpha^5)}(Q_{1,2}) = 0. \ \Delta_3 = D_{(0,1)}^{(1,\alpha^5)}(Q_{1,3}) = \alpha^3$$

$$\Delta_4 = D_{(0,1)}^{(1,\alpha^5)}(Q_{1,4}) = 0. \ \Delta_5 = D_{(0,1)}^{(1,\alpha^5)}(Q_{1,5}) = \alpha^6.$$

Find the minimal polynomial with $\Delta_j \neq 0$ as:

$$j' = 1 \quad \text{and} \quad Q' = Q_{1,1}.$$

As $\Delta_0 = \Delta_2 = \Delta_4 = 0$:

$$Q_{2,0} = Q_{1,0} = 1 + x, \text{ and } \text{lod}(Q_{2,0}) = 1$$
$$Q_{2,2} = Q_{1,2} = \sigma^3 + y^2, \text{ and } \text{lod}(Q_{2,2}) = 5$$
$$Q_{2,4} = Q_{1,4} = \sigma^6 + y^4, \text{ and } \text{lod}(Q_{2,4}) = 14.$$

Modify polynomials in G_1 with $\Delta_j \neq 0$ as:

$$Q_{2,1} = \Delta_1(x - x_0)Q' = \alpha^5 + \alpha^5 x + y(1 + x), \ \ \text{lod}(Q_{2,1}) = 4$$
$$Q_{2,3} = \Delta_1 Q_{1,3} - \Delta_3 Q' = \alpha^3 y + y^3, \ \ \text{lod}(Q_{2,3}) = 9$$
$$Q_{2,5} = \Delta_1 Q_{2,5} - \Delta_5 Q' = \alpha^6 y + y^5, \ \ \text{lod}(Q_{2,5}) = 20$$
$$\text{lod}(G_2) = \text{lod}(Q_{2,0}) = 1.$$

When $i = 0$ and $(u, v) = (1, 0)$, $i_k = 2$. No polynomial is eliminated from the group G_2.

Perform Hasse derivative evaluation for each of the polynomial in G_2 as:

$$\Delta_0 = D_{(1,0)}^{(1,\alpha^5)}(Q_{2,0}) = 1. \ \ \Delta_1 = D_{(1,0)}^{(1,\alpha^5)}(Q_{2,1}) = 0$$
$$\Delta_2 = D_{(1,0)}^{(1,\alpha^5)}(Q_{2,2}) = 0. \ \ \Delta_3 = D_{(1,0)}^{(1,\alpha^5)}(Q_{2,3}) = 0$$
$$\Delta_4 = D_{(1,0)}^{(1,\alpha^5)}(Q_{2,4}) = 0. \ \ \Delta_5 = D_{(1,0)}^{(1,\alpha^5)}(Q_{2,5}) = 0.$$

Find the minimal polynomial with $\Delta_j \neq 0$ as:

$$j' = 0 \text{ and } Q' = Q_{2,0}.$$

As $\Delta_1 = \Delta_2 = \Delta_3 = \Delta_4 = \Delta_5 = 0$:

$$Q_{3,1} = Q_{2,1} = \alpha^5 + \alpha^5 x + y(1 + x), \ \ \text{lod}(Q_{3,1}) = 4$$
$$Q_{3,2} = Q_{2,2} = \alpha^3 + y^2, \ \ \text{and } \text{lod}(Q_{3,2}) = 5$$
$$Q_{3,3} = Q_{2,3} = \alpha^3 y + y^3, \ \ \text{lod}(Q_{3,3}) = 9$$
$$Q_{3,4} = Q_{2,4} = \alpha^6 + y^4, \ \ \text{and } \text{lod}(Q_{3,4}) = 14$$
$$Q_{3,5} = Q_{2,5} = \alpha^6 y + y^5, \ \ \text{lod}(Q_{3,5}) = 20.$$

Modify polynomials in G_2 with $\Delta_j \neq 0$ as:

$$Q_{3,0} = \Delta_0(x - x_0)Q' = 1 + x^2, \ \ \text{lod}(Q_{3,0}) = 3$$
$$\text{lod}(G_3) = \text{lod}(Q_{3,0}) = 3.$$

Based on the same process, interpolation is run through all the rest of the points (x_i, r_i) ($i = 1$ to 6). In order to illustrate the complexity reducing modification scheme, Table 5.2 shows the whole iterative process with respect to the polynomials' leading order.

From Table 5.2 we can see that the modified algorithm starts to take action at $i_k = 10$ when there is at least one polynomial with leading order over 21 and eliminating such polynomials will not affect the final outcome. At the end, both the original and the modified GS algorithm produce the same result: $Q(x, y) = \min\{G_{21}\} = Q_{21,2} = 1 + \alpha^4 x^2 + \alpha^2 x^4 + y^2(\alpha^5 + \alpha^4 x^2)$. From this example we can see that more computation can be reduced if the modified algorithm starts to take action at earlier steps.

Table 5.2 Iterative process of Example 5.2.

i (i_k)	0 (0)	0 (1)	0 (2)	1 (3)	1 (4)	1 (5)	2 (6)	2 (7)	2 (8)	3 (9)
$lod(Q_{i_k,0})$	0	1	1	3	6	6	10	15	15	21
$lod(Q_{i_k,1})$	2	2	4	4	4	7	7	7	11	11
$lod(Q_{i_k,2})$	5	5	5	5	5	5	5	5	5	5
$lod(Q_{i_k,3})$	9	9	9	9	9	9	9	9	9	9
$lod(Q_{i_k,4})$	14	14	14	14	14	14	14	14	14	14
$lod(Q_{i_k,5})$	20	20	20	20	20	20	20	20	20	20
$lod(G_{i_k})$	0	1	1	3	4	5	5	5	5	5

Original GS

i (i_k)	3 (10)	3 (11)	4 (12)	4 (13)	4 (14)	5 (15)	5 (16)	5 (17)	6 (18)	6 (19)	6 (20)	7 (21)
$lod(Q_{i_k,0})$	28	28	36	45	45	55	55	55	55	66	66	78
$lod(Q_{i_k,1})$	11	16	16	16	22	22	22	22	22	22	29	29
$lod(Q_{i_k,2})$	5	5	5	5	5	5	8	8	12	12	12	$\boxed{12}$
$lod(Q_{i_k,3})$	9	9	9	9	9	9	9	13	13	13	13	13
$lod(Q_{i_k,4})$	14	14	14	14	14	14	14	14	14	14	14	14
$lod(Q_{i_k,5})$	20	20	20	20	20	20	20	20	20	20	20	20
$lod(G_{i_k})$	5	5	5	5	5	5	8	8	12	12	12	12

Modified GS

i (i_k)	3 (10)	3 (11)	4 (12)	4 (13)	4 (14)	5 (15)	5 (16)	5 (17)	6 (18)	6 (19)	6 (20)	7 (21)
$lod(Q_{i_k,0})$	—	—	—	—	—	—	—	—	—	—	—	—
$lod(Q_{i_k,1})$	11	16	16	16	—	—	—	—	—	—	—	—
$lod(Q_{i_k,2})$	5	5	5	5	5	5	8	8	12	12	12	$\boxed{12}$
$lod(Q_{i_k,3})$	9	9	9	9	9	9	9	13	13	13	13	13
$lod(Q_{i_k,4})$	14	14	14	14	14	14	14	14	14	14	14	14
$lod(Q_{i_k,5})$	20	20	20	20	20	20	20	20	20	20	20	20
$lod(G_{i_k})$	5	5	5	5	5	5	8	8	12	12	12	12

Note: — means the corresponding polynomial is eliminated; □ means the corresponding polynomial is chosen as $Q(x, y)$.

5.2.5 Factorization

In this section, the factorization theorem is explained, followed by a detailed description of an efficient algorithm known as Roth–Ruckenstein's algorithm [15].

As mentioned in Section 5.2.2, given the interpolated polynomial $Q(x, y)$, the transmitted message polynomial $f(x)$ can be found by determining $Q(x, y)$'s y roots.

Lemma 5.1 *If $Q(x, y)$ has a zero of multiplicity at least m over (x_i, r_i) and $p(x)$ is a polynomial in $F_q[x^{k-1}]$ that $p(x_i) = r_i$, then $(x - x_i)^m | Q(x, p(x))$ [6].*

Define $\Lambda(p, R)$ as the number of symbols in received word R that satisfy $p(x_i) = r_i$ as:

$$\Lambda(p, R) = |\{i : p(x_i) = r_i, i = 0, 1, \ldots, n - 1\}|. \tag{5.37}$$

Lemma 5.2 *$p(x)$ is a polynomial in $F_q[x^{k-1}]$ and $p(x_i) = r_i$ for at least $\Lambda(p, R)$ values. If $m\ \Lambda(p, R) > \deg_{1,k-1}(Q(x, y))$ then $y - p(x)|Q(x, y)$, or $Q(x, p(x)) = 0$ [6].*

Based on Lemma 5.1, if $p(x_i) = r_i$ then $(x - x_i)^m | Q(x, y)$. If S is the set of i that satisfies $p(x_i) = r_i$. as $|S| = \Lambda(p, R)$ then $\prod_{i \in S} (x - x_i)^m | Q(x, p(x))$. Assume $g_1(x) = \prod_{i \in S} (x - x_i)^m$ and $g_2(x) = Q(x, p(x))$; therefore $g_1(x)|g_2(x)$. It is obvious that $g_1(x)$ has x-degree $m\ \Lambda(p, R)$ and $g_2(x)$ has x-degree equal to $\deg_{1,k-1}Q(x, y)$. If $m\ \Lambda(p, R) > \deg_{1,k-1} Q(x, y)$ and $g_1(x)|g_2(x)$, the only solution for these two preconditions is $g_2(x) = 0$. Therefore, if $m\ (p, R) > \deg_{1,k-1}(Q(x, y))$, $Q(x, p(x)) = 0$, or equivalently, $y - p(x)|Q(x, y)$.

As $Q(x, y)$ is the interpolated polynomial from the last step, according to (5.35), $\mathrm{lod}(Q) \leq C$. Based on (5.9), $\deg_{1,k-1}(Q(x, y)) \leq S_x(C)$. If $m\ \Lambda(f, R) \geq S_x(C)$ then $m\ \Lambda(f, R) \geq \deg_{1,k-1}(Q(x, y))$. Based on Lemma 5.2, if $\Lambda(f, R) \geq 1 + \lfloor \frac{S_x(C)}{m} \rfloor$ then the transmitted message polynomial $f(x)$ can be found by factorizing $Q(x, y)$. As $\Lambda(f, R)$ represents the number of points that satisfy $r_i = f(x_i) = c_i$, those points that do not satisfy this equation are where the errors are located.

Therefore, the error-correction capability of the GS algorithm is $\tau_m = n - \lfloor \frac{S_x(C)}{m} \rfloor - 1$, which is defined by (5.10). Under $(1, k - 1)$-lexicographic order, $x^0 y^j$ is the maximal monomial with weighted degree $(k - 1)j$. In polynomial $Q(x, y)$ there should not be any monomials with y-degree over $S_y(C)$, otherwise $\mathrm{lod}(Q) > C$. As a result, $\max\{\deg_y Q(x, y)\} \leq S_y(C)$. As the factorization output list contains the y-roots of $Q(x, y)$, and the number of y-roots of $Q(x, y)$ should not exceed its y-degree, the maximal number of candidate messages in the output list is $l_m = S_y(C)$, which is defined by (5.11).

5.2.6 Recursive Coefficient Search

To find the y-roots of the interpolated polynomial $Q(x, y)$, Roth and Ruckenstein [15] introduced an efficient algorithm for factorizing these bivariate polynomials.

In general, the factorization output $p(x) \in F_q[x^{k-1}]$ can be expressed in the form of:

$$p(x) = p_0 + p_1 x + \cdots + p_{k-1} x^{k-1}, \tag{5.38}$$

where $p_0, p_1, \ldots, p_{k-1} \in GF(q)$. In order to find the polynomials $p(x)$, we must determine their coefficients $p_0, p_1, \ldots, p_{k-1}$, respectively. The idea of Roth–Ruckenstein's algorithm is to recursively deduce $p_0, p_1, \ldots, p_{k-1}$ one at a time.

For any bivariate polynomial, if h is the highest degree such that $x^h | Q(x, y)$, we can define [15]:

$$Q^*(x, y) = \frac{Q(x, y)}{x^h}. \tag{5.39}$$

If we denote $p_0 = p(x)$ and $Q_0(x, y) = Q^*(x, y)$, where $Q(x, y)$ is the new interpolated polynomial (5.30), we can define the recursive updated polynomials $p_s(x)$ and $Q_s(x, y)$, where $s \geq 1$, as [15]:

$$p_s(x) = \frac{p_{s-1}(x) - p_{s-1}(0)}{x} = p_s + \cdots + p_{k-1}x^{k-1-s}, \quad (s = k - 1). \quad (5.40)$$

$$Q_s(x, y) = Q_{s-1}^*(x, xy + p_{s-1}). \quad (5.41)$$

Lemma 5.3 *With* $p_s(x)$ *and* $Q_s(x, y)$ *defined by (5.40) and (5.41), when* $s \geq 1$, $(y - p(x))|Q(x, y)$ *if and only if* $(y - p_s(x))|Q_s(x, y)$ *[4].*

This means that if polynomial $p_s(x)$ is a y-root of $Q_s(x, y)$, we can trace back to find the coefficients $p_{s-1}, \ldots, p_1, p_0$ to reconstruct the polynomial $p(x)$, which is the y-root of polynomial $Q(x, y)$.

The first coefficient p_0 can be determined by finding the roots of $Q_0(0, y) = 0$. If we assume that $Q(x, p(x)) = 0$, $p_0(x)$ should satisfy $Q_0(x, p_0(x)) = 0$. When $x = 0$, $Q_0(0, p_0(0)) = 0$. According to (5.38), $p_0(0) = p_0$, therefore p_0 is the root of $Q_0(0, y) = 0$. By finding the roots of $Q_0(0, y) = 0$, a number of different p_0 can be determined. For each p_0, we can deduce further to find the rest of p_s ($s = 1, \ldots, k - 1$) based on the recursive transformation (5.40) and (5.41).

Assume that after $s - 1$ deductions, polynomial $p_{s-1}(x)$ is the y-root of $Q_{s-1}(x, y)$. Based on (5.40), $p_{s-1}(0) = p_{s-1}$ and a number of p_{s-1} can be determined by finding the roots of $Q_{s-1}(0, y) = 0$. For each p_{s-1}, we can find p_s. As $Q_{s-1}(x, p_{s-1}(x)) = 0$, $(y - p_{s-1}(x))|Q_{s-1}(x, y)$. If we define $y = xy + p_{s-1}$, then $(xy + p_{s-1} - p_{s-1}(x))|Q_{s-1}(x, xy + p_{s-1})$. Based on (5.40), $xy + p_{s-1} - p_{s-1}(x) = xy - xp_s(x)$. As $Q_s(x, y) = Q_{s-1}^*(x, xy + p_{s-1})$, $(xy - xp_s(x))|Q_{s-1}(x, xy + p_{s-1})$ and $(y - p_s(x))|Q_s(x, y)$. Therefore, p_s can again be determined by finding the roots of $Q_s(0, y) = 0$. This root-finding algorithm can be explained as a tree-growing process, as shown in Figure 5.2. There can be an exponential number of routes for choosing coefficients p_s ($s = 0, 1, \ldots, k - 1$) to construct $p(x)$. However, the intended $p(x)$ should satisfy $\deg(p(x)) < k$ and $(y - p(x))|Q(x, y)$. Based on (5.40), when $s = k$, $p_k(x) = 0$. Therefore if $Q_k(x, 0) = 0$, or equivalently $Q_k(x, p_k(x)) = 0$, $(y - p_k(x))|Q_k(x, y)$. According to Lemma 5.3, $(y - p(x))|Q(x, y)$ and $p(x)$ is found.

Based on the above analysis, the factorization process can be summarized as [4, 15]:

Algorithm 5.2: Factorization of list decoding a (n, k) Reed–Solomon code [15]

1. Initialize $Q_0(x, y) = Q^*(x, y)$, $s = 0$.
2. Find roots p_s of $Q_s(0, y) = 0$.
3. For each p_s, perform Q transformation (5.41) to calculate $Q_{s+1}(x, y)$.
4. $s = s + 1$.
5. If $s < k$, go to (ii). If $s = k$ and $Q_s(x, 0) \neq 0$, stop this deduction route. If $s = k$ and $Q_s(x, 0) = 0$, trace the deduction route to find $p_{s-1}, \ldots, p_1, p_0$.

Example 5.3 demonstrates Roth–Ruckenstein's algorithm.

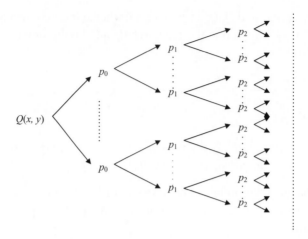

Figure 5.2   Coefficients deduction in Roth–Ruckenstein's algorithm.

Example 5.3: Based on polynomial $Q(x, y) = 1 + \alpha^4 x^2 + \alpha^2 x^4 + y^2(\alpha^5 + \alpha^4 x^2)$, which is the interpolation result of Example 5.2, determine the factorization output list L using Roth–Ruckenstein's algorithm.

   Initialize $Q_0(x, y) = Q^*(x, y) = 1 + \alpha^4 x^2 + \alpha^2 x^4 + y^2(\alpha^5 + \alpha^4 x^2)$ and $s = 0$.
$Q_0(0, y) = 1 + \alpha^5 y^2$ and $p_0 = \alpha$ is the root of $Q_0(0, y) = 0$.
For $p_0 = \alpha$, generate $Q_1(x, y) = Q_0(x, xy + p_0)$:

$$
\begin{aligned}
Q_0(x, xy + \alpha) &= 1 + \alpha^4 x^2 + \alpha^2 x^4 + (xy + \alpha)^2(\alpha^5 + \alpha^4 x^2) \\
&= 1 + \alpha^4 x^2 + \alpha^2 x^4 + (x^2 y^2 + \alpha^2)(\alpha^5 + \alpha^4 x^2) \\
&= 1 + \alpha^4 x^2 + \alpha^2 x^4 + \alpha^5 x^2 y^2 + \alpha^4 x^4 y^2 + 1 + \alpha^6 x^2 \\
&= \alpha^3 x^2 + \alpha^2 x^4 + y^2(\alpha^5 x^2 + \alpha^4 x^4).
\end{aligned}
$$

Now, from (5.39), we can see that x^2 is the highest power of x that divides $Q_0(x, xy + \alpha)$, so:

$$
Q_0^*(x, xy + \alpha) = \frac{Q_0(x, xy + \alpha)}{x^h} = \frac{Q_0(x, xy + \alpha)}{x^2} = \alpha^3 + \alpha^2 x^2 + y^2(\alpha^5 + \alpha^4 x^2).
$$

$s = s + 1 = 1$. As $s < k$, go to step 2 of Algorithm 5.2.
$Q_1(0, y) = \alpha^3 + \alpha^5 y^2$ and $p_1 = \alpha^6$ is a root of $Q_1(0, y) = 0$.
   For $p_1 = \alpha^6$, generate $Q_2(x, y) = Q_1^*(x, xy + p_1) = y^2(\alpha^5 + \alpha^4 x^2)$. $s = s + 1 = 2$. As $s = k$ and $Q_2(x, 0) = 0$, trace this route to find its output $p_0 = \alpha$ and $p_1 = \alpha^6$.
   As a result, factorization output list $L = \{p(x) = \alpha + \alpha^6 x\}$. From Example 5.2, $p(x)$ matches the transmitted message polynomial $f(x)$.

Example 5.4: Factorizing an interpolation polynomial containing two messages Using the same $(7, 2)$ Reed–Solomon code as in Example 5.3, assume that the interpolated polynomial is $Q(x, y) = \alpha x + \alpha^6 x^2 + y(\alpha^3 + \alpha^3 x) + \alpha^2 y^2$.

For $s = 0$, $Q_0(x, y) = Q^*(x, y) = \alpha x + \alpha^6 x^2 + y(\alpha^3 + \alpha^3 x) + \alpha^2 y^2$.

Setting $x = 0$, $Q_0(0, y) = \alpha^3 y + \alpha^2 y^2$ and has two roots, $p_0 = 0$ and α.

Taking $p_0 = 0$:

$$Q_1(x, y) = Q_0^*(x, xy + 0) = \alpha + \alpha^6 x + y(\alpha^3 + \alpha^3 x) + \alpha^2 x y^2.$$

$s = s + 1 = 1$. As $s < k = 2$ we go back to step 2 of Algorithm 5.2.

For $s = 1$, $Q_1(0, y) = \alpha + \alpha^3 y$, which has one root, $p_1 = \alpha^5$.

Taking $p_1 = \alpha^5$:

$$Q_2(x, y) = Q_1^*(x, xy + \alpha^5) = y(\alpha^3 + \alpha^3 x) + \alpha^2 x^2 y^2$$

$s = s + 1 = 2$. Now $s = k$ and $Q_2(x, 0)$, so this route is terminated and the message is $p(x) = 0 + \alpha^5 x = \alpha^5 x$.

However, we must now determine whether there is another message contained within the interpolation polynomial and so we now take the second root of $Q_0(0, y)$.

Taking $p_0 = \alpha$:

$$Q_1(x, y) = Q_0^*(x, xy + \alpha) = \alpha^2 + \alpha^6 x + y(\alpha^3 + \alpha^3 x) + \alpha^2 x y^2$$

$s = s + 1 = 1$. As $s < k = 2$ we go back to step 2 of Algorithm 5.2.

For $s = 1$, $Q_1(0, y) = \alpha^2 + \alpha^3 y$, which has one root, $p_1 = \alpha^6$.

Taking $p_1 = \alpha^6$:

$$Q_2(x, y) = Q_1^*(x, xy + \alpha^6) = y(\alpha^3 + \alpha^3 x) + \alpha^2 x^2 y^2$$

$s = s + 1 = 2$. Now $s = k$ and $Q_2(x, 0)$, so this route is terminated and the message is $p(x) = \alpha + \alpha^6 x$.

The roots of $Q(x, y)$ are shown graphically in Figure 5.3. To complete the decoding procedure, each message would be re-encoded and the code word with the minimum Hamming distance from the received word would be chosen, along with its corresponding message.

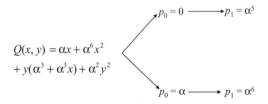

Figure 5.3 Two messages from Roth–Ruckenstein's algorithm for Example 5.3.

5.3 Soft-Decision List Decoding of Reed–Solomon Codes Using the Kötter–Vardy Algorithm

Increases in the performance of a Reed–Solomon code can be achieved by taking into consideration the soft values from the output of the demodulator. These can be used to give a measure of the reliability of each symbol in the received word. To modify the Guruswami–Sudan algorithm to use reliability values instead of hard values, there needs to be a method of mapping reliability values to multiplicity values. Kotter and Vardy presented a seminal paper in 2003 [5] which allowed the algebraic soft-decision decoding of Reed–Solomon codes. In this paper, they also gave an algorithm to convert the reliability of each symbol in the received word to a multiplicity value of each point (x_j, ρ_i) for the interpolation process, where ρ_i can be one of q finite field elements in $\mathrm{GF}(q) = \{\rho_0, \rho_1, \rho_2, \ldots, \rho_{q-1}\}$. The reliability values of each received symbol are arranged in a reliability matrix $\mathbf{\Pi}$ and converted into a multiplicity matrix $\mathbf{M}$. The interpolation and factorization processes then follow in the same way as described previously. The whole system model for soft-decision list decoding is illustrated in Figure 5.4.

5.3.1 Mapping Reliability Values into Multiplicity Values

Instead of making a hard decision on the received symbols, the soft values of each symbol are used to give a measure of reliability. Therefore, the received vector $\mathbf{R} = (r_0, r_1, \ldots, r_{n-1})$ now contains soft values and not finite field elements. The reliability of each received symbol is denoted as $\pi_{i,j}$, which gives the probability of the jth transmitted coded symbol c_j being the $j = i$th element ρ_i in $\mathrm{GF}(q)$, given the soft value of the jth received symbol r_j.

$$p_{i,j} = \mathrm{P}(c_j = \rho_i | r_j)(i = 0, 1, \ldots, q-1 \text{ and } j = 0, 1, \ldots, n-1). \quad (5.42)$$

These reliabilities are entered into a $q \times n$ reliability matrix $\mathbf{\Pi}$.

$$\mathbf{\Pi} = \begin{bmatrix} \pi_{0,0} & \pi_{0,1} & \cdots & \cdots & \cdots & \pi_{0,n-1} \\ \pi_{1,0} & \pi_{1,1} & & & & \pi_{1,n-1} \\ \vdots & & \ddots & & & \vdots \\ \vdots & & & \pi_{i,j} & & \vdots \\ \vdots & & & & \ddots & \vdots \\ \pi_{q-1,0} & \pi_{q-1,1} & \cdots & \cdots & \cdots & \pi_{q-1,n-1} \end{bmatrix}. \quad (5.43)$$

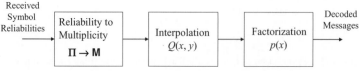

Figure 5.4 System model for soft-decision list decoding.

Referring to Figure 5.1, the matrix $\mathbf{\Pi}$ is taken as an input to the soft-decision decoder and converted to a multiplicity matrix $\mathbf{M}$, followed by the interpolation and factorization processes. An algorithm presented in [5] to convert the reliability matrix $\mathbf{\Pi}$ to a multiplicity matrix $\mathbf{M}$ is now given.

Algorithm 5.3: Convert reliability matrix $\mathbf{\Pi}$ to multiplicity matrix M [5]   Input: Reliability matrix $\mathbf{\Pi}$ and a desired value of the sum of multiplicities in matrix M as:

$$s = \sum_{i=0}^{q-1} \sum_{j=0}^{n-1} m_{i,j}.$$

Initialization: Set $\mathbf{\Pi}^* = \mathbf{\Pi}$ and $q \times n$ all-zero multiplicity matrix $\mathbf{M}$:

1. While $(s > 0)$
   {
2. Find the maximal entry $\pi_{i,j}^*$ in $\mathbf{\Pi}^*$ with position (i, j).
3. Update $\pi_{i,j}^*$ in $\mathbf{\Pi}^*$ as $\pi_{i,j}^* = \frac{\pi_{i,j}}{m_{i,j}+2}$.
4. Update $m_{i,j}$ in $\mathbf{M}$ as $m_{i,j} = m_{i,j} + 1$.
5. $s = s - 1$.
   }

Algorithm 5.3 results in a $q \times n$ multiplicity matrix $\mathbf{M}$, which can be written as:

$$\mathbf{M} = \begin{bmatrix} m_{0,0} & m_{0,1} & \cdots & \cdots & \cdots & m_{0,n-1} \\ m_{1,0} & m_{1,1} & & & & m_{1,n-1} \\ \vdots & & \ddots & & & \vdots \\ \vdots & & & m_{i,j} & & \vdots \\ \vdots & & & & \ddots & \vdots \\ m_{q-1,0} & m_{q-1,1} & \cdots & \cdots & \cdots & m_{q-1,n-1} \end{bmatrix}, \tag{5.44}$$

The entry $m_{i,j}$ represents the multiplicity value of interpolated point (x_j, ρ_i) ($j = 0, 1, \ldots, n - 1$ and $i = 0, 1, \ldots, q - 1$). x_j are the finite field elements used in the encoding process described in Chapter 3. In Algorithm 5.3, the desired value s indicates the total value of multiplicity of all interpolated points. This algorithm gives priority to those interpolated points which correspond to a higher reliability value $\pi_{i,j}$, to be assigned with a higher multiplicity value $m_{i,j}$. For an illustration of the algorithm, see Example 5.5.

Example 5.5: For soft-decision list decoding of the $(7, 2)$ Reed–Solomon code defined in GF(8), the following 8×7 reliability matrix $\mathbf{\Pi}$ is obtained by the receiver:

$$
\mathbf{\Pi} = \begin{bmatrix}
0.959796 & 0.214170 & 0.005453 & 0.461070 & 0.001125 & 0.000505 & 0.691729 \\
0.001749 & 0.005760 & 0.000000 & 0.525038 & 0.897551 & 0.025948 & 0.000209 \\
0.028559 & 0.005205 & 0.000148 & 0.003293 & 0.000126 & 0.018571 & 0.020798 \\
0.000052 & 0.000140 & 0.000000 & 0.003750 & 0.100855 & 0.954880 & 0.000006 \\
0.009543 & 0.736533 & \underline{0.968097} & 0.003180 & 0.000000 & 0.000000 & 0.278789 \\
0.000017 & 0.019810 & 0.000006 & 0.003621 & 0.000307 & 0.000003 & 0.000084 \\
0.000284 & 0.017900 & 0.026295 & 0.000023 & 0.000000 & 0.000002 & 0.008382 \\
0.000001 & 0.000481 & 0.000000 & 0.000026 & 0.000035 & 0.000092 & 0.000003
\end{bmatrix}.
$$

(Note: in the matrix $\mathbf{\Pi}$ ($\mathbf{\Pi}^*$), the maximal entry is underlined).

Apply Algorithm 5.3 with a desired value $s = 20$.

Initialization: Set $\mathbf{\Pi}^* = \mathbf{\Pi}$ and $\mathbf{M} = 0$.

As $s = 20 > 0$, find the maximal entry $\pi^*_{i,j} = 0.968097$ in $\mathbf{\Pi}^*$ with position $(i, j) = (4, 2)$.

Update $\pi^*_{4,2}$ as $\pi^*_{4,2} = \dfrac{\pi_{4,2}}{m_{4,2} + 2} = \dfrac{0.968097}{0 + 2} = 0.484048$.

Update $m_{4,2}$ in $\mathbf{M}$ as $m_{4,2} = 0 + 1 = 1$

$$s = s - 1 = 19.$$

Now the updated $\mathbf{\Pi}^*$ is:

$$
\mathbf{\Pi}^* = \begin{bmatrix}
\underline{0.959796} & 0.214170 & 0.005453 & 0.461070 & 0.001125 & 0.000505 & 0.691729 \\
0.001749 & 0.005760 & 0.000000 & 0.525038 & 0.897551 & 0.025948 & 0.000209 \\
0.028559 & 0.005205 & 0.000148 & 0.003293 & 0.000126 & 0.018571 & 0.020798 \\
0.000052 & 0.000140 & 0.000000 & 0.003750 & 0.100855 & 0.954880 & 0.000006 \\
0.009543 & 0.736533 & 0.484048 & 0.003180 & 0.000000 & 0.000000 & 0.278789 \\
0.000017 & 0.019810 & 0.000006 & 0.003621 & 0.000307 & 0.000003 & 0.000084 \\
0.000284 & 0.017900 & 0.026295 & 0.000023 & 0.000000 & 0.000002 & 0.008382 \\
0.000001 & 0.000481 & 0.000000 & 0.000026 & 0.000035 & 0.000092 & 0.000003
\end{bmatrix}
$$

and the updated $\mathbf{M}$ is:

$$
\mathbf{M} = \begin{bmatrix}
0 & 0 & 0 & 0 & 0 & 0 & 0 \\
0 & 0 & 0 & 0 & 0 & 0 & 0 \\
0 & 0 & 0 & 0 & 0 & 0 & 0 \\
0 & 0 & 0 & 0 & 0 & 0 & 0 \\
0 & 0 & 1 & 0 & 0 & 0 & 0 \\
0 & 0 & 0 & 0 & 0 & 0 & 0 \\
0 & 0 & 0 & 0 & 0 & 0 & 0 \\
0 & 0 & 0 & 0 & 0 & 0 & 0
\end{bmatrix}.
$$

In the next iteration, as $s = 19 > 0$, find the maximal entry $\pi^*_{i,j} = 0.959696$ in $\mathbf{\Pi}^*$ with position $(i, j) = (0, 0)$.

Update $\pi_{0,0}^*$ as $\pi_{0,0}^* = \dfrac{\pi_{0,0}}{m_{0,0} + 2} = \dfrac{0.959796}{0 + 2} = 0.479898.$

Update $m_{0,0}$ in $\mathbf{M}$ as $m_{0,0} = 0 + 1 = 1.$

$$s = s - 1 = 18.$$

Now the updated $\mathbf{\Pi}^*$ is:

$$\mathbf{\Pi}^* = \begin{bmatrix} 0.479898 & 0.214170 & 0.005453 & 0.461070 & 0.001125 & 0.000505 & 0.691729 \\ 0.001749 & 0.005760 & 0.000000 & 0.525038 & 0.897551 & 0.025948 & 0.000209 \\ 0.028559 & 0.005205 & 0.000148 & 0.003293 & 0.000126 & 0.018571 & 0.020798 \\ 0.000052 & 0.000140 & 0.000000 & 0.003750 & 0.100855 & \underline{0.954880} & 0.000006 \\ 0.009543 & 0.736533 & 0.484048 & 0.003180 & 0.000000 & 0.000000 & 0.278789 \\ 0.000017 & 0.019810 & 0.000006 & 0.003621 & 0.000307 & 0.000003 & 0.000084 \\ 0.000284 & 0.017900 & 0.026295 & 0.000023 & 0.000000 & 0.000002 & 0.008382 \\ 0.000001 & 0.000481 & 0.000000 & 0.000026 & 0.000035 & 0.000092 & 0.000003 \end{bmatrix}$$

and the updated $\mathbf{M}$ is:

$$\mathbf{M} = \begin{bmatrix} 1 & 0 & 0 & 0 & 0 & 0 & 0 \\ 0 & 0 & 0 & 0 & 0 & 0 & 0 \\ 0 & 0 & 0 & 0 & 0 & 0 & 0 \\ 0 & 0 & 0 & 0 & 0 & 0 & 0 \\ 0 & 0 & 1 & 0 & 0 & 0 & 0 \\ 0 & 0 & 0 & 0 & 0 & 0 & 0 \\ 0 & 0 & 0 & 0 & 0 & 0 & 0 \\ 0 & 0 & 0 & 0 & 0 & 0 & 0 \end{bmatrix}.$$

Following the same process until $s = 0$, the updated $\mathbf{\Pi}^*$ is:

$$\mathbf{\Pi}^* = \begin{bmatrix} 0.239949 & 0.214170 & 0.005453 & 0.230535 & 0.001125 & 0.000505 & 0.230576 \\ 0.001749 & 0.005760 & 0.000000 & 0.175013 & 0.224388 & 0.025948 & 0.000209 \\ 0.028559 & 0.005205 & 0.000148 & 0.003293 & 0.000126 & 0.018571 & 0.020798 \\ 0.000052 & 0.000140 & 0.000000 & 0.003750 & 0.100855 & 0.238720 & 0.000006 \\ 0.009543 & \underline{0.245511} & 0.242024 & 0.003180 & 0.000000 & 0.000000 & 0.139395 \\ 0.000017 & 0.019810 & 0.000006 & 0.003621 & 0.000307 & 0.000003 & 0.000084 \\ 0.000284 & 0.017900 & 0.026295 & 0.000023 & 0.000000 & 0.000002 & 0.008382 \\ 0.000001 & 0.000481 & 0.000000 & 0.000026 & 0.000035 & 0.000092 & 0.000003 \end{bmatrix}$$

and the updated $\mathbf{M}$ is:

$$\mathbf{M} = \begin{bmatrix} 3 & 0 & 0 & 1 & 0 & 0 & 2 \\ 0 & 0 & 0 & 2 & 3 & 0 & 0 \\ 0 & 0 & 0 & 0 & 0 & 0 & 0 \\ 0 & 0 & 0 & 0 & 0 & 3 & 0 \\ 0 & 2 & 3 & 0 & 0 & 0 & 1 \\ 0 & 0 & 0 & 0 & 0 & 0 & 0 \\ 0 & 0 & 0 & 0 & 0 & 0 & 0 \\ 0 & 0 & 0 & 0 & 0 & 0 & 0 \end{bmatrix}.$$

In the resulting multiplicity matrix $\mathbf{M}$, it can be seen that the sum of its entries $\sum_{i=0}^{7}\sum_{j=0}^{6} m_{i,j} = 20$, which is the desired value s set at the beginning. From $\mathbf{M}$ we see that there are nine nonzero entries, implying that there are now nine points used to generate the interpolation polynomial $Q(x, y)$. These points are:

(x_0, r_0) with $m_{0,0} = 3$, (x_1, ρ_4) with $m_{4,1} = 2$, (x_2, r_4) with $m_{4,2} = 3$,
(x_3, r_0) with $m_{0,3} = 1$, (x_3, r_1) with $m_{1,3} = 2$, (x_4, r_1) with $m_{1,4} = 3$,
(x_5, r_3) with $m_{3,5} = 3$, (x_6, r_0) with $m_{0,6} = 2$, (x_6, r_4) with $m_{4,6} = 1$.

5.3.2 Solution Analysis for Soft-Decision List Decoding

Based on Section 5.2.2, to have a multiplicity of m_{ij} over interpolated point (x_j, ρ_i), the Hasse derivative evaluation of an interpolation polynomial is now defined as:

$$D_{uv} Q(x_j, \rho_i) = \sum_{a \geq u, b \geq v} \binom{a}{u}\binom{b}{v} Q_{ab} x_j^{a-u} \rho_i^{b-v}, u + v < m_{i,j} \tag{5.45}$$

where r_i in (5.17) is replaced with ρ_i. The total number of iterations for all points is:

$$C_M = \frac{1}{2}\sum_{i=0}^{q-1}\sum_{j=0}^{n-1} m_{i,j}(m_{i,j} + 1). \tag{5.46}$$

C_M is called the 'cost' of multiplicity matrix $\mathbf{M}$, which also denotes the number of iterations in the interpolation process.

Based on Lemma 5.1, if $f(x)$ is the message polynomial that satisfies $f(x_j) = c_j$ $(j = 0, 1, \ldots, n - 1)$, polynomial $Q(x, f(x))$ should satisfy:

$$(x - x_0)^{m_0}(x - x_1)^{m_1} \cdots (x - x_{n-1})^{m_{n-1}} | Q(x, f(x)). \tag{5.47}$$

Again, if we let $g_1(x) = (x - x_0)^{m_0}(x - x_1)^{m_1} \cdots (x - x_{n-1})^{m_{n-1}}$ and $g_2(x) = Q(x, f(x))$, based on (5.12), $g_1(x)|g_2(x)$. $g_1(x)$ has x-degree $\deg_x(g_1(x)) = m_0 + m_1 + \cdots + m_{n-1}$. The x-degree of $g_1(x)$ is defined as the code word score S_M with respect to multiplicity matrix $\mathbf{M}$:

$$S_M(\overline{c}) = \deg_x(g_1(x)) = m_0 + m_1 + \cdots + m_{n-1}$$
$$= \sum_{j=0}^{n-1}\{m_{i,j}|\rho_i = c_j, i = 0, 1, \ldots, q - 1\}. \tag{5.48}$$

The x-degree of $g_2(x)$ is bounded by $\deg_x(g_2(x)) \leq \deg_{1,k-1} Q(x, y)$. Therefore, if $S_M(\overline{c}) > \deg_{1,k-1} Q(x, y)$ then $\deg_x(g_1(x)) > \deg_x(g_2(x))$. To satisfy both $\deg_x(g_1(x)) > \deg_x(g_2(x))$ and $g_1(x)|g_2(x)$, the only solution is $g_2(x) = 0$, which indicates

$Q(x, f(x)) = 0$, or equivalently $y - f(x)|Q(x, y)$, and the message polynomial $f(x)$ can be found by determining $Q(x, y)$'s y-roots. As a result, if the code word score with respect to multiplicity matrix $\mathbf{M}$ is greater than the interpolated polynomial $Q(x, y)$'s $(1, k - 1)$-weighted degree

$$S_M(\bar{c}) > \deg_{1,k-1} Q(x, y) \tag{5.49}$$

then $Q(x, f(x)) = 0$, or equivalently $y - f(x)|Q(x, y)$. Message polynomial $f(x)$ can be found by determining the y roots of $Q(x, y)$.

Based on the $(1, k - 1)$-weighted degree definition of monomial $x^a y^b$ given in Section 5.2.1, let us define the following two parameters:

$$N_{1,k-1}(\delta) = \left|\{x^a y^b : a, b \geq 0 \text{ and } \deg_{1,k-1}(x^a y^b) \leq \delta, \delta \in \mathbb{N}\}\right|, \tag{5.50}$$

which represents the number of bivariate monomial $x^a y^b$ with $(1, k - 1)$-weighted degree not greater than a nonnegative integer δ [5]; and:

$$\Delta_{1,k-1}(v) = \min\{\delta : N_{1,k-1}(\delta) > v, v \in \mathbb{N}\}, \tag{5.51}$$

which denotes the minimal value of δ that guarantees $N_{1,k-1}(\delta)$ is greater than a nonnegative integer v [5].

If the $(1, k - 1)$-weighted degree of interpolated polynomial Q is δ^*, based on (5.50), Q has at most $N_{1,k-1}(\delta^*)$ nonzero coefficients. The interpolation procedure generates a system of C_M linear equations of type (5.45). The system will be solvable if [5]:

$$N_{1,k-1}(\delta^*) > C_M. \tag{5.52}$$

Based on (5.51), in order to guarantee the solution, the $(1, k - 1)$-weighted degree δ^* of the interpolated polynomial Q should be large enough that:

$$\deg_{1,k-1}(Q(x, y)) = \delta^* = \Delta_{1,k-1}(C_M). \tag{5.53}$$

Therefore, according to (5.49), given the soft-decision code word score (5.48) and the $(1, k - 1)$-weighted degree of the interpolated polynomial Q (5.53), the message polynomial f can be found if:

$$S_M(\bar{c}) > \Delta_{1,k-1}(C_M). \tag{5.54}$$

As the $(1, k - 1)$-weighted degree of the interpolated polynomial $Q(x, y)$ can be determined by (5.53), while $\Delta_{1,k-1}(C_M)$ can be realized by $\Delta_{1,k-1}(C_M) = \deg_{1,k-1}(x^a y^b | \text{ord}(x^a y^b) = C_M)$, a stopping rule for Algorithm 5.3 based on the designed length of output list l can be imposed. This is more realistic for assessing the performance soft-decision list decoding. As the factorization outputs are the y-roots of the interpolated polynomial Q, the maximal number of outputs l_M based on the interpolated polynomial Q is:

$$l_M = \deg_{0,1} Q(x, y) = \left\lfloor \frac{\deg_{1,k-1} Q(x, y)}{k - 1} \right\rfloor = \left\lfloor \frac{\Delta_{1,k-1}(C_M)}{k - 1} \right\rfloor. \tag{5.55}$$

Therefore, after step 5 of Algorithm 5.3, the updated cost C_M of the multiplicity matrix **M** can be determined using (5.46). As C_M has been determined, the interpolated polynomial $Q(x, y)$'s $(1, k - 1)$-weighted degree can be determined by (5.53). (5.55) can then be applied to calculate the maximal number of factorization outputs, l_M. Based on a designed length of output list l, Algorithm 5.3 is stopped once l_M is greater than l.

In practice, due to the decoding complexity restriction, soft-decision list decoding can only be performed based on a designed length of output list l. This output length restriction in fact leads to practical decoding performance degradation. This phenomenon will be seen later when the simulation results are discussed.

As mentioned in Section 5.2.5, to build interpolated polynomial $Q(x, y)$ there are in total C_M (5.46) iterations. Therefore, the iteration index i_k used in Algorithm 5.1 is: $i_k = 0, 1, \ldots, C_M$. Based on a designed length of output list l, the initialization at step 1 of Algorithm 5.1 can be modified as:

$$G_0 = \{Q_{0,j} = y^j, j = 0, 1, \ldots, l\}, \tag{5.56}$$

where l is the designed length of the output list. As there are in total C_M iterations, based on the complexity reducing scheme's description given in Section 5.2.4, the interpolated polynomial Q's leading order is less than or equal to the total number of iterations C_M:

$$\text{lod}(Q(x, y)) \leq C_M. \tag{5.57}$$

This indicates the fact that (5.53) is an upper bound for the interpolated polynomial's $(1, k - 1)$-weighted degree:

$$\deg_{1,k-1} Q(x, y) \leq (C_M). \tag{5.58}$$

Based on (5.56), those polynomials with leading order greater than C_M will neither be chosen as the interpolated polynomial nor be modified with the interpolated polynomial. Therefore they can be eliminated from the polynomial group and the modification at step 2 can be rewritten as:

$$G_{i_k} = \{Q_{i_k,j} | \text{lod}(Q_{i_k,j}) = C_M\}. \tag{5.59}$$

With respect to interpolated point (x_j, ρ_i) and Hasse derivative parameter (u, v), where $u + v < m_{i,j}$, the Hasse derivative evaluation performed at step 3 of Algorithm 5.1 can be modified and determined by (5.45). The resulting process is the same as Algorithm 5.1, with the exception that for polynomial modification in (5.59) the interpolated point's x-coordinate x_i should be replaced by the x_j which is the current interpolated point's x-coordinate. Also, the index j should be used for the interpolated point's x-coordinate x_j and polynomials in the group $Q_{i_k, j}$.

5.3.3 Simulation Results

This section presents both hard-decision and soft-decision list decoding results for two Reed–Solomon codes: (63, 15) and (63, 31). They are simulated on both the AWGN and Rayleigh fading channels. The Rayleigh fading channel is frequency nonselective with Doppler frequency [16] 126.67 Hz and data rate 30 kb/s. The fading profile is generated using Jakes' method [16]. The fading coefficients have mean value 1.55 and variance 0.60. On the Rayleigh fading channel, a block interleaver of size 63 × 63 is used to combat the fading effect. QPSK modulation is used and simulations are run using the C programming language.

Comparisons between hard-decision and soft-decision are made based on output length l. For an output length l, there are $l + 1$ polynomials taking part in the iterative interpolation process. The total number of iterations (C_m in (5.11) for hard-decision and C_M in (5.46) for soft-decision) also grow with length l. The number of polynomials $l + 1$ and the number of iterations (C_m, C_M) are the important parameters that determine the decoding complexity. Based on the same designed length, from Figures 5.5 and 5.6 it can be seen that soft-decision can achieve significant coding gains over hard-decision list decoding, especially on the Rayleigh fading channel. For example, with a designed length $l = 2$, soft-decision list decoding of the (63, 15) Reed–Solomon code can achieve about a 5.8 dB coding gain at BER $= 10^{-5}$ over hard-decision decoding.

According to the analysis in [14], the performance improvement of soft-decision list decoding over hard-decision list decoding is achieved with an insignificant complexity penalty. This is because for the list decoding algorithm, the complexity is mainly dominated by the interpolation process, and the complexity introduced by the a priori process (Algorithm 5.3) is marginal. For the interpolation process, the important parameter that determines its complexity is the iteration number. As the iteration number of soft-decision does not vary much from hard-decision based on the same designed output length, the complexity of soft-decision list decoding is not much higher than that of hard-decision list decoding.

5.4 List Decoding of Algebraic–Geometric Codes

The GS algorithm consists of two processes: interpolation and factorization. Given a received word $R = (r_0, r_1, \ldots, r_{n-1})$ ($r_i \in \mathrm{GF}(q)$, $i = 0, 1, \ldots, n - 1$), n interpolated

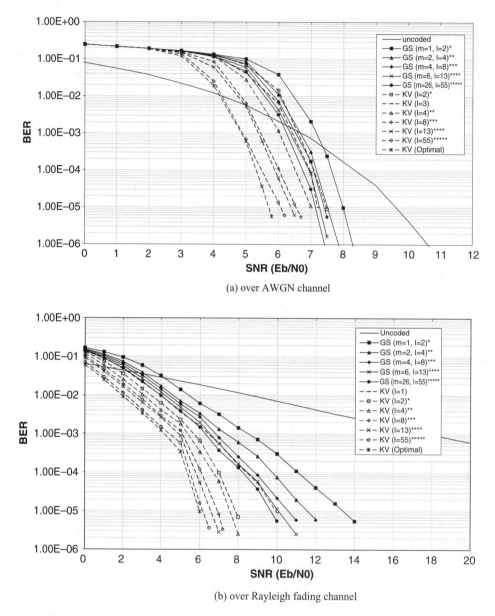

(a) over AWGN channel

(b) over Rayleigh fading channel

Figure 5.5   Hard-decision and soft-decision list decoding Reed–Solomon code (63, 15).

units can be formed by combining each received symbol with its respective affine point used in encoding, as: $(p_0, r_0), (p_1, r_1), \ldots, (p_{n-1}, r_{n-1})$. Interpolation builds the minimal polynomial $Q \in F_q[x, y, z]$, which has a zero of multiplicity of at least m over the n interpolated units. Q can be written as: $Q = \sum_{a,b} Q_{ab} \phi_a z^b$, where $Q_{ab} \in \mathrm{GF}(q)$ and ϕ_a is a set of rational functions with pole orders up to a [17–19]. If (p_i, r_i)

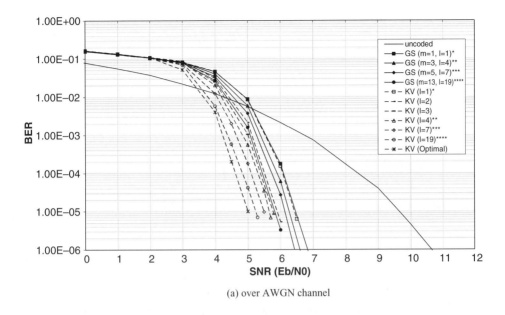

(a) over AWGN channel

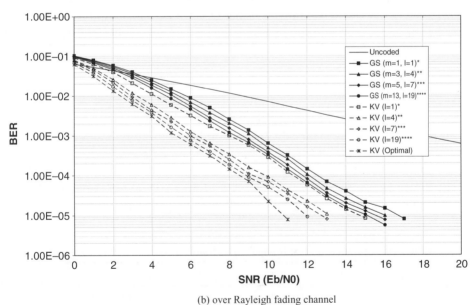

(b) over Rayleigh fading channel

Figure 5.6   Hard-decision and soft-decision list decoding Reed–Solomon code (63, 31).

is the intended interpolated unit, it can also be written with respect to the zero basis functions in Z_{w,p_i} defined in Chapter 2 as [9]:

$$Q = \sum_{u,v} Q_{uv}^{(p_i,r_i)} \psi_{p_i,u}(z - r_i)^v, \tag{5.60}$$

where $Q_{uv}^{(p_i,r_i)} \in GF(q)$. If $Q_{uv}^{(p_i,r_i)} = 0$ for $u + v < m$, polynomial Q has a zero of multiplicity at least m at unit (p_i, r_i) [9, 11]. As $z^b = (z - r_i + r_i)^b = \sum_{v \leq b} \binom{b}{v} r_i^{b-v} (z - r_i)^v$ and $\phi_a = \sum_u \gamma_{a,p_i,u} \psi_{p_i,u}$, substitute them into (5.15):

$$
\begin{aligned}
Q &= \sum_{a,b} Q_{ab} \left(\sum_u \gamma_{a,p_i,u} \psi_{p_i,u} \right) \left(\sum_{v \leq b} \binom{b}{v} r_i^{b-v} (z - r_i)^v \right) \\
&= \sum_{u,v} \left(\sum_{a,b \geq v} Q_{ab} \binom{b}{v} \gamma_{a,p_i,u} r_i^{b-v} \right) \psi_{p_i,u} (z - r_i)^v
\end{aligned}
\tag{5.61}
$$

Therefore, the coefficients $Q_{uv}^{(p_i,r_i)}$ of (5.61) can be written as:

$$
Q_{uv}^{(p_i,r_i)} = \sum_{a,b \geq v} Q_{ab} \binom{b}{v} \gamma_{a,p_i,u} r_i^{b-v}.
\tag{5.62}
$$

(5.62) defines the zero condition constraints on the coefficients Q_{ab} of polynomial Q analogous to the Hasse derivative, so that Q has a zero of multiplicity at least m over unit (p_i, r_i). Example 5.6 shows how to define the zero condition of a polynomial in $F_q[x, y, z]$ using (5.62).

Example 5.6: Given the polynomial $Q(x, y, z) = 1 + \alpha y + \alpha x^2 + z^2 (1 + \alpha^2 y)$ defined in GF(4)$[x, y, z]$ justify the fact that it has a zero of multiplicity at least 2 over the unit $(p, r) = ((1, \alpha), \alpha)$.

Polynomial $Q(x, y, z) = 1 + \alpha y + \alpha x^2 + z^2(1 + \alpha^2 y) = Q_{00}\phi_0 z^0 + Q_{20}\phi_2 z^0 + Q_{30}\phi_3 z^0 + Q_{02}\phi_0 z^2 + Q_{22}\phi_2 z^2$. Supporting the zero condition calculations, the corresponding coefficients $\gamma_{a,p,u}$ are shown in Table 5.3.

Table 5.3 Corresponding coefficients $\gamma_{a,p,u}$ given $p = (1, \alpha)$.

$a \setminus u$	0	1	2	3	...
0	1	1	α	1	...
1	0	1	1	0	...
$\vdots$	$\vdots$	$\vdots$	$\vdots$	$\vdots$	

Based on the above description, to justify that Q has a zero of multiplicity m over unit (p, r), its coefficients Q_{ab} should satisfy $Q_{uv}^{(p,r)} = 0$ for $u + v < 2$ as: $Q_{00}^{(p,r)} = 0$, $Q_{01}^{(p,r)} = 0$ and $Q_{10}^{(p,r)} = 0$.

Based on definition (5.62):

$$Q_{00}^{(p,r)} = Q_{00}\begin{pmatrix}0\\0\end{pmatrix}\gamma_{0,p,0}\alpha^{0-0} + Q_{20}\begin{pmatrix}0\\0\end{pmatrix}\gamma_{2,p,0}\alpha^{0-0} + Q_{30}\begin{pmatrix}0\\0\end{pmatrix}\gamma_{3,p,0}\alpha^{0-0}$$

$$+ Q_{02}\begin{pmatrix}2\\0\end{pmatrix}\gamma_{0,p,0}\alpha^{2-0} + Q_{22}\begin{pmatrix}2\\0\end{pmatrix}\gamma_{2,p,0}\alpha^{2-0}$$

$$= 1 + \alpha^2 + \alpha + \alpha^2 + \alpha^2 = 0$$

$$Q_{01}^{(p,r)} = Q_{02}\begin{pmatrix}2\\1\end{pmatrix}\gamma_{0,p,0}\alpha^{2-1} + Q_{22}\begin{pmatrix}2\\1\end{pmatrix}\gamma_{2,p,0}\alpha^{2-1} = 0 + 0 = 0$$

$$Q_{10}^{(p,r)} = Q_{00}\begin{pmatrix}0\\0\end{pmatrix}\gamma_{0,p,1}\alpha^{0-0} + Q_{20}\begin{pmatrix}0\\0\end{pmatrix}\gamma_{2,p,1}\alpha^{0-0}$$

$$+ Q_{30}\begin{pmatrix}0\\0\end{pmatrix}\gamma_{3,p,1}\alpha^{0-0} + Q_{02}\begin{pmatrix}2\\0\end{pmatrix}\gamma_{0,p,1}\alpha^{2-0} + Q_{22}\begin{pmatrix}2\\0\end{pmatrix}\gamma_{2,p,1}\alpha^{2-0}$$

$$= \alpha + \alpha = 0.$$

Therefore, polynomial Q has a zero of multiplicity at least 2 over unit $(p, r) = (1, \alpha), \alpha)$.

If constraint (5.62) for the coefficients of polynomial Q is denoted as $D_{uv}^{(p_i,r_i)}(Q)$, such that:

$$D_{uv}^{(p_i,r_i)}(Q) = Q_{uv}^{(p_i,r_i)} = \sum_{a,b \geq v} Q_{ab}\begin{pmatrix}b\\v\end{pmatrix}\gamma_{a,p_i,u} r_i^{b-v} \tag{5.63}$$

then interpolation builds a polynomial Q defined as:

$$Q = \min_{\mathrm{lod}(Q)} \left\{ Q \in F_q[x, y, z] | D_{uv}^{(p_i,r_i)}(Q) = 0 \quad \text{for} \quad i = 0, 1, \dots, n-1 \wedge u \right.$$

$$\left. + v < m \ (u, v \in \mathbb{N}) \right\}. \tag{5.64}$$

As there are $\binom{m+1}{2}$ permutations of (u, v) for $u + v < m$, there are in total:

$$C = n\begin{pmatrix}m+1\\2\end{pmatrix} \tag{5.65}$$

zero condition constraints that the coefficients Q_{ab} of polynomial Q need to satisfy. C also represents the number of iterations in the interpolation algorithm [9, 11], in which each iteration imposes a zero condition constraint to Q_{ab}.

Definition: For monomial $\phi_a z^b$, where $\phi_a \in L_w$ and L_w is the Hermitian curve's pole basis defined in $GF(w^2)$, its $(1, w_z)$-weighted degree is defined as:

$$\deg_{1,w_z}(\phi_a z^b) = v_{p_\infty}(\phi_a^{-1}) + b \cdot w_z,$$

where w_z is the weighted degree for variable z, and defined as: $w_z = v_{p_\infty}(z^{-1}) = v_{p_\infty}(\phi_{k-1}^{-1})$. ϕ_{k-1} is the maximal term in the message polynomial. A $(1, w_z)$-lexicographic order (ord) can be defined to arrange monomials $\phi_a z^b$:

$$\phi_{a_1} z^{b_1} < \phi_{a_2} z^{b_2},$$

if $\deg_{1,w_z}(\phi_{a_1} z^{b_1}) < \deg_{1,w_z}(\phi_{a_2} z^{b_2})$, or $\deg_{1,w_z}(\phi_{a_1} z^{b_1}) = \deg_{1,w_z}(\phi_{a_2} z^{b_2})$ and $b_1 < b_2$ [9]. If $\phi_{a'} z^{b'}$ is the maximal monomial in polynomial $Q = \sum_{a,b} Q_{ab} \phi_a z^b$ as:

$$\phi_{a'} z^{b'} = \max\{\phi_a z^b | Q_{ab} \neq 0\}.$$

$\phi_{a'} z^{b'}$ is called Q's leading monomial (LM) and its coefficient $Q_{a'b'}$ is called f's leading coefficient (LC), denoted as: $\text{LM}(f) = \phi_{a'} z^{b'}$ and $\text{LC}(f) = f_{a'b'}$. Polynomial Q's $(1, w_z)$-weighted degree ($\deg_{1,w_z}(Q)$) and leading order (lod(Q)) are defined as:

$$\deg_{1,w_z}(Q) = \deg_{1,w_z}(\phi_{a'} z^{b'}), \quad \text{and} \quad \text{lod}(Q) = \text{ord}(\phi_{a'} z^{b'}).$$

The $(1, w_z)$-weighted degree upper bound of polynomial Q is defined as [9, 11]:

$$\max\{\deg_{1,w_z} Q\} = l_m v_{p_\infty}(z - 1) + t_m, \tag{5.66}$$

where l_m is the maximal number of output candidates from factorization, defined as:

$$l_m = \max\left\{u | \binom{u}{2} v_{p_\infty}(z - 1) - (u - 1)g \leq C\right\} - 1, \tag{5.67}$$

and parameter t_m is defined as:

$$t_m = \max\left\{u | (l_m + 1)u - \Gamma(u) + \binom{l_m + 1}{2} v_{p_\infty}(z - 1) - l_m g \leq C\right\}, \tag{5.68}$$

where g is the genus of the Hermitian curve, $u \in \mathbb{N}$ and $\Gamma(u)$ denotes the number of gaps that are less than or equal to the nonnegative integer u [11].

If there exists a polynomial $h \in F_q^{u_z}[x, y]$ such that:

$$\Lambda(h, R) = |\{i | h(p_i) = r_i, i = 0, 1, \ldots, n - 1\}| \tag{5.69}$$

then the total zero orders of polynomial $Q(x, y, h)$ over all the interpolated units is:

$$\sum_{i=0}^{n-1} v_{p_i}(Q(x, y, h)) = m\Lambda(h, R). \tag{5.70}$$

To define the total zero order of polynomial $Q(x, y, h)$, the following lemma is applied:

Lemma 5.4 *If $Q(x, y, z)$ has a zero of multiplicity m over unit (p_i, r_i) and h is a polynomial in $F_q^{u_z}[x, y]$ that satisfies $h(p_i) = r_i$ then $Q(x, y, h)$ has a zero order of at least m at p_i, as $v_{p_i}(Q(x, y, h)) \geq m$ [9, 11].*

(5.69) defines the total number of affine points that satisfy $h(p_i) = r_i$, and therefore the total zero order of polynomial $Q(x, y, h)$ over all the affine points is defined by (5.70).

Theorem 5.1 *If polynomial $Q(x, y, h)$'s total zero order is greater than its pole order, as:*

$$\sum_{i=0}^{n-1} v_{p_i}(Q(x, y, h)) > v_{p_\infty}(Q(x, y, h)^{-1}), \tag{5.71}$$

then h is the z root of Q: $Q(x, y, h) = 0$, or equivalently $z - h | Q(x, y, z)$ [6, 9, 11].
As $h \in F_q^{u_z}[x, y]$, $v_{p_\infty}(Q(x, y, h)^{-1}) = v_{p_\infty}(Q(x, y, z)^{-1}) = \deg_{1,w_z}(Q(x, y, z))$.
Therefore, based on (5.69) and (5.70), Theorem 5.1 results in the following corollary:

Corollary 5.1 *If there exists a polynomial $h \in F_q^{u_z}[x, y]$ such that:*

$$m \Lambda(h, R) > \deg_{1,w_z}(Q(x, y, z)) \tag{5.72}$$

then the list decoding outputs h can be found by factorizing the interpolated polynomial $Q(x, y, z)$ as: $z - h | Q(x, y, z)$ [18].

If $h = f$, (5.69) defines the number of uncorrupted received symbols. Therefore, the GS algorithm's error-correction capability τ_m is:

$$t_m = n - \Lambda(h, R) = n - \left\lfloor \frac{\deg_{1,w_z} Q}{m} \right\rfloor - 1. \tag{5.73}$$

Since the upper bound of $\deg_{1,w_z} Q$ is defined by (5.70):

$$t_m \geq n - \left\lfloor \frac{l_m v_{p_\infty}(z^{-1}) + t_m}{m} \right\rfloor - 1. \tag{5.74}$$

The GS algorithm's error-correction capability upper bound for a (n, k) Hermitian code is defined by:

$$t_{GS} = n - \left\lfloor \sqrt{n(n - d^*)} \right\rfloor - 1.$$

5.5 Determining the Corresponding Coefficients

Based on (5.62), the corresponding coefficients $\gamma_{a,p_i,u}$ are critical for defining the zero condition of a polynomial in $F_q[x, y, z]$. Without knowing them, we have to transfer a general polynomial written with respect to the zero basis functions and find the coefficients $Q_{uv}^{(p_i,r_i)}$, which is not efficient during the iterative interpolation. In fact, the corresponding coefficients $\gamma_{a,p_i,u}$ can be determined independently of the received word. Therefore, if they can be determined beforehand and applied during the iterations, the interpolation efficiency can be greatly improved. This section proposes an algorithm to determine them.

The problem we intend to solve can be simply stated as: given an affine point $p_i = (x_i, y_i)$ of curve H_w and a pole basis monomial ϕ_a, determine the corresponding coefficients $\gamma_{a,p_i,u}$ so that ϕ_a can be written as a sum of the zero basis functions $\psi_{p_i,u} : \phi_a = \sum_u \gamma_{a,p_i,u} \psi_{p_i,u}$. For any two pole basis monomials ϕ_{a_1} and ϕ_{a_2} in L_w, $\phi_{a_1}\phi_{a_2} = \sum_{a \in N} \phi_a$ and the zero basis function $\psi_{p_i,u}$ (2.13) can be written as a sum of pole basis monomials ϕ_a [9]:

$$\psi_{p_i,u} = \sum_a \zeta_a \phi_a, \tag{5.75}$$

where coefficients $\zeta_a \in \mathrm{GF}(q)$. Partition $\psi_{p_i,u}(x, y)$ as:

$$\psi_{p_i,u} = \psi_{p_i,u}^A \cdot \psi_{p_i,u}^B, \tag{5.76}$$

where $\psi_{p_i,u}^A = (x - x_i)^\lambda$ and $\psi_{p_i,u}^B = [(y - y_i) - x_i^w(x - x_i)]^\delta = [y - x_i^w x - (y_i - x_i^{w+1})]^\delta$. It is easy to recognize that $\psi_{p_i,u}^A$ has leading monomial $\mathrm{LM}(\psi_{p_i,u}^A) = x^\delta$ and leading coefficient $\mathrm{LC}(\psi_{p_i,u}^A) = 1$. As $v_{p_\infty}(y^{-1}) > v_{p_\infty}(x^{-1})$, $\psi_{p_i,u}^B$ has leading monomial $\mathrm{LM}(\psi_{p_i,u}^B) = y^\delta$ and leading coefficient $\mathrm{LC}(\psi_{p_i,u}^B) = 1$. Based on (5.76), $\psi_{p_i,u}$ has leading monomial $\mathrm{LM}(\psi_{p_i,u}^A) \cdot \mathrm{LM}(\psi_{p_i,u}^B) = x^\lambda y^\delta$ and leading coefficient $\mathrm{LC}(\psi_{p_i,u}^A) \cdot \mathrm{LC}(\psi_{p_i,u}^B) = 1$. As $0 \leq \lambda \leq w$ and $\delta \geq 0$, the set of leading monomials of zero basis functions in Z_{w,p_i} contains all the monomials defined in pole basis L_w. Summarizing the above analysis, Corollary 5.2 is proposed as follows:

Corollary 5.2 *If ϕ_L is the leading monomial of zero basis function $\psi_{p_i,u}$ as* $\mathrm{LM}(\psi_{p_i,u}) = \phi_L$*, the leading coefficient of $\psi_{p_i,u}$ equals 1 and (5.75) can be written as [18]:*

$$\psi_{p_i,u} = \sum_{a<L} \zeta_a \phi_a + \phi_L. \tag{5.77}$$

The set of leading monomials of zero basis functions in Z_{w,p_i} contains all the monomials in L_w:

$$\{\mathrm{LM}(\psi_{p_i,u}) = \phi_L, \psi_{p_i,u} \in Z_{w,p_i}\} \subseteq L_w. \tag{5.78}$$

Following on, by identifying the second-largest pole basis monomial ϕ_{L-1} with coefficient $\zeta_{L-1} \in \mathrm{GF}(q)$ in $\psi_{p_i,u}$, (5.77) can also be written as [18]:

$$\psi_{p_i,u} = \sum_{a < L-1} \zeta_a \phi_a + \zeta_{L-1} \phi_{L-1} + \phi_L. \tag{5.79}$$

Now it is sufficient to propose the new efficient algorithm in order to determine the corresponding coefficients $\gamma_{a,p_i,u}$.

Algorithm 5.4: Determine the corresponding coefficients $\gamma_{a,p_i,u}$ between a pole basis monomial and zero basis functions [14, 18]

1. Initialize all corresponding coefficients $\gamma_{a,p_i,u} = 0$.
2. Find the zero basis function $\psi_{p_i,u}$ with $\mathrm{LM}(\psi_{p_i,u}) = \phi_a$, and let $\gamma_{a,p_i,u} = 1$.
3. Initialize function $\hat{\psi} = \psi_{p_i,u}$.
4. While ($\hat{\psi} \neq \phi_a$)
   {
5. Find the second-largest pole basis monomial ϕ_{L-1} with coefficient ζ_{L-1} in $\hat{\psi}$.
6. In Z_{w,p_i}, find a zero basis function $\psi_{p_i,\alpha}$ whose leading monomial $\mathrm{LM}(\psi_{p_i,u}) = \phi_{L-1}$, and let the corresponding coefficient $\gamma_{a,p_i,u} = \zeta_{L-1}$.
7. Update $\hat{\psi} = \hat{\psi} + \gamma_{a,p_i,u} \psi_{p_i,u}$.
   }

Proof: Notice that functions $\psi_{p_i,\alpha}$ with $\mathrm{LM}(\psi_{p_i,u}) > \phi_a$ will not contribute to the sum calculation of (5.62) and their corresponding coefficients $\gamma_{a,p_i,u} = 0$. The zero basis function $\psi_{p_i,u}$ found at step 2 has leading monomial $\phi_L = \phi_a$. Based on (5.79), it can be written as [18]:

$$\psi_{p_i,u} = \sum_{a' < L-1} \zeta_{a'} \phi_{a'} + \zeta_{L-1} \phi_{L-1} + \phi_a \tag{5.80}$$

(5.80) indicates that the corresponding coefficient between ϕ_a and $\psi_{p_i,u}$ is 1; $\gamma_{a,p_i,u} = 1$. Polynomial $\hat{\psi}$, initialized by step 3, is an accumulated polynomial resulting in ϕ_a. While $\hat{\psi} \neq \phi_a$, in (5.80), the second-largest monomial ϕ_{L-1} with coefficient ζ_{L-1} is identified by step 5. Next find another zero basis function $\psi_{p_i,u}$ in Z_{w,p_i} such that $\mathrm{LM}(\psi_{p_i,u}) = \phi_{L-1}$. According to Corollary 5.2, this zero basis function always exists and it can be written as: $\psi_{p_i,u} = \sum_{a' < L-1} \zeta_{a'} \phi_{a'} + \phi_{L-1}$. At step 6, the corresponding coefficient between monomial ϕ_a and the found zero basis function $\psi_{p_i,u}$ can be determined as: $\gamma_{a,p_i,u} = \zeta_{L-1}$. As a result, the accumulated calculation of step 7 can be written as:

$$\hat{\psi} = \sum_{a' < L-1} \zeta_{a'} \phi_{a'} + \zeta_{L-1} \phi_{L-1} + \phi_a + \gamma_{a,p_i,u} \psi_{p_i,u}$$

$$= \sum_{a' < L-1} \zeta_{a'} \phi_{a'} + \zeta_{L-1} \phi_{L-1} + \phi_a + \sum_{a' < L-1} \zeta_{L-1} \zeta_{a'} \phi_{a'} + \zeta_{L-1} \phi_{L-1}. \tag{5.81}$$

Therefore, in the new accumulated $\hat{\psi}$, $\zeta_{L-1}\phi_{L-1}$ is eliminated, while the leading monomial ϕ_a is preserved. If the updated $\hat{\psi} \neq \phi_a$, its second-largest monomial ϕ_{L-1} is again eliminated, while ϕ_a is always preserved as a leading monomial by the same process. The algorithm terminates after all monomials that are smaller than ϕ_a have been eliminated and results in $\hat{\psi} = \phi_a$. This process is equivalent to the sum calculation of (5.62). Example 5.7 illustrates Algorithm 5.4.

Example 5.7: Given $p_i = (\alpha^2, \alpha^2)$ is an affine point on curve H_2 and a pole basis (L_2) monomial $\phi_5 = y^2$, determine the corresponding coefficients $\gamma_{5,p_i,u}$ so that ϕ_5 can be written as $\phi_5 = \sum_u \gamma_{5,p_i,u}\psi_{p_i,u}$.

Based on (2.13), the first eight zero basis functions in Z_{2,p_i} can be listed as:

$$\psi_{p_i,0} = (x - \alpha^2)^0 = 1 \quad \psi_{p_i,1} = (x - \alpha^2)^1 = \alpha^2 + x$$
$$\psi_{p_i,2} = (x - \alpha^2)^2 = \alpha + x^2 \quad \psi_{p_i,3} = (y - \alpha^2) - \alpha(x - \alpha^2) = \alpha + \alpha x + y$$
$$\psi_{p_i,4} = (x - \alpha^2)[(y - \alpha^2) - \alpha(x - \alpha^2)] = 1 + \alpha^2 x + \alpha^2 y + \alpha x^2 + xy$$
$$\psi_{p_i,5} = (x - \alpha^2)^2[(y - \alpha^2) - \alpha(x - \alpha^2)] = \alpha^2 + \alpha^2 x + \alpha x^2 + \alpha y^2 + x^2 y$$
$$\psi_{p_i,6} = [(y - \alpha^2) - \alpha(x - \alpha^2)]^2 = \alpha^2 + \alpha^2 x^2 + y^2$$
$$\psi_{p_i,7} = (x - \alpha^2)[(y - \alpha^2) - \alpha(x - \alpha^2)]^2 = \alpha + \alpha^2 x + \alpha^2 y + \alpha x^2 + xy^2.$$

Initialize all $\gamma_{5,p_i,u} = 0$. In Z_{2,p_i}, as $\text{LM}(\psi_{p_i,6}) = \phi_5$, we let $\gamma_{5,p_i,6} = 1$ and initialize the accumulated polynomial $\hat{\psi} = \psi_{p_i,6} = \alpha^2 + \alpha^2 x^2 + y^2$.

As $\hat{\psi} \neq \phi_5$, its second-largest monomial $\phi_{L-1} = x^2$ with coefficient $\zeta_{L-1} = \alpha^2$ is identified. Among the zero basis functions in Z_{2,p_i}, we find $\psi_{p_i,2}$ with $\text{LM}(\psi_{p_i,2}) = \phi_{L-1} = x^2$, and let $\gamma_{5,p_i,2} = \zeta_{L-1} = \alpha^2$. Update $\hat{\psi} = \hat{\psi} + \gamma_{5,p_i,2}\psi_{p_i,2} = \alpha + y^2$.

As $\hat{\psi} \neq \phi_5$, again its second-largest monomial $\phi_{L-1} = 1$ with coefficient $\zeta_{L-1} = \alpha$ is identified. Among the zero basis functions in Z_{2,p_i}, we find $\psi_{p_i,0}$ with $\text{LM}(\psi_{p_i,0}) = \phi_{L-1} = 1$, and let $\gamma_{5,p_i,0} = \zeta_{L-1} = \alpha$. Update $\hat{\psi} = \hat{\psi} + \gamma_{5,p_i,0}\psi_{p_i,0} = y^2$.

Now, $\hat{\psi} = \phi_5$, we can stop the algorithm and output $\gamma_{5,p_i,0} = \alpha$, $\gamma_{5,p_i,2} = \alpha^2$ and $\gamma_{5,p_i,6} = 1$. The rest of the corresponding coefficients $\gamma_{5,p_i,u} = 0$ ($u \neq 0, 2, 6$).

Before interpolation, monomials ϕ_a that exist in the interpolated polynomial Q are unknown. However, the $(1, w_z)$-weighted degree upper bound of polynomial Q is defined by (5.70), from which the largest pole basis monomial $\phi_{\max}$ that might exist in Q can be predicted by $v_{p_\infty}(\phi_{\max}^{-1}) = \max\{\deg_{1,w_z} Q\}$. Based on interpolation multiplicity m, with parameter $u < m$, the corresponding coefficients that might be used in interpolation are $\gamma_{0,p_i,u} \sim \gamma_{\max,p_i,u}(u < m)$. Therefore Algorithm 5.4 can be used to determine all the corresponding coefficients $\gamma_{0,p_i,u} \sim \gamma_{\max,p_i,u}$ and only $\gamma_{0,p_i,u} \sim \gamma_{\max,p_i,u}$ ($u < m$) are stored for interpolation in order to minimize the memory requirement. For example, to list decode the (8, 4, 4) Hermitian code with multiplicity $m = 2$, $\max\{\deg_{1,w_z} Q\} = 13$. Therefore, the largest pole basis monomial that might exist in Q is $\phi_{max} = \phi_{12} = x^2 y^3$ and Algorithm 5.4 can be applied to calculate all the corresponding coefficients $\gamma_{0,p_i,u} \sim \gamma_{12,p_i,u}$ $(u < 2)$ that are stored.

5.6 Complexity Reduction Interpolation

Interpolation determines a polynomial Q that intersects the points (p_i, r_i) $(i = 0, 1, \ldots, n - 1)$. This can be implemented by an iterative polynomial construction algorithm [4, 9, 11, 13]. At the beginning, a group of polynomials are initialized. During the iterations, they are tested by different zero condition constraints and modified interactively. As with the hard-decision list decoding of Reed–Solomon codes, there are from (5.11) a total of C iterations, after which the minimal polynomial in the group is chosen as the interpolated polynomial Q. According to the iterative process analysis given in Section 5.2.4 and also [13], the interpolated polynomial Q has leading order $\text{lod}(Q) \leq C$. This indicates that those polynomials with leading order greater than C will not be the chosen candidates. Also, if there is a polynomial in the group with leading order greater than C during the iterations, the chosen polynomial Q will not be modified with this polynomial, otherwise $\text{lod}(Q) > C$. Therefore, those polynomials with leading order greater than C can be eliminated from the group during iterations in order to save unnecessary computations.

If $f \in F_q[x, y, z]$ has leading monomial $\text{LM}(f) = \phi_{a'} z^{b'}$, polynomials in $F_q[x, y, z]$ can be partitioned into the following classes according to their leading monomial's z-degree b' and $\phi_{a'}$'s pole order $v_{p_\infty}(\phi_{a'}^{-1})$ [18]:

$$V_{\lambda+w\delta} = \{f \in F_q[x, y, z] | b' = \delta \wedge v_{p_\infty}(\phi_{a'}^{-1}) = uw + \lambda,$$
$$\text{LM}(f) = \phi_{a'} z^{b'}, (\delta, u, \lambda) \in \mathbb{N}, \lambda < w\}, \tag{5.82}$$

such that $F_q[x, y, z] = \bigcup_{\lambda<w} V_{\lambda+w\delta}$. The factorization outputs are the z-roots of Q. Therefore, the z-degree of Q is less than or equal to the maximal number of the output list l_m (5.67) and Q is a polynomial chosen from the following classes:

$$V_j = V_{\lambda+w\delta} \quad (0 \leq \lambda < w, 0 \leq \delta \leq l_m). \tag{5.83}$$

At the beginning of the iterative process, a group of polynomials are initialized to represent each of the polynomial classes defined by (5.83) as:

$$G = \{Q_j = Q_{\lambda+w\delta} = y^\lambda z^\delta, Q_j \in V_j\}. \tag{5.84}$$

During the iterations, each polynomial Q_j in the group G is the minimal polynomial within its class V_j that satisfies all the tested zero conditions. At the beginning of each iteration, the polynomial group G is modified by [18]:

$$G = \{Q_j | \text{lod}(Q_j) \leq C\} \tag{5.85}$$

in order to eliminate those polynomials with leading order greater than C. Then the remaining polynomials in G are tested by the zero condition constraint defined by (5.62) as:

$$\Delta_j = D_{uv}^{(p_i, r_i)}(Q_j). \tag{5.86}$$

The determined corresponding coefficients $\gamma_{a, p_i, u}$ are applied for this calculation. Those polynomials with $\Delta_j = 0$ satisfy the zero condition and do not need to be modified. However, those polynomials with $\Delta_j \neq 0$ need to be modified. Among them, the index of the minimal polynomial is found as j' and the minimal polynomial is recorded as Q':

$$j' = \text{index} \left(\min_{\text{lod}(Q_j)} \{Q_j | \Delta_j \neq 0\} \right) \tag{5.87}$$

$$Q' = Q_{j'}. \tag{5.88}$$

$Q_{j'}$ is modified as:

$$Q_{j'} = (x - x_i)Q', \tag{5.89}$$

where x_i is the x-coordinate of affine point p_i which is included in the current interpolated unit (p_i, r_i). The modified $Q_{j'}$ satisfies $D_{uv}^{(p_i, r_i)}(Q_{j'}) = 0$. Based on Properties 1 and 2 mentioned in Section 5.2.3, $D_{uv}^{(p_i, r_i)}[(x - x_i)Q'] = D_{uv}^{(p_i, r_i)}(xQ') - x_i D_{uv}^{(p_i, r_i)}(Q') = x_i D_{j'} - x_i \Delta_{j'} = 0$. The rest of the polynomials with $\Delta_j \neq 0$ are modified as:

$$Q_j = \Delta_{j'} Q_j - \Delta_j Q'. \tag{5.90}$$

The modified Q_j satisfies $D_{uv}^{(p_i, r_i)}(Q_j) = 0$ because $D_{uv}^{(p_i, r_i)}[\Delta_{j'} Q_j - \Delta_j Q'] = \Delta_{j'} D_{uv}^{(p_i, r_i)}(Q_j) - \Delta_j D_{uv}^{(p_i, r_i)}(Q') = \Delta_{j'} - \Delta_{j'} = 0$. After C iterations, the minimal polynomial in the group G is chosen as the interpolated polynomial Q:

$$Q = \min_{\text{lod}(Q_j)} \{Q_j | Q_j \in G\}. \tag{5.91}$$

From the above description, it can be seen that by applying the complexity reduction scheme in (5.85) the zero condition calculation from (5.86) and modifications from (5.89) and (5.90), for those polynomials Q_j with $\text{lod}(Q_j) > C$, can be avoided and therefore the interpolation efficiency can be improved. According to [13], this complexity reduction scheme is error-dependent, so that it reduces complexity more significantly in low-error-weight situations. This is because the modification scheme of (5.89) takes action in earlier iteration steps for low-error-weight situations, and therefore computation can be reduced. Figure 5.7 shows interpolation complexity reduction (with different multiplicity m) when applying the scheme in (5.85) to decode the (64, 19, 40) Hermitian code. It is shown that complexity can be reduced

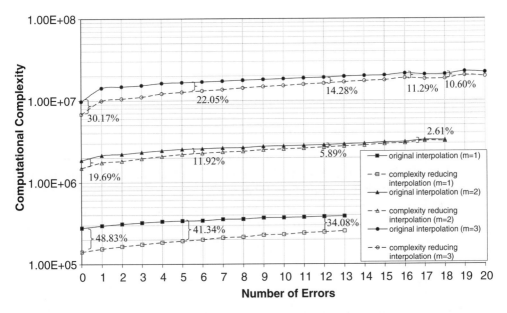

Figure 5.7   Complexity analysis for the interpolation of GS decoding Hermitian code (64, 19, 40).

significantly in low-error-weight situations, especially when $m = 1$ where complexity can be reduced up to 48.83%. However, in high-error-weight situations, complexity reduction is not as significant. Based on Figure 5.7, it can also be observed that the complexity reduction also depends on the interpolation multiplicity m. When $m = 1$, complexity reduction is most significant; when $m = 2$, complexity reduction is the most marginal.

Summarizing Sections 5.5 and 5.6, the modified complexity reduction interpolation process for GS decoding Hermitian codes can be stated as follows:

Initial computation: Apply Algorithm 5.4 to determine all the necessary corresponding coefficients $\gamma_{a,p_i,u}$ and store them for use by the iterative polynomial construction algorithm (Algorithm 5.5).

Algorithm 5.5: Iterative polynomial construction [14, 18]   *Initialization*:: Initialize the group of polynomials G with (5.84).

1. For each interpolated unit (p_i, r_i) $(i = 0, 1, \ldots, n - 1)$
   {
2. For each pair of the zero condition parameters (u, v) $(u + v < m)$
   {
3. Modify polynomial group G by (5.85).
4. Test the zero condition Δ_j of each polynomial in G with (5.86).
5. For polynomials Q_j with $\Delta_j \neq 0$
   {

6. Denote the minimal polynomial's index as j' by (5.87) and record it as Q' by (5.88).
7. If $j = j'$, Q_j is modified by (5.89).
8. If $j \neq j'$, Q_j is modified by (5.90).
   $\}\}\}$

At the end of the iterations, the minimal polynomial Q is chosen from the group G, as in (5.91).

Example 5.8 illustrates this complexity reduction interpolation process.

Example 5.8: Decode the $(8, 4, 4)$ Hermitian code defined in GF(4) using the GS algorithm with interpolation multiplicity $m = 2$.

The Hermitian code word is generated by evaluating the message polynomial over the following affine points: $p_0 = (0, 0)$, $p_1 = (0, 1)$, $p_2 = (1, \alpha)$, $p_3 = (1, \alpha^2)$, $p_4 = (\alpha, \alpha)$, $p_5 = (\alpha, \alpha^2)$, $p_6 = (\alpha^2, \alpha)$, $p_7 = (\alpha^2, \alpha^2)$. The received word is given as $R = (1, \alpha^2, \alpha, \alpha^2, \alpha, \alpha^2, \alpha^2, \alpha)$.

Applying (5.11), the iteration number $C = 8 \binom{3}{2} = 24$. Based on C, the length of the output list can be determined as $l_2 = 3$, and of parameter t_2 as $t_2 = 1$, by using (5.67) and (5.68) respectively. As a result, the $(1, 4)$-weighted degree upper bound for the interpolated polynomial can be determined by (5.66) as $\max\{\deg_{1,4} Q\} = 13$. Therefore, the maximal pole basis (L_2) monomial that might exist in the interpolated polynomial is $\phi_{\max} = \phi_{12} = x^2 y^3$. As the interpolation multiplicity $m = 2$, Algorithm 5.4 is applied to determine the corresponding coefficients $\gamma_{0,p_i,u} - \gamma_{12,p_i,u}$ and $\gamma_{0,p_i,u} - \gamma_{12,p_i,u}$ $(u < 2)$ are stored for the following interpolation process. Table 5.4 lists all the resulted corresponding coefficients $\gamma_{0,p_i,u} \sim \gamma_{12,p_i,u}$ $(u < 2)$.

Following on, Algorithm 5.5 is performed to find the interpolated polynomial $Q(x, y, z)$. At the beginning, a group of polynomials is initialized as: $Q_0 = 1$, $Q_1 = y$, $Q_2 = z$, $Q_3 = yz$, $Q_4 = z^2$, $Q_5 = yz^2$, $Q_6 = z^3$, $Q_7 = yz^3$. Their leading orders are: $\text{lod}(Q_0) = 0$, $\text{lod}(Q_1) = 2$, $\text{lod}(Q_2) = 4$, $\text{lod}(Q_3) = 9$, $\text{lod}(Q_4) = 12$, $\text{lod}(Q_5) = 20$, $\text{lod}(Q_6) = 24$, $\text{lod}(Q_7) = 35$.

For interpolated unit, $(p_0, r_0) = ((0, 0), 1)$.

For the zero parameter, $u = 0$ and $v = 0$.

As $\text{lod}(Q_7) > C$, polynomial Q_7 is eliminated from the group.

Test the zero condition of the remaining polynomials in the group as:

$$\Delta_0 = D_{00}^{(p_0,r_0)}(Q_0) = 1, \quad \Delta_1 = D_{00}^{(p_0,r_0)}(Q_1) = 0, \quad \Delta_2 = D_{00}^{(p_0,r_0)}(Q_2) = 1,$$

$$\Delta_3 = D_{00}^{(p_0,r_0)}(Q_3) = 0,$$

$$\Delta_4 = D_{00}^{(p_0,r_0)}(Q_4) = 1, \quad \Delta_5 = D_{00}^{(p_0,r_0)}(Q_5) = 0, \quad \Delta_6 = D_{00}^{(p_0,r_0)}(Q_6) = 1.$$

Find the minimal polynomial with $\Delta_j \neq 0$ as:

$$j' = 0 \quad \text{and} \quad Q' = Q_0.$$

Table 5.4 Predetermined corresponding coefficients for Example 5.8.

p, γ	a/u	0	1	2	3	4	5	6	7	8	9	10	11	12
$p_0, \gamma_{a,p_0,u}$	0	1	0	0	0	0	0	0	0	0	0	0	0	0
	1	0	1	0	0	0	0	0	0	0	0	0	0	0
$p_1, \gamma_{a,p_1,u}$	0	1	0	1	0	0	1	0	0	1	0	0	1	0
	1	0	1	0	0	1	0	0	1	0	0	1	0	0
$p_2, \gamma_{a,p_2,u}$	0	1	1	α	1	α	α^2	α	α^2	1	α^2	1	α	1
	1	0	1	1	0	α^2	0	1	α^2	α^2	0	α	0	α^2
$p_3, \gamma_{a,p_3,u}$	0	1	1	α^2	1	α^2	α	α^2	α	1	α	1	α^2	1
	1	0	1	1	0	α	0	1	α	α	0	α^2	0	α
$p_4, \gamma_{a,p_4,u}$	0	1	α	α	α^2	α^2	α^2	1	1	1	α	α	α	α^2
	1	0	1	α^2	0	α^2	0	α	α^2	α	0	α	0	1
$p_5, \gamma_{a,p_5,u}$	0	1	α	α^2	α^2	1	α	α	α^2	1	1	α	α^2	α^2
	1	0	1	α^2	0	α	0	α	α	1	0	α^2	0	α^2
$p_6, \gamma_{a,p_6,u}$	0	1	α^2	α	α	1	α^2	α^2	α	1	1	α^2	α	α
	1	0	1	α	0	α^2	0	α^2	α^2	1	0	α	0	α
$p_7, \gamma_{a,p_7,u}$	0	1	α^2	α^2	α	α	α	1	1	1	α^2	α^2	α^2	α
	1	0	1	α	0	α	0	α^2	α	α^2	0	α^2	0	1

As $\Delta_1 = \Delta_3 = \Delta_5 = 0$:

$$Q_1 = Q_1 = y, \text{ and } \text{lod}(Q_1) = 2$$
$$Q_3 = Q_3 = yz, \text{ and } \text{lod}(Q_3) = 9$$
$$Q_5 = Q_5 = yz^2, \text{ and } \text{lod}(Q_5) = 20.$$

Modify polynomials in the group with $\Delta_j \neq 0$ as:

$$Q_0 = (x - 0)Q' = x, \text{ and } \text{lod}(Q_0) = 1$$
$$Q_2 = \Delta_0 Q_2 - \Delta_2 Q' = 1 + z, \text{ and } \text{lod}(Q_2) = 4$$
$$Q_4 = \Delta_0 Q_4 - \Delta_4 Q' = 1 + z_2, \text{ and } \text{lod}(Q_4) = 12$$
$$Q_6 = \Delta_0 Q_6 - \Delta_6 Q' = 1 + z_3 \text{ and } \text{lod}(Q_6) = 24.$$

For the zero parameter, $u = 0$ and $v = 1$.

As there is no polynomial in the group with leading order over C, no polynomial is eliminated in this iteration.

Test the zero condition of the remaining polynomials in the group as:

$$\Delta_0 = D_{01}^{(p_0,r_0)}(Q_0) = 0, \ \Delta_1 = D_{01}^{(p_0,r_0)}(Q_1) = 0, \ \Delta_2 = D_{01}^{(p_0,r_0)}(Q_2) = 1,$$
$$\Delta_3 = D_{01}^{(p_0,r_0)}(Q_3) = 0,$$
$$\Delta_4 = D_{01}^{(p_0,r_0)}(Q_4) = 0, \ \Delta_5 = D_{01}^{(p_0,r_0)}(Q_5) = 0, \ \Delta_6 = D_{01}^{(p_0,r_0)}(Q_6) = 1.$$

Find the minimal polynomial with $\Delta_j \neq 0$ as:

$$j' = 2 \quad \text{and} \quad Q' = Q_2.$$

As $\Delta_0 = \Delta_1 = \Delta_3 = \Delta_4 = \Delta_5 = 0$:

$$Q_0 = Q_0 = x, \text{ and lod}(Q_0) = 1$$
$$Q_1 = Q_1 = y, \text{ and lod}(Q_1) = 2$$
$$Q_3 = Q_3 = y_z, \text{ and lod}(Q_3) = 9$$
$$Q_4 = Q_4 = 1 + z_2, \text{ and lod}(Q_4) = 12$$
$$Q_5 = Q_5 = yz^2 \text{ and lod}(Q_5) = 20.$$

Modify polynomials in the group with $\Delta_j \neq 0$ as:

$$Q_2 = (x - 0)Q' = x + xz \text{ and lod}(Q_2) = 7$$
$$Q_6 = \Delta_2 Q_6 - \Delta_6 Q' = z + z^3 \text{ and lod}(Q_6) = 24.$$

For zero parameter, $u = 1$ and $v = 0$.

As there is no polynomial in the group with leading order over C, no polynomial is eliminated in this iteration.

Test the zero condition of the remaining polynomials in the group as:

$$\Delta_0 = D_{10}^{(p_0,r_0)}(Q_0) = 1, \ \Delta_1 = D_{10}^{(p_0,r_0)}(Q_1) = 0, \ \Delta_2 = D_{10}^{(p_0,r_0)}(Q_2) = 0,$$
$$\Delta_3 = D_{10}^{(p_0,r_0)}(Q_3) = 0,$$
$$\Delta_4 = D_{10}^{(p_0,r_0)}(Q_4) = 0, \ \Delta_5 = D_{10}^{(p_0,r_0)}(Q_5) = 0, \ \Delta_6 = D_{10}^{(p_0,r_0)}(Q_6) = 0.$$

Find the minimal polynomial with $\Delta_j \neq 0$ as:

$$j' = 0 \quad \text{and} \quad Q' = Q_0.$$

As $\Delta_1 = \Delta_2 = \Delta_3 = \Delta_4 = \Delta_5 = \Delta_6 = 0$:

$$Q_1 = Q_1 = y, \text{ and lod}(Q_1) = 2$$
$$Q_2 = Q_2 = x + xz, \text{ and lod}(Q_2) = 7$$
$$Q_3 = Q_3 = yz, \text{ and lod}(Q_3) = 9$$
$$Q_4 = Q_4 = 1 + z^2, \text{ and lod}(Q_4) = 12$$
$$Q_5 = Q_5 = yz^2, \text{ and lod}(Q_5) = 20$$
$$Q_6 = Q_6 = z + z^3, \text{ and lod}(Q_6) = 24.$$

Modify polynomials in the group with $\Delta_j \neq 0$ as:

$$Q_0 = (x - 0)Q' = x^2, \text{ and } \text{lod}(Q_0) = 3.$$

Following the same process, interpolation runs through the rest of the interpolated units $(p_1, r_1) \sim (p_7, r_7)$ and with respect to all zero parameters $(u, v) = (0, 0), (0, 1)$ and $(1, 0)$. After C iterations, the chosen interpolated polynomial is: $Q(x, y, z) = \alpha^2 + \alpha x + y + \alpha x^2 + y^2 + \alpha^2 x^2 y + \alpha^2 xy^2 + \alpha^2 y^3 + \alpha^2 x^2 y^2 + z(x + xy + xy^2) + z^2(\alpha^2 + \alpha x + \alpha^2 y + \alpha x^2)$, and $\text{lod}(Q(x, y, z)) = 23$. Polynomial $Q(x, y, z)$ has a zero of multiplicity at least 2 over the eight interpolated units.

5.7 General Factorization

Based on the interpolated polynomial, factorization finds the polynomial's z-roots in order to determine the output list. Building upon the work of [15, 20], this section presents a generalized factorization algorithm, or the so-called recursive coefficient search algorithm, which can be applied to both Reed–Solomon codes and algebraic–geometric codes. This section's work is based on [21]. In general, the algorithm is described with application to algebraic–geometric codes. Therefore, it has to be stated that when applying this algorithm to Reed–Solomon codes, the rational functions in an algebraic–geometric code's pole basis are simplified to univariate monomials in the Reed–Solomon code's pole basis. As a consequence, polynomials in $F_q[x, y]$ are simplified to univariate polynomials with variable x.

Those polynomials $h \in F_q^{w_z}[x, y]$ will be in the output list if $Q(x, y, h) = 0$. The outcome of the factorization can be written as:

$$\begin{cases} h_1 = h_{1,0}\phi_0 + \cdots + h_{1,k-1}\phi_{k-1} \\ \qquad \vdots \\ h_l = h_{l,0}\phi_0 + \cdots + h_{l,k-1}\phi_{k-1} \end{cases}, \tag{5.92}$$

with $l \leq l_m$. Rational functions $\phi_0, \ldots, \phi_{k-1}$ are predetermined by the decoder; therefore, finding the list of polynomials is equivalent to finding their coefficients, $h_{1,0}, \ldots, h_{1,k-1}, \ldots, h_{l,0}, \ldots, h_{l,k-1}$, respectively. Substituting h into the interpolated polynomial $Q = \sum_{a,b} Q_{ab}\phi_a z^b$, we have:

$$Q(x, y, h) = \sum_{a,b} Q_{ab}\phi_a h^b = \sum_{a,b} Q_{ab}\phi_a(h_0\phi_0 + \cdots + h_{k-1}\phi_{k-1})^b. \tag{5.93}$$

It is important to notice that:

$$(\phi_i\phi_j) \mod \zeta = \sum_v \phi_v, \tag{5.94}$$

where χ is the algebraic curve (e.g. Hermitian curve H_w) and ϕ_i, ϕ_j and ϕ_v are rational functions in the pole basis associated with the curve χ (for example the pole basis L_w associated with curve H_w). Therefore (5.93) can be rewritten as a polynomial in $F_q[x, y]$:

$$Q(x, y, h) = \sum_a Q_a \phi_a, \tag{5.95}$$

where coefficients Q_a are equations with unknowns $h_0, \ldots, h_{k-1}$. If $T = |\{Q_a | Q_a \neq 0\}|$, the rational functions ϕ_a with $Q_a \neq 0$ can be arranged as $\phi_{a_1} < \phi_{a_2} < \cdots < \phi_{a_T}$ and (5.95) can again be rewritten as:

$$Q(x, y, h) = Q_{a_1} \phi_{a_1} + Q_{a_2} \phi_{a_2} + \cdots + Q_{a_T} \phi_{a_T}. \tag{5.96}$$

Again, coefficients $Q_{a_1}, Q_{a_2}, \ldots,$ and Q_{a_T} are equations of unknowns $h_0, \ldots, h_{k-1}$. To have $Q(x, y, h) = 0$, we need $Q_{a_1} = Q_{a_2} = \cdots = Q_{a_T} = 0$. Therefore, $h_0, \ldots, h_{k-1}$ can be determined by solving the following simultaneous set of equations [21]:

$$\begin{cases} Q_{a_1}(h_0, \ldots, h_{k-1}) = 0 \\ Q_{a_2}(h_0, \ldots, h_{k-1}) = 0 \\ \qquad \vdots \\ Q_{a_T}(h_0, \ldots, h_{k-1}) = 0 \end{cases} . \tag{5.97}$$

In order to solve (5.97), a recursive coefficient search algorithm is applied to determine $h_0, \ldots, h_{k-1}$ [20, 22]. Here a more general and efficient factorization algorithm is presented. Let us denote the following polynomials with respect to a recursive index s ($0 \leq s \leq k - 1$):

$$h^{(s)}(x, y) = h_0 \phi_0 + \cdots + h_{k-1-s} k_{-1-s}, \tag{5.98}$$

which is a candidate polynomial with coefficients $h_0, \ldots, h_{k-1-s}$ undetermined. Update $Q(x, y, z)$ recursively as:

$$Q^{(s+1)}(x, y, z) = Q^{(s)}(x, y, z + h_{k-1-s} \phi_{k-1-s}), \tag{5.99}$$

with $Q^{(0)}(x, y, z) = Q(x, y, z)$, which is the interpolated polynomial (5.68). Substituting $h_{k-1-s} \phi_{k-1-s}$ into $Q^{(s)}(x, y, z)$, we have:

$$\tilde{Q}^{(s)}(x, y) = Q^{(s)}(x, y, h_{k-1-s} \phi_{k-1-s}). \tag{5.100}$$

$\tilde{Q}^{(s)} \bmod \chi$ can be transferred into a polynomial in $F_q[x, y]$ with coefficients expressed as $\sum_i \varpi_i h_{k-1-s}^i$ where $\varpi_i \in GF(q)$. Denote $\tilde{Q}^{(s)}$'s leading monomial with its leading coefficient as [21]:

$$\phi_L^{(s)} = LM(\tilde{Q}^{(s)}) \tag{5.101}$$

$$C_L^{(s)}(h_{k-1-s}) = LC(\tilde{Q}^{(s)}). \tag{5.102}$$

Based on (5.102), it can be seen that $LM(h^{(s)}) = \phi_{k-1-s}$ and $LC(h^{(s)}) = h_{k-1-s}$. Therefore, for any recursive polynomial $Q^{(s)}(x, y, z)$, we have:

$$LM(Q^{(s)}(x, y, h^{(s)})) = LM(Q^{(s)}(x, y, h_{k-1-s}\phi_{k-1-s})) = LM(\tilde{Q}^{(s)}) = \phi_L^{(s)} \quad (5.103)$$

$$LC(Q^{(s)}(x, y, h^{(s)})) = LC(Q^{(s)}(x, y, h_{k-1-s}\,\phi_{k-1-s})) = LC(\tilde{Q}^{(s)}) = C_L^{(s)}(h_{k-1-s}). \quad (5.104)$$

As all the candidate outputs should satisfy $Q(x, y, h) = 0$, and from the above definitions it can be seen that $h = h^{(0)}$ and $Q^{(0)}(x, y, z) = Q(x, y, z), Q(x, y, h) = 0$ is equivalent to $Q^{(0)}(x, y, h^{(0)}) = 0$. Based on (5.107) and (5.108), in order to have $Q^{(0)}(x, y, h^{(0)}) = 0$ we need to find its leading monomial $\phi_L^{(0)}$ with leading coefficient $C_L^{(0)}(h_{k-1})$ and determine the values of h_{k-1} that satisfy $C_L^{(0)}(h_{k-1}) = 0$. As a result, the leading monomial of $Q^{(0)}(x, y, h^{(0)})$ is eliminated. Based on each value of h_{k-1} and performing the polynomial update (5.99), $Q^{(1)}(x, y, z)$ is generated, in which $\phi_L^{(0)}$ has been eliminated. Now, $Q(x, y, h) = 0$ is equivalent to $Q^{(1)}(x, y, h^{(1)}) = 0$. Again, to have $Q^{(1)}(x, y, h^{(1)}) = 0$, we need $C_L^{(1)}(h_{k-2}) = 0$. Therefore, h_{k-2} can be determined by solving $C_L^{(1)}(h_{k-2}) = 0$. Based on each value of h_{k-2}, we can trace further to find the rest of the coefficients. In general, after the coefficients h_{k-1-s} $(0 \le s < k-1)$ have been determined by solving $C_L^{(s)}(h_{k-1-s}) = 0$, based on each value of them, perform the polynomial update (5.99) to generate $Q^{(s+1)}(x, y, z)$. From $Q^{(s+1)}(x, y, z)$, $\tilde{Q}^{(s+1)}$ can be calculated and $h_{k-1-(s+1)}$ can be determined by solving $C_L^{(s+1)}(h_{k-1-(s+1)}) = 0$. This process is illustrated in Figure 5.8.

From Figure 5.8 it can be seen that there might be an exponential number of routes to find coefficients $h_{k-1}, \ldots, h_0$. However, not every route will be able to reach h_0, as during the recursive process there may be no solution for $C_L^{(s)}(h_{k-1-s}) = 0 = 0$. If h_0 is produced and $Q^{(k-1)}(x, y, h_0\phi_0) = 0$, this route can be traced to find the rest of the

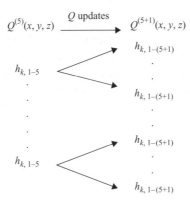

Figure 5.8   Recursive coefficient search.

coefficients $h_1, \ldots, h_{k-1}$ to construct polynomial h which will satisfy $Q(x, y, h) = 0$. The correctness of this judgement will be proven later.

Based on the above description, the generalized factorization algorithm is summarized in Algorithm 5.6.

Algorithm 5.6: Recursive coefficient search [14, 21, 22]   *Initialization*: $Q^{(0)}(x, y, z) = Q(x, y, z)$. The recursive index $s = 0$ and output candidate index $l = 1$.
Perform: Recursive coefficient search (s) $(RCS(s))$.
Recursive coefficient search (RCS):
Input parameter: s $(0 \leq s \leq k - 1)$.

1. Perform (5.100) to calculate $\tilde{Q}^{(s)}(x, y)$.
2. Find out $\phi_L^{(s)}$ with its coefficient $C_L^{(s)}(h_{k-1-s})$.
3. Determine h_{k-1-s} by solving $C_L^{(s)}(h_{k-1-s}) = 0$.
4. For each value of h_{k-1-s}, do
   {
5. $h_{l,k-1-s} = h_{k-1-s}$.
6. If $s = k - 1$, calculate $Q^{(k-1)}(x, y, h_0\phi_0)$ and go to step 7. Else go to step 8.
7. If $Q^{(k-1)}(x, y, h_0\phi_0) = 0$, trace this route to find coefficients $h_{l,k-1}, h_{l,k-2}, \ldots,$ and $h_{l,0}$ to construct the candidate polynomial h_l and $l = l + 1$. Else stop this route.
8. Perform polynomial update (5.99) to generate $Q^{(s+1)}(x, y, z)$.
9. Perform $RCS(s + 1)$.
   }

If a number of h_{k-1-s} have been determined, the algorithm will choose one of them to determine the rest of the coefficients until all the possible routes starting from this h_{k-1-s} have been traced. After this, it will choose another value of h_{k-1-s} and repeat the process. This algorithm will terminate after all the possible routes started from h_{k-1} have been traced. To prove the correctness of this algorithm, we need to justify the fact that the polynomial h_l produced in step 7 satisfies $Q(x, y, h_l) = 0$.

Proof: As $Q^{(k-1)}(x, y, h_0\phi_0) = 0$ and $h^{(k-1)}(x, y) = h_0\phi_0$, we have $Q^{(k-1)}(x, y, h^{(k-1)}) = 0$. Assuming h_1 is the previously-found coefficient, $Q^{(k-1)}(x, y, z)$ is generated by (5.99): $Q^{(k-1)}(x, y, z) = Q^{(k-2)}(x, y, z + h_1\phi_1)$. From $Q^{(k-1)}(x, y, h^{(k-1)}) = 0$ we have $Q^{(k-2)}(x, y, h^{(k-1)} + h_1\phi_1) = 0$. Based on (5.98), it can be seen that $h^{(k-2)} = h_0\phi_0 + h_1\phi_1 = h^{(k-1)} + h_1\phi_1$. Therefore, $Q^{(k-2)}(x, y, h^{(k-2)}) = 0$. It can be deduced further to get $Q^{(k-3)}(x, y, h^{(k-3)}) = 0, \ldots,$ and $Q^{(0)}(x, y, h^{(0)}) = 0$. As $Q^{(0)}(x, y, z) = Q(z)$ and $h(0) = h_0\phi_0 + h_1\phi_1 + \cdots + h_{k-1}\phi_{k-1}$, whose coefficients have been traced as the coefficients of the output candidate h_l, it can be concluded that $Q(x, y, h_l) = 0$.

Here two worked examples are given to illustrate the generalized factorization algorithm applied to both a Hermitian code and a Reed–Solomon code.

Example 5.9: List decoding of a (8, 4, 4) Hermitian code defined in GF(4)
Given the interpolated polynomial is $Q(x, y, z) = \alpha^2 y + \alpha^2 x^2 + \alpha xy + \alpha^2 y^2 + \alpha^2 x^2 y + x^2 y^2 + xy^3 + (\alpha x + \alpha xy + \alpha xy^2)z + (x + x^2)z^2$, apply Algorithm 5.6 to find out its z-roots.

Initialization: $Q^{(0)}(x, y, z) = Q(x, y, z)$, $s = 0$ and $l = 1$.

$RCS(0)$:
$\tilde{Q}^{(0)}(x, y) = Q^{(0)}(x, y, h_3 x^2) = (\alpha^2 + \alpha h_3)y + \alpha^2 x^2 + \alpha xy + (\alpha^2 + h_3^2)y^2 + (\alpha^2 + h_3^2)x^2 y + (1 + h_3^2)x^2 y^2 + xy^3 + (\alpha h_3 + h_3^2)y^4$, with $\phi_L^{(0)} = y^4$ and $C_L^{(0)}(h_3) = \alpha h_3 + h_3^2$. Solving $C_L^{(0)}(h_3) = 0$, we have $h_3 = 0$ or $h_3 = \alpha$.

For $h_3 = 0$, $h_{1,3} = h_3 = 0$. As $s = 0 < 3$, update $Q^{(1)}(x, y, z) = Q^{(0)}(x, y, z + 0x^2) = Q(x, y, z)$, and perform $RCS(1)$...

Based on the same progress, the outcomes from $RCS(1)$, $RCS(2)$ and $RCS(3)$ are summarized in Table 5.5.

Table 5.5 Recursive coefficient search from $h_3 = 0$.

$RCS(s)$	$\phi_L^{(s)}$	$C_L^{(s)}(h_{k-1-s})$	$h_{2,k-1-s} = h_{k-1-s}$
$RCS(1)$	xy^3	$1 + \alpha h_2$	α^2
$RCS(2)$	$x^2 y^2$	$\alpha^2 + \alpha h_1$	α
$RCS(3)$	xy^2	αh_0	0

After $RCS(2)$ we have $Q^{(3)}(x, y, z) = (\alpha x + \alpha xy + \alpha xy^2)z + (x + x^2)z^2$. In $RCS(3)$, by solving $C_L^{(3)}(h_0) = 0$, we have $h_0 = 0$. Therefore, $h_{1,0} = h_0 = 0$. As $s = 3$ and $Q^{(3)}(x, y, h_0\phi_0) = Q^{(3)}(x, y, 0 \cdot 1) = 0$, this route can be traced to construct candidate polynomial $h_1 = \alpha x + \alpha^2 y$ and update the candidate index $l = l + 1 = 2$.

Going back to the closest division point (when $s = 0$), we have:

For $h_3 = \alpha$, $h_{2,3} = h_3 = \alpha$. As $s = 0 < 3$, update $Q^{(1)}(x, y, z) = Q^{(0)}(x, y, z + \alpha x^2) = \alpha^2 x^2 + \alpha xy + \alpha x^2 y^2 + xy^3 + (\alpha x + \alpha xy + \alpha xy^2)z + (x + x^2)z^2$ and perform $RCS(1)$...

Again, the outcomes of $RCS(1)$, $RCS(2)$ and $RCS(3)$ are summarized in Table 5.6.

Table 5.6 Recursive coefficient search from $h_3 = \alpha$.

$RCS(s)$	$\phi_L^{(s)}$	$C_L^{(s)}(h_{k-1-s})$	$h_{2,k-1-s} = h_{k-1-s}$
$RCS(1)$	xy^3	$1 + \alpha h_2$	α^2
$RCS(2)$	$x^2 y^2$	αh_1	0
$RCS(3)$	xy^2	$\alpha^2 + \alpha h_0$	α

After $RCS(2)$ we have $Q^{(3)}(x, y, z) = \alpha^2 x^2 + \alpha^2 xy + \alpha^2 xy^2 + (\alpha x + \alpha xy + \alpha xy^2)z + (x + x^2)z^2$. In $RCS(3)$, by solving $C_L^{(3)}(h_0) = 0$, we have $h_{2,0} = h_0 = \alpha$.

As $s = 3$ and $Q^{(3)}(x, y, h_0\phi_0) = Q^{(3)}(x, y, \alpha \cdot 1) = 0$, this route can be traced to construct the candidate polynomial $h_2 = \alpha + \alpha^2 y + \alpha x^2$. As all the possible routes from h_0 have been traced, the factorization process terminates and outputs: $h_1 = \alpha x + \alpha^2 y$, $h_2 = \alpha + \alpha^2 y + \alpha x^2$.

Example 5.10: List decoding of a $(7, 2, 6)$ Reed–Solomon code defined in GF(8)
α is a primitive element in GF(8) satisfying $\alpha^3 + \alpha + 1 = 0$.

Given the interpolated polynomial $Q(x, z) = \alpha x + \alpha^6 x^2 + (\alpha^3 + \alpha^3 x)z + \alpha^2 z^2$, apply Algorithm 5.6 to determine its z-roots.

Initialization: $Q^{(0)}(x, z) = Q(x, z)$, $s = 0$ and $l = 1$.

$RCS(0)$:
$\tilde{Q}^{(0)}(x, z) = Q^{(0)}(x, h_1 x) = (\alpha + \alpha^3 h_1)x + (\alpha^6 + \alpha^3 h_1 + \alpha^2 h_1^2)x^2$, with $\phi_L^{(0)} = x^2$ and $C_L^{(0)}(h_1) = \alpha^6 + \alpha^3 h_1 + \alpha^2 h_1^2$. Solving $C_L^{(0)}(h_1) = 0$, we have $h_1 = \alpha^5$ or $h_1 = \alpha^6$.

For $h_1 = \alpha^5$, $h_{1,1} = h_1 = \alpha^5$. As $s = 0 < 1$, update $Q^{(1)}(x, z) = Q^{(0)}(x, z + \alpha^5 x)$ $= (\alpha^3 + \alpha^3 x)z + \alpha^2 z^2$ and perform $RCS(1)$.

In $RCS(1)$, following the same progress, we have $\phi_L^{(1)} = x$ and $C_L^{(1)}(h_0) = \alpha^3 h_0$. Solving $C_L^{(1)}(h_0) = 0$, we have $h_0 = 0$.

For $h_0 = 0$, $h_{1,0} = h_0 = 0$. As $s = 1$ and $Q^{(1)}(x, h_0\phi_0) = Q^{(1)}(x, 0 \cdot 1) = 0$, this route can be traced to construct candidate polynomial $h_1 = \alpha^5 x$. Update the output candidate index as $l = l + 1 = 2$.

Going back to the closest division point (when $s = 0$), we have:

For $h_1 = \alpha^6$, $h_{2,1} = h_1 = \alpha^6$. As $s = 0 < 1$, update $Q^{(1)}(x, z) = Q^{(0)}(x, z + \alpha^6 x)$ $= \alpha^4 x + (\alpha^3 + \alpha^3 x)z + \alpha^2 z^2$, and perform $RCS(1)$.

In $RCS(1)$ we have $\phi_L^{(1)} = x$ and $C_L^{(1)}(h_0) = \alpha^4 + \alpha^3 h_0$. Solving $C_L^{(1)}(h_0) = 0$, we have $h_0 = \alpha$.

For $h_0 = \alpha$, $h_{2,0} = h_0 = \alpha$. As $s = 1$ and $Q^{(1)}(x, h_0\phi_0) = Q^{(1)}(x, \alpha \cdot 1) = 0$, this route can be traced to construct candidate polynomial $h_2 = \alpha + \alpha^6 x$. As all the possible routes from h_0 have been traced, the factorization process terminates and outputs: $h_1 = \alpha^5 x$, $h_2 = \alpha + \alpha^6 x$.

5.8 Soft-Decision List Decoding of Hermitian Codes

We can extend the hard-decision GS decoding algorithm for AG codes to soft-decision by extending the Kotter–Vardy algorithm. As with Reed–Solomon codes, soft-decision list decoding is almost the same as hard-decision list decoding, with the addition of a process to convert the reliability values of the received symbols into a multiplicity matrix. In this section, a comparison of the code word scores is made between soft- and hard-decision list decoding and conditions are derived to ensure that the equations from the interpolation process are solvable.

Based on w_z, the two parameters first defined by (5.50) and (5.51) for analyzing bivariate monomial $x^a y^b$ ($a, b \in \mathbb{N}$) can be extended to trivariate monomials $\phi_a z^b$ ($a, b \in \mathbb{N}$):

$$N_{1,w_z}(d) = |\{\phi_a z^b : a, b \geq 0 \text{ and } \deg_{1,w_z}(\phi_a z^b) \leq \delta, \delta \in \mathbb{N}\}|, \qquad (5.105)$$

which represents the number of monomials with $(1, w_z)$-weighted degree not greater than δ, and:

$$\Delta_{1,w_z}(v) = \min \{\delta : N_{1,w_z}(\delta) > v, v \in \mathbb{N}\}, \qquad (5.106)$$

which represents the minimal value of δ that guarantees $N_{1,w_z}(\delta)$ is greater than v.

The difference between soft-decision list decoding of Reed–Solomon codes and of Hermitian codes is in the size of the reliability matrix. For soft-decision list decoding of Reed–Solomon codes, the reliability matrix $\boldsymbol{\Pi}$ has size $q \times n$, where $n = q - 1$. For soft-decision list decoding of Hermitian codes, the reliability matrix $\boldsymbol{\Pi}$ has size $q \times n$, where $n = q^{3/2}$.

5.8.1 System Solution

The reliability matrix $\boldsymbol{\Pi}$ is then converted to the multiplicity matrix $\boldsymbol{M}$, for which Algorithm 5.3 is applied. A version of Algorithm 5.3 introducing a stopping rule based on the designed length of output list is presented later in this subsection, as Algorithm 5.7.

In this subsection, the code word score with respect to matrix $\boldsymbol{M}$ is analyzed so as to present the system solution for this soft-decision list decoder. It is shown that the soft-decision scheme provides a higher code word score than the hard-decision scheme.

The resulting multiplicity matrix $\boldsymbol{M}$ can be written as:

$$M = \begin{bmatrix} m_{0,0} & m_{0,1} & \cdots & \cdots & \cdots & m_{0,n-1} \\ m_{1,0} & m_{1,1} & & & & m_{1,n-1} \\ \vdots & & \ddots & & & \vdots \\ \vdots & & & m_{i,j} & & \vdots \\ \vdots & & & & \ddots & \vdots \\ m_{q-1,0} & m_{q-1,1} & \cdots & \cdots & \cdots & m_{q-1,n-1} \end{bmatrix}, \qquad (5.107)$$

where entry $m_{i,j}$ represents the multiplicity for unit (p_j, ρ_i). Interpolation builds the minimal polynomial $Q_M \in F_q[x, y, z]$, which has a zero of multiplicity at least $m_{i,j}$

$(m_{i,j} \neq 0)$ over all the associated units (p_j, ρ_i). Following on from (5.62), with respect to interpolated unit (p_j, ρ_i), Q_M's coefficients Q_{ab} should satisfy:

$$\sum_{a,b \geq v} Q_{ab} \binom{b}{v} \gamma_{a,p_j,u} \rho_i^{b-v} = 0, \ \forall \, u, v \in N \text{ and } u + v < m_{i,j}. \tag{5.108}$$

For this soft-decision interpolation, the number of interpolated units covered by Q_M is:

$$|\{m_{i,j} \neq 0 | m_{i,j} \in M, i = 0, 1, \ldots, q - 1 \text{ and } j = 0, 1, \ldots, n - 1\}| \tag{5.109}$$

and the cost C_M of multiplicity matrix M is:

$$C_M = \frac{1}{2} \sum_{i=0}^{q-1} \sum_{j=0}^{n-1} m_{i,j}(m_{i,j} + 1), \tag{5.110}$$

which represents the number of constraints in (5.108) to Q_M's coefficients Q_{ab}. These can be imposed by the iterative polynomial construction algorithm [9, 11, 17] in C_M iterations. Notice that (5.109)) and (5.110)) have the same expression as (5.45) and (5.46), respectively, with the exception that $n = q^{3/2}$.

Based on Lemma 5.1, the following units' multiplicities will contribute to the code word score: $(p_0, c_0), (p_1, c_1), \ldots$, and (p_{n-1}, c_{n-1}). Referring to the multiplicity matrix in (5.107), the interpolated polynomial Q_M can be explained as passing through these units with multiplicity at least $m_0 = m_{i,0}$ $(\rho_i = c_0)$, $m_1 = m_{i,1}$ $(\rho_i = c_1), \ldots$, and $m_{n-1} = m_{i,n-1}$ $(\rho_i = c_{n-1})$, respectively. If $f \in F_q^{w_z}[x, y]$ is the transmitted message polynomial such that $f(p_i) = c_i$, the total zero order of $Q_M(x, y, f)$ over units $\{(p_0, c_0), (p_1, c_1), \ldots, (p_{n-1}, c_{n-1})\}$ is at least:

$$m_0 + m_1 + \cdots + m_{n-1} = \sum_{j=0}^{n-1} \{m_{i,j} | \rho_i = c_j, i = 0, 1, \ldots, q - 1\}. \tag{5.111}$$

Therefore, the code word score $S_M(\overline{c})$ with respect to multiplicity matrix M is:

$$S_M(\overline{c}) = \sum_{j=0}^{n-1} \{m_{i,j} | \rho_i = c_j, i = 0, 1, \ldots, q - 1\}. \tag{5.112}$$

If $S_M(\overline{c}) > \deg_{1,w_z}(Q_M(x, y, z))$ then $\sum_{i=0}^{n-1} v_{p_i}(Q_M(x, y, f)) > v_{p_\infty}(Q_M(x, y, f)^{-1})$ and $Q_M(x, y, f) = 0$. f can be found by determining $Q_M(x, y, z)$'s z-roots. It results in the following corollary for successful soft-decision list decoding:

Corollary 5.3 *If the code word score with respect to multiplicity matrix M is greater than the interpolated polynomial Q_M's $(1, w_z)$-weighted degree [23]:*

$$S_M(\bar{c}) > \deg_{1,w_z}(Q_M(x, y, z)) \qquad (5.113)$$

then $Q_M(x, y, f) = 0$ or $z - f | Q_M(x, y, z)$.

To compare the soft-decision's code word score $S_M(\bar{c})$ with the hard-decision's code word score $S_M(\bar{c})$, denote the index of the maximal element in each column of Π as:

$$i_j = \text{index } (\max\{\pi_{i,j} | i = 0, 1, \ldots, q - 1\}), \qquad (5.114)$$

such that $\pi_{i_j,j} > \pi_i$, $j(i \neq i_j)$. The hard-decision received word R can be written as:

$$R = (r_0, r_1, \ldots, r_{n-1}) = (\rho_{i_0}, \rho_{i_1}, \ldots, \rho_{i_{n-1}}). \qquad (5.115)$$

For hard-decision list decoding, only those entries in the multiplicity matrix from (5.107) that correspond to the reliability value $\pi_{i_j,j}$ will be assigned a multiplicity, as $m_{i_j,j} = m$, and therefore the score for hard-decision can also be written with respect to multiplicity matrix M as:

$$S_m(\bar{c}) = S_M(\bar{c}) = \sum_{j=0}^{n-1} \{m_{i_j,j} | \rho_{i_j} = c_j\}. \qquad (5.116)$$

Comparing (5.112) and (5.116), the soft-decision list decoder gains its improvements by increasing its code word score. This is done by increasing the total number of interpolated units (5.111) so that the possibility of covering more interpolated units, including the corresponding code word symbols, is also increased.

If the $(1, w_z)$-weighted degree of interpolated polynomial Q_M is δ^*, based on (5.106), Q_M has at most $N_{1,w_z}(\delta^*)$ nonzero coefficients. The interpolation procedure generates a system of C_M linear equations of type (5.108). The system will be solvable if [5]:

$$N_{1,w_z}(\delta^*) > C_M. \qquad (5.117)$$

Based on (5.106), in order to guarantee the solution, the $(1, w_z)$-weighted degree δ^* of the interpolated polynomial Q_M should be large enough that:

$$\deg_{1,w_z}(Q_M(x, y, z)) = d^* = \Delta_{1,w_z}(C_M). \qquad (5.118)$$

Therefore, based on Corollary 5.3, given the soft-decision code word score (5.116) and the $(1, w_z)$-weighted degree of the interpolated polynomial Q_M (5.118), the message polynomial f can be found if [23]:

$$S_M(\bar{c}) > \Delta_{1,w_z}(C_M). \qquad (5.119)$$

The factorization output list contains the z-roots of polynomial Q_M. Therefore, the maximal length of output list l_M should be equal to polynomial Q_M's z-degree $(\deg_z Q_M)$, as:

$$l_M = \deg_z(Q_M(x, y, z)) = \left\lfloor \frac{\deg_{1,w_z}(Q_M(x, y, z))}{w_z} \right\rfloor = \left\lfloor \frac{\Delta_{1,w_z}(C_M)}{w_z} \right\rfloor. \quad (5.120)$$

When converting matrix Π to matrix M, based on a designed length of output list l, Algorithm 5.1 will stop once l_M is greater than l. $\Delta_{1,w_z}(C_M)$ can be determined by finding the monomial with $(1, w_z)$-lexicographic order C_M, as:

$$\Delta_{1,w_z}(C_M) = \deg_{1,w_z}(\phi_a z^b | \mathrm{ord}(\phi_a z^b) = C_M). \quad (5.121)$$

Therefore, in order to assess the soft-decision list decoding algorithm's performance with a designed length of output list l, a large enough value must be set for s when initializing Algorithm 5.3. In the algorithm, after step 5, we can determine the cost C_M (5.110) of the updated matrix M and apply (5.121) to determine $\Delta_{1,w_z}(C_M)$. The maximal length of output list l_M can then be determined by (5.120). Stop Algorithm 5.3 once l_M is greater than l and output the updated matrix M.

Algorithm 5.7 is based on Algorithm 5.3 and includes this stopping rule. This algorithm is used to obtain the simulation results shown later.

Algorithm 5.7: Convert reliability matrix Π to multiplicity matrix M [14, 23]
Input: Reliability matrix Π, a high enough desired value of the sum of multiplicities in matrix M:

$s = \sum_{i=0}^{q-1} \sum_{j=0}^{n-1} m_{i,j}$, and designed output length l.

Initialization: Set $\Pi^* = \Pi$ and $q \times n$ all-zero multiplicity matrix M.

1. While (s > 0 or $\lfloor l_M \rfloor < l$)
   {
2. Find the maximal entry $\pi_{i,j}^*$ in Π^* with position (i, j).
3. Update $\pi_{i,j}^*$ in Π^* as $\pi_{i,j}^* = \frac{\pi_{i,j}}{m_{i,j}+2}$.
4. Update $m_{i,j}$ in M as $m_{i,j} = m_{i,j} + 1$.
5. $s = s - 1$.
6. For the updated M, calculate its interpolation cost C_M by (5.110).
7. Determine $\Delta_{1,w_z}(C_M)$ by (5.121).
8. Calculate l_M by (5.120).
   }

Again, this algorithm gives priority to those interpolated points which correspond to a higher reliability value $\pi_{i,j}$, to be assigned with a higher multiplicity value $m_{i,j}$. For example, if $\pi_{i_1 j_1} < \pi_{i_2 j_2}$ then $m_{i_1 j_1} \leq m_{i_2 j_2}$.

The interpolation and factorization methods given in Algorithms 5.5 and 5.6 are unchanged for soft-decision list decoding of AG codes. Again, as the total number of

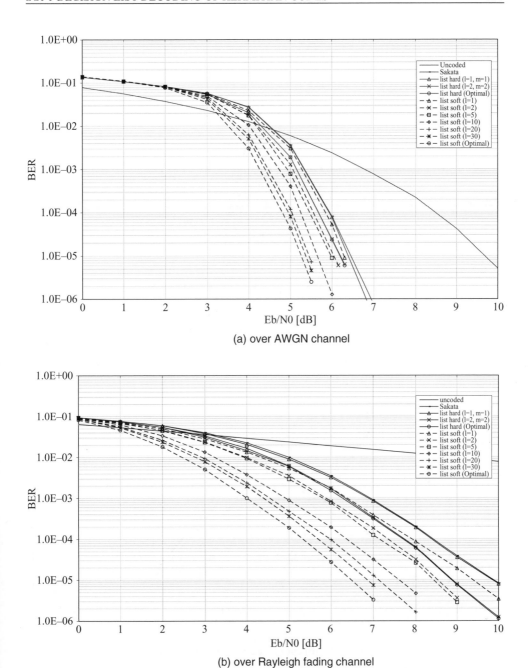

(a) over AWGN channel

(b) over Rayleigh fading channel

Figure 5.9   Soft-decision list decoding of Hermitian code (64, 39, 20).

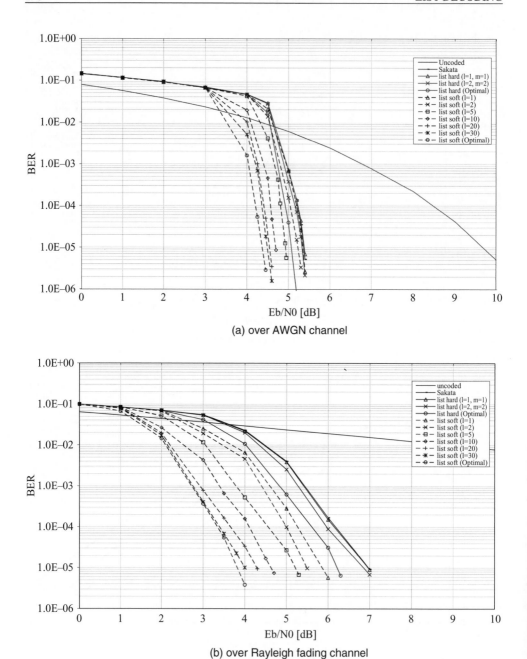

(a) over AWGN channel

(b) over Rayleigh fading channel

Figure 5.10   Soft-decision list decoding of Hermitian code (512, 289, 196).

iterations C_M has been determined before running the interpolation, those polynomials with leading order greater than C_M can be eliminated from the group during the iterations [23].

5.8.2 Simulation Results

The performance of hard- and soft-decision list decoding for Hermitian codes is evaluated by simulation results on the AWGN channel and a frequency nonselective Rayleigh fading channel with Doppler frequency 126.67 Hz. The fading profile is generated using Jakes' method [16] and the fading coefficients have a mean value of 1.55 and variance 0.60. During simulation, quasi-static fading is assumed in which the fading amplitude changes for each code word block. For combating the fading effect, 64×64 and 100×512 block interleavers are employed for codes defined in GF(16) and GF(64) respectively.

In Figure 5.9a the soft-decision list decoding performance of the (64, 39, 20) Hermitian code over GF(16) is shown on the AWGN channel for different list lengths. This is compared with the hard-decision list decoding performance with multiplicity values $m = 1$ and 2, and the performance of the Sakata algorithm is included too. We can see that there are small coding gains for soft-decision list decoding over hard-decision list decoding up to approximately 1.3 dB at a BER of 10^{-4}. In Figure 5.9b, more significant coding gains are achieved on the fading channel, with a coding gain greater than 3 dB over hard-decision list decoding.

In Figure 5.10a the soft-decision list decoding performance of the (512, 289, 196) Hermitian code defined over GF(64) is shown on the AWGN channel for different output list lengths. Again, small coding gains are achieved over hard-decision decoding. In Figure 5.10b the soft-decision list decoding performance on the fading channel is shown, with more significant coding gains over hard-decision decoding.

5.9 Conclusions

In this chapter we have introduced the concept of list decoding for Reed–Solomon and AG codes. We can see that list decoding allows more errors to be corrected than with conventional algebraic decoding algorithms, but at a cost of higher complexity. However, a method to reduce this complexity without degrading the performance is given for both Reed–Solomon and AG codes. The Kotter–Vardy algorithm for the soft-decision list decoding of Reed–Solomon codes was presented; this is almost identical to the Guruswami–Sudan algorithm but with an extra process to convert the reliability of the received symbols into a multiplicity matrix. The Kotter–Vardy algorithm was also extended to soft decode AG codes. One interesting observation from the soft-decision list decoding algorithm was that it could achieve the same performance as the hard-decision list decoder but at a *lower* complexity. More significant coding gains can be achieved on a slow fading channel, particularly for AG codes, suggesting that soft-decision list decoding is very suitable for channels where bursts of errors are common.

References

[1] Elias, P. (1957) *List Decoding for Noisy Channels*, Res. Lab. Electron, MIT, Cambridge, MA.

[2] Elias, P. (1991) Error-correcting codes for list decoding. *Information Theory, IEEE Transactions*, **37**, 5–12.

[3] Wozencraft, J.M. (1958) *List Decoding*, Res. Lab. Electron, MIT, Cambridge, MA.

[4] McEliece, R.J. (2003) The Guruswami–Sudan Decoding Algorithm for Reed–Solomon Codes, California Institute. Tech, Pasadena, California, IPN Progress Rep, pp. 42–153.

[5] Koetter, R. and Vardy, A. (2003) Algebraic soft-decision decoding of Reed–Solomon codes. *IEEE Trans. Inform. Theory*, **49**, 2809–25.

[6] Guruswami, V. and Sudan, M. (1999) Improved decoding of Reed–Solomon and algebraic–geometric codes. *IEEE Trans. Inform. Theory*, **45**, 1757–67.

[7] Hasse, H. (1936) Theorie der hoheren differentiale in einem algebraishen funcktionenkorper mit vollkommenem konstantenkorper nei beliebeger charakteristic. *J. Reine. Aug. Math*, **175**, 50–4.

[8] Koetter, R. (1996) On algebraic decoding of algebraic–geometric and cyclic codes. Linkoping, Sweden: University Linkoping.

[9] Nielsen, R.R. (2001) List decoding of linear block codes. Lyngby, Denmark: Tech. Univ. Denmark.

[10] Moon, T.K. (2005) *Error Correction Coding – Mathematical and Algorithms*, Wiley Interscience.

[11] Høholdt, T. and Nielsen, R.R. (1999) Decoding Hermitian codes with Sudan's algorithm, in *Applied Algebra, Algebraic Algorithms and Error-Correcting Codes* (eds H.I.N. Fossorier, S. Lin, and A. Pole), (Lecture Notes in Computer Science), Vol. **1719**, Springer-Verlag, Berlin, Germany, pp. 260–70.

[12] Feng, G.-L. and Tzeng, K.K. (1991) A generalization of the Berlekamp-Massey algorithm for multisequence shift-register synthesis with application to decoding cyclic codes. *IEEE Trans. Inform. Theory*, **37**, 1274–87.

[13] Chen, L., Carrasco, R.A. and Chester, E.G. (2007) Performance of Reed–Solomon codes using the Guruswami–Sudan algorithm with improved interpolation efficiency. *IET Commun*, **1**, 241–50.

[14] Chen, L. (2007) Design of an efficient list decoding system for Reed–Solomon and algebraic–geometric codes, PhD Thesis *School of electrical, electronic and computer engineering*. Newcastle-upon-Tyne: Newcastle University.

[15] Roth, R. and Ruckenstein, G. (2000) Efficient decoding of Reed–Solomon codes beyond half the minimum distance. *IEEE Trans. Inform. Theory*, **46**, 246–57.

[16] Proakis, J.G. (2000) *Digital Communications*, 4th edn, McGraw-Hill International.

[17] Chen, L., Carrasco, R.A. and Johnston, M. (2006) List decoding performance of algebraic geometric codes. *IET Electronic Letters*, **42**, 986–7.

[18] Chen, L., Carrasco, R.A. and Johnston, M. (XXXX) Reduced complexity interpolation for list decoding Hermitian codes. *IEEE Trans. Wireless Commun*, Accepted for publication.

[19] Chen, L. and Carrasco, R.A. (2007) Efficient list decoder for algebraic–geometric codes. Presented at 9th International Symposium on Communication Theory and Application (ISCTA'07), Ambleside, Lake district, UK.

[20] Wu, X.-W. and Siegel, P. (2001) Efficient root-finding algorithm with application to list decoding of algebraic–geometric codes. *IEEE Trans. Inform. Theory*, **47**, 2579–87.

[21] Chen, L., Carrasco, R.A., Johnston, M. and Chester, E.G. (2007) Efficient factorisation algorithm for list decoding algebraic–geometric and Reed–Solomon codes. Presented at ICC 2007, Glasgow, UK.

[22] Wu, X.-W. (2002) An algorithm for finding the roots of the polynomials over order domains. Presented at ISIT 2002, Lausanne, Switzerland.

[23] Chen, L., Carrasco, R.A. and Johnston, M. (XXXX) Soft-decision list decoding of Hermitian codes. *IEEE Trans. Commun*, Submitted for publication.

6

Non-Binary Low-Density Parity Check Codes

6.1 Introduction

The previous chapters have described non-binary block codes that have algebraic decoders, but now a new type of block code is introduced that is decoded using a graph. This is the well-known low-density parity check (LDPC) code, which was first presented by Gallager in 1962 [1] but only became popular when Mackay rediscovered it in 1995 [2]. It has been shown that with iterative soft-decision decoding LDPC codes can perform as well as or even better than turbo codes [3]. It is claimed that LDPC codes will be chosen in future standards, such as 4G, later versions of WiMax and magnetic storage devices.

In this chapter, binary LDPC codes are first introduced with a discussion on random and structured construction methods. The Belief Propagation algorithm is then presented in detail to decode LDPC codes. We then expand binary LDPC codes to non-binary LDPC codes defined over a finite field GF(q) and again discuss different construction methods. Finally the Belief Propagation algorithm is extended to non-binary symbols, which increases its complexity, but a method to reduce complexity is introduced using fast Fourier transforms (FFT), allowing us to decode non-binary LDPC codes defined over large finite fields.

6.2 Construction of Binary LDPC Codes – Random and Structured Methods

A low-density parity check (LDPC) code is characterized by its sparse parity check matrix, that is a matrix containing mostly zero elements and few nonzero elements. Three important parameters are its code word length, n, its dimension, k, and its number of parity bits, $m = n - k$. The number of nonzero elements in a row of the

Non-Binary Error Control Coding for Wireless Communication and Data Storage   Rolando Antonio Carrasco and Martin Johnston
© 2008 John Wiley & Sons, Ltd

parity check matrix is called the *row weight*, denoted by ρ. The number of nonzero elements in a column of the parity check matrix is called the *column weight*, denoted by γ. There are two general classes of LDPC code:

- If the row weights and column weights are constant for each row and column in the parity check matrix then the LDPC code is said to be *regular*.
- If the row and column weights vary for each row and column in the parity check matrix then the LDPC code is said to be *irregular*.

In general, irregular LDPC codes have a better performance than regular LDPC codes. An example of a parity check matrix **H** for a small binary LDPC code is given in (6.1):

$$\mathbf{H} = \begin{bmatrix} 1 & 1 & 0 & 1 & 0 & 0 \\ 0 & 1 & 1 & 0 & 1 & 0 \\ 1 & 0 & 0 & 0 & 1 & 1 \\ 0 & 0 & 1 & 1 & 0 & 1 \end{bmatrix} \tag{6.1}$$

We can see that **H** has a constant row weight $\rho = 3$ and constant column weight $\gamma = 2$, implying that this is a regular LDPC code. Originally, LDPC codes were constructed by first choosing the row and column weights and then randomly determining where the nonzero elements were located.

6.2.1 Tanner Graph Representation

A parity check matrix can be expressed in the form of a bipartite graph known as a Tanner graph. Each row in the parity check matrix represents a parity check equation z_i, $1 \le i \le m$, and each column represents a coded bit c_j, $1 \le j \le n$. For example, the first row of **H** in (6.1) is the parity check equation $z_1 = c_1 \oplus c_2 \oplus c_4$. The Tanner graph connects the coded bits to each parity check equation. For each row of the parity check matrix, if the coded bit is a 1 then there is a connection between that coded bit and the corresponding parity check equation. The Tanner graph for **H** in (6.1) is given in Figure 6.1.

The Tanner graph reveals the presence of *cycles*, that is a path starting and ending at the same coded bit. In Figure 6.1 it can be seen that the smallest cycle has a length of 6. One such cycle is $c_1 \rightarrow z_1 \rightarrow c_4 \rightarrow z_4 \rightarrow c_6 \rightarrow z_3 \rightarrow c_1$. The length of the smallest cycle in a Tanner graph is known as its *girth*. It is desirable to avoid LDPC codes with Tanner graphs with a girth of 4 as this degrades the performance, particularly for short LDPC codes. From the Tanner graph, we introduce two sets:

- M_n is the set of indices of the parity checks connected to coded bit c_n.
- N_m is the set of indices the coded bits that are connected to parity check z_m.

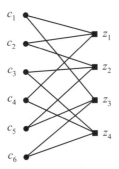

Figure 6.1 Tanner graph for the parity check matrix of (6.1).

So from the Tanner graph of Figure 6.1, the set $M_1 = \{1, 3\}$, since parity checks z_1 and z_3 are connected to coded bit c_1. The set $N_1 = \{1, 2, 4\}$, since coded bits c_1, c_2 and c_4 are connected to parity check z_1. We also introduce the notation M_n/m, which is the set M_n excluding the parity check z_m that is connected to the coded bit c_n. Similarly, N_m/n is the set N_m excluding the coded bit c_n that is connected to the parity check z_m. Therefore, $M_1/1 = \{3\}$ and $N_1/1 = \{2, 4\}$. This notation will be useful when describing the Belief Propagation algorithm for decoding LDPC codes.

6.2.2 Structured Construction Methods

There are many different methods for constructing a binary LDPC code [5–6], but in this chapter we will discuss just one. This method is known as the Balanced Incomplete Block Design (BIBD), which can be used to construct high-rate LDPC codes [4]. Let $X = \{x_1, x_2, \ldots, x_v\}$ be a set of v objects. A BIBD of X is a collection of n γ-subsets of X, denoted by $B_1, B_2, \ldots, B_n$, called blocks, such that the following conditions are satisfied:

1. Each object appears in exactly ρ of the n blocks.
2. Every two objects appear together in exactly λ of the n blocks.
3. The number γ of objects in each block is small compared to the total number of objects in X.

Instead of a list of the blocks, a BIBD can be efficiently described by a $v \times n$ matrix **Q** over GF(2), as follows:

1. The rows of **Q** correspond to the v objects in X.
2. The columns of **Q** correspond to the n blocks of the design.

3. Each element $Q_{i,j}$ in $\mathbf{Q}$ is a 1 if and only if the ith object x_i is contained in the jth block B_j of the design; otherwise it is 0. This matrix is called the *incidence matrix* of the design.

It follows from the structural properties of a BIBD that the incidence matrix $\mathbf{Q}$ has constant row weight ρ and constant column weight γ and any two rows of $\mathbf{Q}$ have exactly λ '1-components' in common.

Example 6.1: Construction of an incidence matrix using BIBD: Let $X = \{x_1,$ $x_2, x_3, x_4, x_5, x_6, x_7\}$ be a set of seven objects. The blocks $\{x_1, x_2, x_4\}$, $\{x_2, x_3,$ $x_5\}$, $\{x_3, x_4, x_6\}$, $\{x_4, x_5, x_7\}$, $\{x_5, x_6, x_1\}$, $\{x_6, x_7, x_2\}$,$\{x_7, x_1, x_3\}$ form a BIBD for the set X. Every block consists of three objects, each object appears in three blocks, and every two objects appear together in exactly one block, that is $\lambda = 1$. The incidence matrix $\mathbf{Q}$ is given below:

$$\mathbf{Q} = \begin{pmatrix} 1 & 0 & 0 & 0 & 1 & 0 & 1 \\ 1 & 1 & 0 & 0 & 0 & 1 & 0 \\ 0 & 1 & 1 & 0 & 0 & 0 & 1 \\ 1 & 0 & 1 & 1 & 0 & 0 & 0 \\ 0 & 1 & 0 & & 1 & 0 & 0 \\ 0 & 0 & 1 & 0 & 1 & 1 & 0 \\ 0 & 0 & 0 & 1 & 0 & 1 & 1 \end{pmatrix}.$$

Note that each row of $\mathbf{Q}$ is a cyclic right-shift of the row above it and the first row is the cyclic shift of the last row. We also note that each column is a downward cyclic shift of the column on its left and the first column is the downward cyclic shift of the last column. Therefore $\mathbf{Q}$ is a 7×7 square *circulant matrix*. Therefore, the null space of $\mathbf{Q}$ gives a (γ, ρ)-regular LDPC code, called a *BIBD-LDPC code*, of length n. The Tanner graph of a BIBD-LDPC code is free of cycles of length 4 and hence its girth is at least 6. In fact, the second property of a BIBD with $\lambda = 1$ ensures that the girth of the code's Tanner graph is exactly 6. If $\lambda = 2$ then the incidence matrix will have a Tanner graph full of short cycles of length 4, which is undesirable.

6.2.3 LDPC Codes from Class 1 Bose BIBD

There are four classes of Bose BIBD, each with different types producing many different binary LDPC codes. However, in this book we only deal with LDPC constructed from Class 1 Bose BIBDs of type 1 [4]. Choose a finite field GF$(2t + 1)$, where t is selected so that GF$(2t + 1)$ is prime, with a primate element x that satisfies $x^{4t} - 1 = x^c$, where c is an odd integer less than $2t + 1$. The resulting LDPC code parameters will have the following values: $n = t(12t + 1)$, $v = 12t + 1$, $\gamma = 4$, $\rho = 4t$, $\lambda = 1$. The base blocks of this BIBD are:

$$\left\{0, x^{2i}, x^{2i+4t}, x^{2i+8t}\right\}, \tag{6.2}$$

for $0 \le i < t$. From each base block B_i, we can form $12t + 1$ blocks by adding each element of the field GF($12t + 1$) in turn to the elements in B_i. This results in $t(12t + 1)$ blocks. The incidence matrix $\mathbf{Q}$ of this BIBD is a $(12t + 1) \times t(12t + 1)$ matrix. It can be written in cyclic form, which consists of t $(12t + 1) \times (12t + 1)$ circulant submatrices in a row, as shown below:

$$\mathbf{Q} = \left(\mathbf{Q}_1 \, \mathbf{Q}_2 \cdots \mathbf{Q}_t\right), \tag{6.3}$$

where the ith circulant $\mathbf{Q}_i$ is the incidence matrix formed by adding each element in GF($12t + 1$) in turn to the elements of the ith base block B_i. A circulant is a square matrix which has the following structure:

1. Each row is a right-cyclic shift of the row above it and the first row is the right-cyclic shift of the last row.
2. Each column is a downward cyclic shift of the column on its left and the first column is the downward cyclic shift of the last column. The set of columns is the same as the set of rows, except that each row reads from right to left.

Both the row and the column weight of each circulant $\mathbf{Q}_i$ are 4. Therefore, the row and column weights of $\mathbf{Q}$ are $4t$ and 4 respectively. The rank of $\mathbf{Q}$ (or a circulant $\mathbf{Q}_i$) is observed to be $12t$.

Example 6.2: Construction of an LDPC code using a class 1 Bose BIBD: Let $t = 1$. Then $12t + 1 = 13$. The element 6 of GF(13) = $\{0, 1, 2, \ldots, 12\}$ is a primitive element so with $c = 3$ we can easily check that $x^4 - 1 = 64 - 1 = 9 - 1$ $= 8 = 63$, which is less than 13. Therefore, there is a BIBD with parameters, $v = 13$, $n = 13$, $\gamma = 4$, $\rho = 4$ and $\lambda = 1$. The base block for this design is $\{0, 6^0, 6^4, 6^8\} = \{0, 1, 9, 3\}$. By adding each element in GF(13) to the base block we get the following 13 blocks:

$B_0 = \{0, 1, 9, 3\}$, $B_1 = \{1, 2, 10, 4\}$, $B_2 = \{2, 3, 11, 5\}$, $B_3 = \{3, 4, 12, 6\}$, $B_4 = \{4, 5, 0, 7\}$,
$B_5 = \{5, 6, 1, 8\}$, $B_6 = \{6, 7, 2, 9\}$, $B_7 = \{7, 9, 3, 10\}$, $B_8 = \{8, 9, 4, 11\}$, $B_9 = \{9, 10, 5, 12\}$,
$B_{10} = \{10, 11, 6, 0\}$, $B_{11} = \{11, 12, 7, 1\}$, $B_{12} = \{12, 0, 8, 2\}$.

Hence, the circulant matrix $\mathbf{Q}$ is:

$$\mathbf{Q} = \begin{bmatrix}
1 & 0 & 0 & 0 & 1 & 0 & 0 & 0 & 0 & 0 & 1 & 0 & 1 \\
1 & 1 & 0 & 0 & 0 & 1 & 0 & 0 & 0 & 0 & 0 & 1 & 0 \\
0 & 1 & 1 & 0 & 0 & 0 & 1 & 0 & 0 & 0 & 0 & 0 & 1 \\
1 & 0 & 1 & 1 & 0 & 0 & 0 & 1 & 0 & 0 & 0 & 0 & 0 \\
0 & 1 & 0 & 1 & 1 & 0 & 0 & 0 & 1 & 0 & 0 & 0 & 0 \\
0 & 0 & 1 & 0 & 1 & 1 & 0 & 0 & 0 & 1 & 0 & 0 & 0 \\
0 & 0 & 0 & 1 & 0 & 1 & 1 & 0 & 0 & 0 & 1 & 0 & 0 \\
0 & 0 & 0 & 0 & 1 & 0 & 1 & 1 & 0 & 0 & 0 & 1 & 0 \\
0 & 0 & 0 & 0 & 0 & 1 & 0 & 1 & 1 & 0 & 0 & 0 & 1 \\
1 & 0 & 0 & 0 & 0 & 0 & 1 & 0 & 1 & 1 & 0 & 0 & 0 \\
0 & 1 & 0 & 0 & 0 & 0 & 0 & 1 & 0 & 1 & 1 & 0 & 0 \\
0 & 0 & 1 & 0 & 0 & 0 & 0 & 0 & 1 & 0 & 1 & 1 & 0 \\
0 & 0 & 0 & 1 & 0 & 0 & 0 & 0 & 0 & 1 & 0 & 1 & 1
\end{bmatrix}.$$

This example only produces a single circulant matrix, but the next value of t that ensures the finite field is prime is $t = 6$, giving GF(73). This will result in an LDPC code with parameters $v = 73$, $n = 438$, $\gamma = 4$, $\rho = 292$ and $\lambda = 1$. In this case, there will be six base blocks each containing four elements, and six circulant matrices can be formed by adding every element in GF(73) to each of the base blocks. Each circulant matrix will have dimensions 73×73, with $\gamma = 4$, $\rho = 4$. The six circulant matrices are then concatenated to form the final parity check matrix, H, with dimensions 73×438. The null space of this matrix will be the (438, 365) LDPC code, assuming each row in H is independent.

6.2.4 Constructing the Generator Matrix of a Binary LDPC Code

The encoding of a binary message to form an LDPC code word is achieved by multiplying the binary message vector by a systematic generator matrix, $\mathbf{G}$. The systematic generator matrix is obtained by first performing Gauss–Jordan elimination on the parity check matrix so that it is of the form $\mathbf{H} = [\mathbf{I_m}|\mathbf{P}]$, where $\mathbf{I_m}$ is the $m \times m$ identity matrix and $\mathbf{P}$ is a parity matrix. The generator matrix is then $\mathbf{G} = [\mathbf{P^T}|\mathbf{I_k}]$, where $\mathbf{P^T}$ is the matrix transpose of $\mathbf{P}$, and $\mathbf{I_k}$ is the $k \times k$ identity matrix. Gauss–Jordan elimination is related to Gaussian elimination in that it eliminates all elements below and also above each pivot element. Gaussian elimination results in a matrix in row echelon form, whereas Gauss–Jordan elimination results in a matrix in reduced row echelon form. It can therefore be seen as a two-stage process, where Gaussian elimination is first performed to transform the matrix into row echelon form, followed by elimination of elements above the pivot elements.

The elimination procedure on a matrix containing finite field elements is performed in the same way as if the matrix contained real numbers. The only difference is that elements are eliminated with modulo-2 addition instead of subtraction.

It is not unusual for a parity check matrix to have dependent rows. When performing Gaussian elimination on the matrix, all dependent rows will appear as rows containing all zeroes located at the bottom of the new matrix. This is the case if we apply Gaussian elimination to the parity check matrix of (6.1):

$$\mathbf{H} = \begin{bmatrix} 1 & 1 & 0 & 1 & 0 & 0 \\ 0 & 1 & 1 & 0 & 1 & 0 \\ 1 & 0 & 0 & 0 & 1 & 1 \\ 0 & 0 & 1 & 1 & 0 & 1 \end{bmatrix} \rightarrow \mathbf{H}' \begin{bmatrix} 1 & 0 & 0 & 0 & 1 & 1 \\ 0 & 1 & 1 & 0 & 1 & 0 \\ 0 & 0 & 1 & 1 & 0 & 1 \\ 0 & 0 & 0 & 0 & 0 & 0 \end{bmatrix}.$$

We can see that row 4 in $\mathbf{H}$ is the modulo-2 sum of row 1, row 2 and row 3 and hence it is a dependent row. Therefore, performing Gaussian elimination results in an all-zero row in the new matrix, $\mathbf{H}'$. However, a generator matrix can still be obtained by removing the all-zero rows and then eliminating elements above the pivot elements in the resulting submatrix:

$$\mathbf{H}' = \begin{bmatrix} 1 & 0 & 0 & 0 & 1 & 1 \\ 0 & 1 & 1 & 0 & 1 & 0 \\ 0 & 0 & 1 & 1 & 0 & 1 \end{bmatrix} \rightarrow \mathbf{H}'' = \begin{bmatrix} 1 & 0 & 0 & 0 & 1 & 1 \\ 0 & 1 & 0 & 1 & 1 & 1 \\ 0 & 0 & 1 & 1 & 0 & 1 \end{bmatrix}.$$

The systematic parity check matrix $\mathbf{H}''$ is simply obtained by adding row 3 to row 2 in $\mathbf{H}'$. Finally, the generator matrix is:

$$\mathbf{G} = \begin{bmatrix} 0 & 1 & 1 & 1 & 0 & 0 \\ 1 & 1 & 0 & 0 & 1 & 0 \\ 1 & 1 & 1 & 0 & 0 & 1 \end{bmatrix}.$$

We can see that dependent rows in the parity check matrix will increase the code rate of the LDPC code. If all the rows were independent then the null space of the parity check matrix of (6.1) would be the (6, 2) LDPC code with code rate of 1/3. However, the dependent row means the null space is the (6, 3) LDPC code with code rate of 1/2.

For larger parity check matrices, the swapping of columns is usually required during Gaussian elimination to ensure that the pivot elements are on the main diagonal of the matrix. However, this will obviously change the parity check equations and the resulting systematic generator matrix will not be the null space of the original parity check matrix. In this case, any column that is swapped must also be swapped on the original parity check matrix, but this will not alter its column and row weights.

6.3 Decoding of Binary LDPC Codes Using the Belief Propagation Algorithm

The decoding of LDPC codes using the sum product algorithm involves finding a code word of length n where each coded bit c_n maximizes the probability of c_n conditioned on the parity check equations associated with c_n being satisfied [7].

$$P(c_n|\{z_m = 0, \ m \in M_n\}). \tag{6.4}$$

The sum product algorithm involves calculating two probabilities. The first is the probability q_{mn}, which is the probability of the nth code bit conditioned on its associated parity checks being satisfied with the exception of the mth parity check:

$$q_{mn}(x) = P(c_n = x \left|\{z_m = 0, m' \in M_n/m\}\right.. \tag{6.5}$$

The second probability is denoted r_{mn}, which is the probability that the mth parity check is satisfied conditioned on all the possible values of the coded bits $\mathbf{c}$:

$$r_{mn}(x) = P(z_m = 0|\mathbf{c}). \tag{6.6}$$

Figure 6.2 shows how these probabilities are exchanged in the Tanner graph of Figure 6.1 for the connections between c_1 and z_1, and c_6 and z_4.

6.3.1 Initialization of the Belief Propagation Algorithm

The probabilities $q_{mn}(x)$ are initialized to the probability $f_n^{(x)}$ that the nth received bit is x, that is $P(c_n = x)$. For the additive white Gaussian noise (AWGN) channel the probability $g_n^{(x)}$ is [7]:

$$g_n^{(0)} \propto \frac{1}{\sigma\sqrt{2\pi}}e^{-(r_n-1)^2/2\sigma^2} \quad \text{and} \quad g_n^{(1)} \propto \frac{1}{\sigma\sqrt{2\pi}}e^{-(r_n+1)^2/2\sigma^2}. \tag{6.7}$$

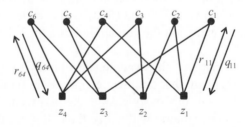

Figure 6.2 Tanner graph showing the exchange of information between the coded bits and their parity checks.

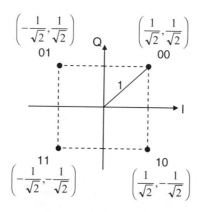

Figure 6.3 QPSK constellation.

In general, when the modulation scheme has an in-phase and quadrature component the probabilities are determined by calculating the Euclidean distances for the in-phase component from each constellation point. The coordinates of each constellation point in the QPSK constellation are shown in Figure 6.3.

Therefore, the four probabilities $g_n^{(00)}$, $g_n^{(01)}$, $g_n^{(11)}$ and $g_n^{(10)}$ for the QPSK constellation are:

$$g_n^{(00)} \propto \frac{1}{\sigma\sqrt{2\pi}}e^{-\left[\left(r_n^{(I)}-\frac{1}{\sqrt{2}}\right)^2+\left(r_n^{(Q)}-\frac{1}{\sqrt{2}}\right)^2\right]/2\sigma^2},$$

$$g_n^{(01)} \propto \frac{1}{\sigma\sqrt{2\pi}}e^{-\left[\left(r_n^{(I)}-\frac{1}{\sqrt{2}}\right)^2+\left(r_n^{(Q)}+\frac{1}{\sqrt{2}}\right)^2\right]/2\sigma^2}$$

$$g_n^{(11)} \propto \frac{1}{\sigma\sqrt{2\pi}}e^{-\left[\left(r_n^{(I)}+\frac{1}{\sqrt{2}}\right)^2+\left(r_n^{(Q)}+\frac{1}{\sqrt{2}}\right)^2\right]/2\sigma^2},$$

$$g_n^{(10)} \propto \frac{1}{\sigma\sqrt{2\pi}}e^{-\left[\left(r_n^{(I)}+\frac{1}{\sqrt{2}}\right)^2+\left(r_n^{(Q)}-\frac{1}{\sqrt{2}}\right)^2\right]/2\sigma^2}$$

(6.8)

where $r_n^{(I)}$ and $r_n^{(Q)}$ are the in-phase and quadrature components of the received symbol. The likelihoods of the received bits are then placed in a vector **f**:

$$\mathbf{f} = \begin{bmatrix} g_1^{(0)} & g_2^{(0)} & \cdots & g_{n-1}^{(0)} & g_n^{(0)} \\ g_1^{(1)} & g_2^{(1)} & \cdots & g_{n-1}^{(1)} & g_n^{(1)} \end{bmatrix}.$$

(6.9)

To initialize the sum–product algorithm, the likelihoods in $\mathbf{f}$ are used to initialize a matrix $\mathbf{Q}$ defined as:

$$
\mathbf{Q} =
\left[
\begin{array}{ccccc|cc}
g_1^{(0)} & g_2^{(0)} & g_3^{(0)} & \cdots & g_{n-1}^{(0)} & g_n^{(0)} \\
g_1^{(1)} & g_2^{(1)} & g_3^{(1)} & \cdots & g_{n-1}^{(1)} & g_n^{(1)} \\[4pt]
\vdots & \vdots & \vdots & \ddots & \vdots & \vdots \\
\vdots & \vdots & \vdots & \ddots & \vdots & \vdots \\[4pt]
g_1^{(0)} & g_2^{(0)} & g_3^{(0)} & \cdots & g_{n-1}^{(0)} & g_n^{(0)} \\
g_1^{(1)} & g_2^{(1)} & g_3^{(1)} & \cdots & g_{n-1}^{(1)} & g_n^{(1)}
\end{array}
\right].
\tag{6.10}
$$

The matrix $\mathbf{Q}$ has partitions containing the likelihoods that the ith received bit is a 0, $g_i^{(0)}$ or a 1, $g_i^{(1)}$.

6.3.2 The Horizontal Step

The horizontal step of the sum–product algorithm determines the probability r_{mn}, defined as [7]:

$$
r_{mn}(x) = \sum_{\mathbf{c}:c_n=x} P(z_m = 0|\mathbf{c})P(\mathbf{c}|c_n = x),
\tag{6.11}
$$

where, given that the nth coded bit $c_n = x$, $\mathbf{c}$ is a binary vector of the remaining coded bits of length $\rho - 1$, containing every possible combination of binary sequences. There are therefore $2^{\rho-1}$ binary sequences that (6.9) is summed over. However, not all these sequences will satisfy $z_m = 0$ so we only consider those sequences that do satisfy the parity check. For example, if $x = 0$ then only those binary sequences with an even number of 1s are able to satisfy $z_m = 0$. Therefore, the probability of a sequence containing an odd number of 1s satisfying $z_m = 0$ is zero. Therefore, $r_{mn}(x)$ is the sum of all the probabilities of the binary sequences of length $\rho - 1$ that when summed together equal x:

$$
r_{mn}(x) = \sum_{\mathbf{c}:c_n=x} P(z_m = 0|\mathbf{c}) \prod_{n' \in N_m \setminus n} q_{mn'}(x),
\tag{6.12}
$$

where $P(z_m = 0 \mid \mathbf{c})$ is either zero or one. The probabilities $r_{mn}(x)$ are placed in a matrix
$\mathbf{R}$:

$$\mathbf{R} = \begin{bmatrix} r_{11}(0) & r_{12}(0) & r_{13}(0) & \cdots & r_{1(n-1)}(0) & r_{1n}(0) \\ r_{11}(1) & r_{12}(1) & r_{13}(1) & \cdots & r_{1(n-1)}(1) & r_{1n}(1) \\ \vdots & \vdots & \vdots & \ddots & \vdots & \vdots \\ \vdots & \vdots & \vdots & \ddots & \vdots & \vdots \\ r_{m1}(0) & r_{m2}(0) & r_{m3}(0) & \cdots & r_{m(n-1)}(0) & r_{mn}(0) \\ r_{m1}(1) & r_{m2}(1) & r_{m3}(1) & \cdots & r_{m(n-1)}(1) & r_{mn}(1) \end{bmatrix}, \tag{6.13}$$

where each element in the parity check matrix has a corresponding 1×2 column
vector containing the probabilities $r_{mn}(0)$ and $r_{mn}(1)$ (or a $1 \times q$ column vector for an
LDPC code defined over GF(q)).

6.3.3 The Vertical Step

The vertical step updates the probabilities q_{mn} and is much simpler than the horizontal
step. From (6.5), q_{mn} can be written using Baye's Rule as [7]:

$$\begin{aligned} q_{mn}(x) &= P(c_n = x \mid \{z_m = 0, m' \in M_n/m\} \\ &= \frac{P(c_n = x)P\left(\{z_m = 0, m' \in M_n/m\} \mid c_n = x\right)}{P\left(\{z_m = 0, m' \in M_n/m\}\right)}. \end{aligned} \tag{6.14}$$

Using (6.6) and letting $f_n^x = P(c_n = x)$, (6.10) can be written as [7]:

$$q_{mn}(x) = \beta_{mn} f_n^x \prod_{m' \in M_n/m} r_{m'n}(x), \tag{6.15}$$

where β_{mn} is a normalizing constant such that $\sum q_{mn}(x) = 1$, that is:

$$\beta_{mn} = \frac{1}{\sum\limits_{x} f_n^x \prod\limits_{m' \in M_n \backslash m} r_{m'n}(x)}. \tag{6.16}$$

The $q_{mn}(x)$ are placed in the matrix $\mathbf{Q}$ as shown in (6.17):

$$\mathbf{Q} = \begin{bmatrix} q_{11}(0) & q_{12}(0) & q_{13}(0) & \cdots & q_{1(n-1)}(0) & q_{1n}(0) \\ q_{11}(1) & q_{12}(1) & q_{13}(1) & \cdots & q_{1(n-1)}(1) & q_{1n}(1) \\ \vdots & \vdots & \vdots & \ddots & \vdots & \vdots \\ \vdots & \vdots & \vdots & \ddots & \vdots & \vdots \\ q_{m1}(0) & q_{m2}(0) & q_{m3}(0) & \cdots & q_{m(n-1)}(0) & q_{mn}(0) \\ q_{m1}(1) & q_{m2}(1) & q_{m3}(1) & \cdots & q_{m(n-1)}(1) & q_{mn}(1) \end{bmatrix}, \qquad (6.17)$$

where each element in the parity check matrix has a corresponding 1×2 column vector containing the probabilities $q_{mn}(0)$ and $q_{mn}(1)$ (or a $1 \times q$ column vector for an LDPC code defined over GF(q)).

In this step, the pseudo posterior probabilities $q_n(x)$ are also determined by [7]:

$$q_n(x) = \beta_n f_n^x \prod_{m \in M_n} r_{mn}(x), \qquad (6.18)$$

where again β_n is a normalizing constant such that $\sum q_n(x) = 1$. The pseudo posterior probabilities are place in a matrix $\mathbf{Q}'$:

$$\mathbf{Q}' = \begin{bmatrix} q_1(0) & q_2(0) & q_3(0) & \cdots & q_{n-1}(0) & q_n(0) \\ q_1(1) & q_2(1) & q_3(1) & \cdots & q_{n-1}(1) & q_n(1) \end{bmatrix}. \qquad (6.19)$$

From these pseudo posterior probabilities estimates of the transmitted code word can be found by:

$$\hat{c}_n = \arg \max_x \beta_n f_n^x \prod_{m \in N_m} r_{mn}(x). \qquad (6.20)$$

Example 6.3: Decoding a binary LDPC code with the Belief Propagation algorithm: In this example, we use the regular $(10, 5)$ binary LDPC code with row weight $\rho = 6$ and column weight $\gamma = 3$. The parity check matrix $\mathbf{H}$ is given as:

$$\mathbf{H} = \begin{bmatrix} 1 & 1 & 1 & 0 & 0 & 1 & 1 & 0 & 0 & 1 \\ 1 & 0 & 1 & 0 & 1 & 1 & 0 & 1 & 1 & 0 \\ 0 & 0 & 1 & 1 & 1 & 0 & 1 & 0 & 1 & 1 \\ 0 & 1 & 0 & 1 & 1 & 1 & 0 & 1 & 0 & 1 \\ 1 & 1 & 0 & 1 & 0 & 0 & 1 & 1 & 1 & 0 \end{bmatrix}.$$

Let us assume that the transmitted code word is $\mathbf{c} = [0 \ \ 0 \ \ 0 \ \ 1 \ \ 0 \ \ 1 \ \ 0 \ \ 1 \ \ 0 \ \ 1]$ and the received word is $\mathbf{r} = [0 \ \ 0 \ \ 0 \ \ 1 \ \ 1 \ \ 1 \ \ 0 \ \ 1 \ \ 0 \ \ 1]$, with a single error in the fifth position. The associated likelihoods are:

$$
\mathbf{f} = \begin{bmatrix} 0.78 & 0.84 & 0.81 & 0.52 & 0.45 & 0.13 & 0.82 & 0.21 & 0.75 & 0.24 \\ 0.22 & 0.16 & 0.19 & 0.48 & 0.55 & 0.87 & 0.18 & 0.79 & 0.25 & 0.76 \end{bmatrix}.
$$

We initialize the matrix $\mathbf{Q}$ with these likelihoods as follows:

$$
\mathbf{Q} = \begin{bmatrix}
0.78 & 0.84 & 0.81 & 0 & 0 & 0.13 & 0.82 & 0 & 0 & 0.24 \\
0.22 & 0.16 & 0.19 & 0 & 0 & 0.87 & 0.18 & 0 & 0 & 0.76 \\
0.78 & 0 & 0.81 & 0 & 0.45 & 0.13 & 0 & 0.21 & 0.75 & 0 \\
0.22 & 0 & 0.19 & 0 & 0.55 & 0.87 & 0 & 0.79 & 0.25 & 0 \\
0 & 0 & 0.81 & 0.52 & 0.45 & 0 & 0.82 & 0 & 0.75 & 0.24 \\
0 & 0 & 0.19 & 0.48 & 0.55 & 0 & 0.18 & 0 & 0.25 & 0.76 \\
0 & 0.84 & 0 & 0.52 & 0.45 & 0.13 & 0 & 0.21 & 0 & 0.24 \\
0 & 0.16 & 0 & 0.48 & 0.55 & 0.87 & 0 & 0.79 & 0 & 0.76 \\
0.78 & 0.84 & 0 & 0.52 & 0 & 0 & 0.82 & 0.21 & 0.75 & 0 \\
0.22 & 0.16 & 0 & 0.48 & 0 & 0 & 0.18 & 0.79 & 0.25 & 0
\end{bmatrix}.
$$

The matrix $\mathbf{Q}$ containing all the probabilities $q_{mn}(x)$ is then used for the horizontal step to determine the matrix $\mathbf{R}$ containing the probabilities $r_{mn}(x)$. To determine the probability $r_{11}(1)$ we must calculate the probabilities of all the possible binary sequences that satisfy z_1 when $c_1 = 1$. Hence, we must first find all possible binary values of c_2, c_3, c_6, c_7 and c_{10} that satisfy:

$$
c_2 + c_3 + c_6 + c_7 + c_{10} = c_1 = 1.
$$

Since we are only working with binary values, the above equation can only be satisfied when an odd number of the coded bits are equal to 1. The 16 possible bit sequences and their probabilities are given in Table 6.1.

$r_{11}(1) = \big(q_{12}(1)q_{13}(0)q_{16}(0)q_{17}(0)q_{1,10}(0)\big) + \big(q_{12}(0)q_{13}(1)q_{16}(0)q_{17}(0)q_{1,10}(0)\big) +$
$+ \big(q_{12}(0)q_{13}(0)q_{16}(1)q_{17}(0)q_{1,10}(0)\big) + \big(q_{12}(0)q_{13}(0)q_{16}(0)q_{17}(1)q_{1,10}(0)\big)$
$+ \big(q_{12}(0)q_{13}(0)q_{16}(0)q_{17}(0)q_{1,10}(1)\big) + \big(q_{12}(1)q_{13}(1)q_{16}(1)q_{17}(0)q_{1,10}(0)\big)$
$+ \big(q_{12}(0)q_{13}(1)q_{16}(1)q_{17}(1)q_{1,10}(0)\big) + \big(q_{12}(0)q_{13}(0)q_{16}(1)q_{17}(1)q_{1,10}(1)\big)$
$+ \big(q_{12}(1)q_{13}(1)q_{16}(0)q_{17}(0)q_{1,10}(1)\big) + \big(q_{12}(1)q_{13}(1)q_{16}(0)q_{17}(1)q_{1,10}(0)\big)$
$+ \big(q_{12}(0)q_{13}(1)q_{16}(1)q_{17}(0)q_{1,10}(1)\big) + \big(q_{12}(1)q_{13}(0)q_{16}(1)q_{17}(0)q_{1,10}(1)\big)$
$+ \big(q_{12}(1)q_{13}(0)q_{16}(0)q_{17}(1)q_{1,10}(1)\big) + \big(q_{12}(0)q_{13}(1)q_{16}(0)q_{17}(1)q_{1,10}(1)\big)$
$+ \big(q_{12}(1)q_{13}(0)q_{16}(1)q_{17}(1)q_{1,10}(0)\big) + \big(q_{12}(1)q_{13}(1)q_{16}(1)q_{17}(1)q_{1,10}(1)\big)$
$= 0.00316 + 0.004038 + 0.116495 + 0.003821 + 0.055123 + 0.005205$
$+ 0.005998 + 0.080978 + 0.002463 + 0.000171 + 0.086533 + 0.070267$
$+ 0.002305 + 0.002838 + 0.004871 + 0.0036618 r_{11}(1) = 0.448086.$

Table 6.1 Bit sequences and their associated probabilities of determining $r_{11}(1)$.

Bit sequence	Probability
10 000	$q_{12}(1)\, q_{13}(0)\, q_{16}(0)\, q_{17}(0)\, q_{1,10}(0) = 0.16 \times 0.81 \times 0.13 \times 0.82 \times 0.24 = 0.003316$
01 000	$q_{12}(0)\, q_{13}(1)\, q_{16}(0)\, q_{17}(0)\, q_{1,10}(0) = 0.84 \times 0.19 \times 0.13 \times 0.82 \times 0.24 = 0.004083$
00 100	$q_{12}(0)\, q_{13}(0)\, q_{16}(1)\, q_{17}(0)\, q_{1,10}(0) = 0.84 \times 0.81 \times 0.87 \times 0.82 \times 0.24 = 0.116495$
00 010	$q_{12}(0)\, q_{13}(0)\, q_{16}(0)\, q_{17}(1)\, q_{1,10}(0) = 0.84 \times 0.81 \times 0.13 \times 0.18 \times 0.24 = 0.003821$
00 001	$q_{12}(0)\, q_{13}(0)\, q_{16}(0)\, q_{17}(0)\, q_{1,10}(1) = 0.84 \times 0.81 \times 0.13 \times 0.82 \times 0.76 = 0.055123$
11 100	$q_{12}(1)\, q_{13}(1)\, q_{16}(1)\, q_{17}(0)\, q_{1,10}(0) = 0.16 \times 0.19 \times 0.87 \times 0.82 \times 0.24 = 0.005205$
01 110	$q_{12}(0)\, q_{13}(1)\, q_{16}(1)\, q_{17}(1)\, q_{1,10}(0) = 0.84 \times 0.19 \times 0.87 \times 0.18 \times 0.24 = 0.005998$
00 111	$q_{12}(0)\, q_{13}(0)\, q_{16}(1)\, q_{17}(1)\, q_{1,10}(1) = 0.84 \times 0.81 \times 0.87 \times 0.18 \times 0.76 = 0.080978$
11 001	$q_{12}(1)\, q_{13}(1)\, q_{16}(0)\, q_{17}(0)\, q_{1,10}(1) = 0.16 \times 0.19 \times 0.13 \times 0.82 \times 0.76 = 0.002463$
11 010	$q_{12}(1)\, q_{13}(1)\, q_{16}(0)\, q_{17}(1)\, q_{1,10}(0) = 0.16 \times 0.19 \times 0.13 \times 0.18 \times 0.24 = 0.000171$
01 101	$q_{12}(0)\, q_{13}(1)\, q_{16}(1)\, q_{17}(0)\, q_{1,10}(1) = 0.84 \times 0.19 \times 0.87 \times 0.82 \times 0.76 = 0.086533$
10 101	$q_{12}(1)\, q_{13}(0)\, q_{16}(1)\, q_{17}(0)\, q_{1,10}(1) = 0.16 \times 0.81 \times 0.87 \times 0.82 \times 0.76 = 0.070267$
10 011	$q_{12}(1)\, q_{13}(0)\, q_{16}(0)\, q_{17}(1)\, q_{1,10}(1) = 0.16 \times 0.81 \times 0.13 \times 0.18 \times 0.76 = 0.002305$
01 011	$q_{12}(0)\, q_{13}(1)\, q_{16}(0)\, q_{17}(1)\, q_{1,10}(1) = 0.84 \times 0.19 \times 0.13 \times 0.18 \times 0.76 = 0.002838$
10 110	$q_{12}(1)\, q_{13}(0)\, q_{16}(1)\, q_{17}(1)\, q_{1,10}(0) = 0.16 \times 0.81 \times 0.87 \times 0.18 \times 0.24 = 0.004871$
11 111	$q_{12}(1)\, q_{13}(1)\, q_{16}(1)\, q_{17}(1)\, q_{1,10}(1) = 0.16 \times 0.19 \times 0.87 \times 0.18 \times 0.76 = 0.003618$

By adding these probabilities we get $r_{11}(1) = 0.448086$ and therefore $r_{11}(0) = 1 - r_{11}(1) = 0.551914$. The complete matrix **R** is then:

$$\mathbf{R} =$$

0.551914	0.542753	0.546890	0	0	0.460714	0.545425	0	0	0.444092
0.448086	0.457247	0.453110	0	0	0.539286	0.454575	0	0	0.555908
0.493347	0	0.493991	0	0.537255	0.505034	0	0.506423	0.493347	0
0506653	0	0.506009	0	0.462745	0.494966	0	0.493577	0.507451	0
0	0	0.500333	0.505158	0.497937	0	0.500322	0	0.500413	0.499603
0	0	0.499667	0.494842	0.502063	0	0.499678	0	0.499587	0.500397
0	0.500446	0	0.507588	0.496965	0.499590	0	0.499477	0	0.499416
0	0.499554	0	0.492412	0.503035	0.500410	0	0.500523	0	0.500584
0.497476	0.497921	0	0.464662	0	0	0.497791	0.502437	0.497173	0
0.502524	0.502079	0	0.535338	0	0	0.502209	0.497563	0.502827	0

The vertical step then updates the matrix **Q** using (6.15). To determine the probability $q_{11}(x)$:

$$q_{11}(0) = \beta_{11} f_1^0 \prod_{m' \in M_1 \setminus 1} r_{m'1}(0)$$
$$= \beta_{11} \times 0.78 \times (0.493347 \times 0.497476) = 0.191434 \beta_{11}$$

$$q_{11}(1) = \beta_{11} f_1^1 \prod_{m' \in M_1 \setminus 1} r_{m'1}(1)$$
$$= \beta_{11} \times 0.22 \times (0.506653 \times 0.502524) = 0.056013 \beta_{11}$$

Since $0.191334\beta_{11} + 0.05601\beta_{11} = 1$, $\beta_{11} = \frac{1}{0.191334+0.056013} = 4.042903$.
So:

$$q_{11}(0) = 4.042903 \times 0.78 \times (0.493347 \times 0.497476) = 0.773636$$
$$q_{11}(1) = 4.042903 \times 0.22 \times (0.506653 \times 0.502524) = 0.226364$$

The remaining q_{mn} are shown in **Q**:

$$\mathbf{Q} =$$

0.773636	0.839121	0.806481	0	0	0.132106	0.818884	0	0	0.239285
0.226364	0.160879	0.193519	0	0	0.867894	0.181116	0	0	0.760715
0.812140	0	0.837461	0	0.444958	0.113039	0	0.211273	0.748185	0
0.187860	0	0.162539	0	0.555042	0.886961	0	0.788727	0.251815	0
0	0	0.833978	0.492203	0.484126	0	0.844187	0	0.742212	0.201076
0	0	0.166022	0.507797	0.515874	0	0.155813	0	0.257788	0.798924
0	0.860727	0	0.489773	0.485097	0.115241	0	0.215940	0	0.201196
0	0.139273	0	0.510227	0.514903	0.884759	0	0.784060	0	0.798804
0.809608	0.861934	0	0.532711	0	0	0.845514	0.213942	0.744684	0
0.190392	0.138066	0	0.467289	0	0	0.154486	0.786058	0.255316	0

Finally the pseudo posterior probabilities q_n are determined using (6.18). To find $q_1(x)$:

$$q_1(0) = \beta_1 f_1^0 \prod_{m \in M_1} r_{m1}(0)$$
$$= \beta_1 \times 0.78 \times (0.551914 \times 0.493347 \times 0.500413) = 0.116898\beta_1$$

$$q_1(1) = \beta_1 f_1^1 \prod_{m \in M_1} r_{m1}(1)$$
$$= \beta_1 \times 0.22 \times (0.448086 \times 0.506653 \times 0.502524) = 0.025099\beta_1$$

$$\therefore \beta_1 = \frac{1}{0.116898 + 0.025099} = 7.042402$$

$$q_1(0) = 7.042402 \times 0.78 \times (0.551914 \times 0.493347 \times 0.500413) = 0.808046$$
$$q_1(1) = 7.042402 \times 0.22 \times (0.448086 \times 0.506653 \times 0.502524) = 0.191954$$

Applying (6.19), the decoded symbol $\hat{c}_1 = 0$ since $q_1(0)$ has the higher probability. The pseudo posterior probabilities of the received word are:

$$Q' = \begin{bmatrix} 0.808046 & 0.860941 & 0.834162 & 0.497361 & 0.482065 & 0.115074 & 0.844356 & 0.215586 & 0.742528 & 0.200821 \\ 0.191954 & 0.139059 & 0.165838 & 0.502639 & 0.517935 & 0.884926 & 0.155644 & 0.784414 & 0.257472 & 0.799179 \end{bmatrix}$$

and the corresponding decoded code word is $\hat{c} = [0 \quad 0 \quad 0 \quad 1 \quad 1 \quad 1 \quad 0 \quad 1 \quad 0 \quad 1]$. We can see that this does not match the original transmitted code word and so further iterations are required. It turns out that the decoded code word is correct after the third iteration and the final matrices $\mathbf{R}$, $\mathbf{Q}$ and $\mathbf{Q}'$ are given along with the correct decoded code word:

$$\mathbf{R} = \begin{bmatrix}
0.549960 & 0.540086 & 0.544369 & 0 & 0 & 0.463092 & 0.542650 & 0 & 0 & 0.447890 \\
0.450040 & 0.459914 & 0.455631 & 0 & 0 & 0.536908 & 0.457350 & 0 & 0 & 0.552110 \\
0.493114 & 0 & 0.493650 & 0 & 0.545393 & 0.505532 & 0 & 0.507453 & 0.491301 & 0 \\
0.506886 & 0 & 0.506350 & 0 & 0.454607 & 0.494468 & 0 & 0.492547 & 0.508699 & 0 \\
0 & 0 & 0.499989 & 0.500176 & 0.502649 & 0 & 0.499989 & 0 & 0.499985 & 0.500012 \\
0 & 0 & 0.500011 & 0.499824 & 0.497351 & 0 & 0.500011 & 0 & 0.500015 & 0.499988 \\
0 & 0.499975 & 0 & 0.500415 & 0.503915 & 0.500023 & 0 & 0.500032 & 0 & 0.500030 \\
0 & 0.500025 & 0 & 0.499585 & 0.496085 & 0.499977 & 0 & 0.499968 & 0 & 0.499970 \\
0.496595 & 0.497094 & 0 & 0.457904 & 0 & 0 & 0.496955 & 0.503693 & 0.495668 & 0 \\
0.503405 & 0.502906 & 0 & 0.542096 & 0 & 0 & 0.503045 & 0.496307 & 0.504332 & 0
\end{bmatrix}$$

$$\mathbf{Q} = \begin{bmatrix}
0.772854 & 0.838418 & 0.806053 & 0 & 0 & 0.132534 & 0.818189 & 0 & 0 & 0.240031 \\
0.227146 & 0.161582 & 0.193947 & 0 & 0 & 0.867466 & 0.181811 & 0 & 0 & 0.759969 \\
0.810391 & 0 & 0.835883 & 0 & 0.456507 & 0.114177 & 0 & 0.212482 & 0.746725 & 0 \\
0.189609 & 0 & 0.164117 & 0 & 0.543493 & 0.885823 & 0 & 0.787518 & 0.253275 & 0 \\
0 & 0 & 0.832375 & 0.478244 & 0.499266 & 0 & 0.842265 & 0 & 0.740099 & 0.203955 \\
0 & 0 & 0.167625 & 0.521756 & 0.500734 & 0 & 0.157735 & 0 & 0.259901 & 0.796045 \\
0 & 0.859034 & 0 & 0.478005 & 0.498000 & 0.116425 & 0 & 0.217493 & 0 & 0.203943 \\
0 & 140966 & 0 & 0.521995 & 0.502000 & 0.883575 & 0 & 0.782507 & 0 & 0.796057 \\
0.808242 & 0.860424 & 0 & 0.520590 & 0 & 0 & 0.843871 & 0.215010 & 0.743407 & 0 \\
0.191758 & 0.139576 & 0 & 0.479410 & 0 & 0 & 0.156129 & 0.784990 & 0.256593 & 0
\end{bmatrix}$$

$$\mathbf{Q}' = \begin{bmatrix}
0.806122 & 0.859023 & 0.832369 & 0.478419 & 0.501915 & 0.116434 & 0.842260 & 0.217514 & 0.740088 & 0.203963 \\
0.193878 & 0.140977 & 0.167631 & 0.521581 & 0.498085 & 0.883566 & 0.157740 & 0.782486 & 0.259913 & 0.796037
\end{bmatrix}$$

$$\hat{\mathbf{c}} = [0 \quad 0 \quad 0 \quad 1 \quad 0 \quad 1 \quad 0 \quad 1 \quad 0 \quad 1].$$

The majority of the complexity of the sum–product algorithm is due to the horizontal step, since the number of possible non-binary sequences can become very large. However, it has been shown that fast Fourier transforms (FFTs) can be used to replace this step, resulting in a significant reduction in complexity [8].

6.3.4 Reducing the Decoding Complexity Using Fast Fourier Transforms

The horizontal step described previously involves finding all possible binary sequences that satisfy a parity check equation, determining the probability of each sequence and adding them all together, as defined in (6.10).

If we take the simple parity check equation $z_1 = c_1 \oplus c_2 \oplus c_3 = 0$ then to determine $r_{11}(x)$ we first need to find all the solutions of $c_2 \oplus c_3 = c_1 = x$. For $x = 0$ the solutions are $c_2 = 0$, $c_3 = 0$ and $c_2 = 1$, $c_3 = 1$. Therefore:

$$r_{11}(0) = q_{12}(0)q_{13}(0) + q_{12}(1)q_{13}(1).$$

For $x = 1$ the solutions are $c_2 = 0$, $c_3 = 1$ and $c_2 = 1$, $c_3 = 0$. Therefore:

$$r_{11}(1) = q_{12}(0)q_{13}(1) + q_{12}(1)q_{13}(0).$$

In general, we can write this as the *convolution* operation:

$$r_{11}(x) = \sum_{v=0}^{1} q_{12}(v)q_{13}(x - v),$$

where $v \in \text{GF}(2)$.

This implies that the same result can be achieved by replacing the convolution operation with Fourier transforms. Therefore, to determine r_{mn} we must first calculate the product of the Fourier transforms of the other $q_{mn'}$ and then apply the inverse Fourier transform:

$$r_{mn}(x) = F^{-1}\left(\prod_{n' \in N_m/n} F\left(q_{mn'}(x)\right) \right), \tag{6.21}$$

where $F(\ )$ is the Fourier transform and F^{-1} is the inverse Fourier transform. Since all elements belong to the additive group $\mathbb{Z}_2$, the Fast Fourier transform reduces to the Hadamard transform [9] $\mathbf{W}_2$, where:

$$\mathbf{W}_2 = \frac{1}{\sqrt{2}} \begin{bmatrix} 1 & 1 \\ 1 & -1 \end{bmatrix}. \tag{6.22}$$

A property of the Hadamard matrix is that its inverse is also the Hadamard matrix, that is:

$$\mathbf{W}_2\mathbf{W}_2 = \frac{1}{2} \begin{bmatrix} 1 & 1 \\ 1 & -1 \end{bmatrix} \begin{bmatrix} 1 & 1 \\ 1 & -1 \end{bmatrix} = \frac{1}{2} \begin{bmatrix} 2 & 0 \\ 0 & 2 \end{bmatrix} = \mathbf{I},$$

where $\mathbf{I}$ is the identity matrix. Consequently, the inverse Fourier transform can be obtained by also multiplying by the Hadamard matrix. We will now use FFTs to determine $r_{11}(0)$ and $r_{11}(1)$ for the first iteration of Example 3.2. From (6.20):

$$r_{11}(x) = F^{-1}\left(\prod_{n' \in N_0 \backslash 0} F\left(q_{1n'}(x)\right) \right)$$

$$= F^{-1}\left(\frac{1}{\sqrt{2}} \begin{bmatrix} 1 & 1 \\ 1 & -1 \end{bmatrix} \begin{bmatrix} q_{13}(0) \\ q_{13}(1) \end{bmatrix} \times \frac{1}{\sqrt{2}} \begin{bmatrix} 1 & 1 \\ 1 & -1 \end{bmatrix} \begin{bmatrix} q_{14}(0) \\ q_{14}(1) \end{bmatrix} \right)$$

$$\times \frac{1}{\sqrt{2}}\begin{bmatrix} 1 & 1 \\ 1 & -1 \end{bmatrix}\begin{bmatrix} q_{16}(0) \\ q_{16}(1) \end{bmatrix} \times \frac{1}{\sqrt{2}}\begin{bmatrix} 1 & 1 \\ 1 & -1 \end{bmatrix}\begin{bmatrix} q_{17}(0) \\ q_{17}(1) \end{bmatrix}$$

$$\times \frac{1}{\sqrt{2}}\begin{bmatrix} 1 & 1 \\ 1 & -1 \end{bmatrix}\begin{bmatrix} q_{1,10}(0) \\ q_{1,10}(1) \end{bmatrix} \Bigg)$$

$$= F^{-1}\Bigg(\frac{1}{\sqrt{2}}\begin{bmatrix} 1 & 1 \\ 1 & -1 \end{bmatrix}\begin{bmatrix} 0.84 \\ 0.16 \end{bmatrix} \times \frac{1}{\sqrt{2}}\begin{bmatrix} 1 & 1 \\ 1 & -1 \end{bmatrix}\begin{bmatrix} 0.81 \\ 0.19 \end{bmatrix}$$

$$\times \frac{1}{\sqrt{2}}\begin{bmatrix} 1 & 1 \\ 1 & -1 \end{bmatrix}\begin{bmatrix} 0.13 \\ 0.87 \end{bmatrix} \times \frac{1}{\sqrt{2}}\begin{bmatrix} 1 & 1 \\ 1 & -1 \end{bmatrix}\begin{bmatrix} 0.82 \\ 0.18 \end{bmatrix}$$

$$\times \frac{1}{\sqrt{2}}\begin{bmatrix} 1 & 1 \\ 1 & -1 \end{bmatrix}\begin{bmatrix} 0.24 \\ 0.76 \end{bmatrix}\Bigg)$$

$$= F^{-1}\Bigg(\frac{1}{\sqrt{2}}\begin{bmatrix} 1 \\ 0.68 \end{bmatrix} \times \frac{1}{\sqrt{2}}\begin{bmatrix} 1 \\ 0.62 \end{bmatrix} \times \frac{1}{\sqrt{2}}\begin{bmatrix} 1 \\ -0.74 \end{bmatrix}$$

$$\times \frac{1}{\sqrt{2}}\begin{bmatrix} 1 \\ 0.64 \end{bmatrix} \times \frac{1}{\sqrt{2}}\begin{bmatrix} 1 \\ -0.52 \end{bmatrix}\Bigg).$$

$$= \frac{1}{2}\begin{bmatrix} 1 & 1 \\ 1 & -1 \end{bmatrix}\begin{bmatrix} 1 \\ 0.103828 \end{bmatrix}$$

$$= \begin{bmatrix} 0.551914 \\ 0.448086 \end{bmatrix}$$

Therefore, $r_{11}(0) = 0.551914$ and $r_{11}(1) = 0.448086$, which is identical to the result obtained in the previous example. It is quite obvious by comparing these two methods that FFTs are much simpler to perform. FFTs will also be applied to the decoding of non-binary LDPC codes, explained later in this chapter.

6.4 Construction of Non-Binary LDPC Codes Defined Over Finite Fields

A non-binary LDPC code is simply an LDPC code with a sparse parity check matrix containing elements that could be defined over groups, rings or fields. In this book, we concentrate solely on LDPC codes defined over finite fields GF(2^i), where i is a positive integer greater than 1. In 1998, Mackay presented the idea of LDPC codes over finite fields [10], proving that they can achieve increases in performance over their binary counterparts with increasing finite field size. Mackay also showed how the sum–product algorithm could be extended to decode non-binary LDPC codes, but the overall complexity was much higher. At that time it was therefore only feasible to decode non-binary LDPC codes over small finite fields.

There are only a few papers in the literature on non-binary LDPC codes and most of these codes are constructed by taking a known binary LDPC code and replacing its nonzero elements with randomly-generated finite field elements. Shu Lin has presented several structured methods to construct good non-binary LDPC codes using a technique known as *array dispersion* [11], one of which we now describe.

6.4.1 Construction of Non-Binary LDPC Codes from Reed–Solomon Codes

Shu Lin has demonstrated several array dispersion methods using tools such as Euclidean and finite geometries and also using a single code word from a very low-rate Reed–Solomon code [11]. Since this book is concerned with non-binary codes, this section explains how to construct non-binary LPDC codes from a Reed–Solomon code word.

In the context of a non-binary code defined over finite fields $GF(q)$, array dispersion is an operation applied to each nonzero element in a matrix whereby each element is transformed to a location vector of length $q - 1$. For an element $\alpha^i \in GF(q)$, $0 \leq i \leq q - 2$, it will be placed in the ith index of the location vector. For example, the element α^5 in $GF(8)$ would be placed in the fifth index of the location vector. The element in this location vector is used to build an array with each row defined as the previous row cyclically shifted to the right and multiplied by the primitive element in $GF(q)$, resulting in a $(q - 1) \times (q - 1)$ array. Performing array dispersion on a matrix with dimensions $a \times b$ would result in a larger matrix with dimensions $a(q - 1) \times b(q - 1)$. Therefore, after array dispersion the element α^5 would become:

$$\begin{pmatrix} 0 & 0 & 0 & 0 & 0 & \alpha^5 & 0 \\ 0 & 0 & 0 & 0 & 0 & 0 & \alpha^6 \\ 1 & 0 & 0 & 0 & 0 & 0 & 0 \\ 0 & \alpha & 0 & 0 & 0 & 0 & 0 \\ 0 & 0 & \alpha^2 & 0 & 0 & 0 & 0 \\ 0 & 0 & 0 & \alpha^3 & 0 & 0 & 0 \\ 0 & 0 & 0 & 0 & \alpha^4 & 0 & 0 \end{pmatrix}$$

where the top row is the original location vector. For the zero element, array dispersion will result in a $(q - 1) \times (q - 1)$ zero matrix.

These non-binary LDPC codes are constructed from Reed–Solomon codes with message length $k = 2$. Since these code are maximum distance separable (MDS) their minimum Hamming distance is $d = n - k + 1 = n - 1$. Since a Reed–Solomon code defined over $GF(q)$ has a block length of $n = q - 1$, the Reed–Solomon code will be a $(q - 1, 2, q - 2)$ Reed–Solomon code. Since $d = q - 2$, this means that the minimum weight of the code word is also $q - 2$, implying that the code word will contain $q - 2$ nonzero elements and a single zero. It can be shown that the vector containing all 1s, $[1, 1, 1, \ldots, 1]$, and the vector $[1, \alpha, \alpha^2, \ldots, \alpha^{q-2}]$ are both valid code words. As a Reed–Solomon code is linear, adding these two code words will result in another code word:

$$\begin{bmatrix} 1 & 1 & 1 & \cdots & 1 \end{bmatrix} + \begin{bmatrix} 1 & \alpha & \alpha^2 & \cdots & \alpha^{q-2} \end{bmatrix} = \begin{bmatrix} 0 & 1+\alpha & 1+\alpha^2 & \cdots & 1+\alpha^{q-2} \end{bmatrix},$$

which has a weight of $q - 2$. This code word is then used to build a $(q - 1) \times (q - 1)$ array by cyclically shifting it to the right to form the next row of the array. Observe that the main diagonal of this array will be zeroes.

Example 6.4: Array dispersion of a (7, 2, 6) Reed–Solomon code word over GF(8): The code word of weight $q - 2 = 6$ is obtained by adding the two vectors:

$$\begin{bmatrix} 1 & 1 & 1 & 1 & 1 & 1 & 1 \end{bmatrix} + \begin{bmatrix} 1 & \alpha & \alpha^2 & \alpha^3 & \alpha^4 & \alpha^5 & \alpha^6 \end{bmatrix}$$
$$= \begin{bmatrix} 0 & \alpha^3 & \alpha^6 & \alpha & \alpha^5 & \alpha^4 & \alpha^2 \end{bmatrix}.$$

This then is used to build the 7×7 array:

$$\begin{pmatrix} 0 & \alpha^3 & \alpha^6 & \alpha & \alpha^5 & \alpha^4 & \alpha^2 \\ \alpha^2 & 0 & \alpha^3 & \alpha^6 & \alpha & \alpha^5 & \alpha^4 \\ \alpha^4 & \alpha^2 & 0 & \alpha^3 & \alpha^6 & \alpha & \alpha^5 \\ \alpha^5 & \alpha^4 & \alpha^2 & 0 & \alpha^3 & \alpha^6 & \alpha \\ \alpha & \alpha^5 & \alpha^4 & \alpha^2 & 0 & \alpha^3 & \alpha^6 \\ \alpha^6 & \alpha & \alpha^5 & \alpha^4 & \alpha^2 & 0 & \alpha^3 \\ \alpha^3 & \alpha^6 & \alpha & \alpha^5 & \alpha^4 & \alpha^2 & 0 \end{pmatrix}$$

Since Reed–Solomon codes are cyclic, each row is a code word and each column, when read from bottom to top, is also a code word, all with weights equal to 6. After applying array dispersion we will obtain a much larger 49×49 array. This is too large to reproduce here but a small part of it is shown below, corresponding to the 2×2 subarray circled.

$$\begin{bmatrix}
0 & 0 & 0 & 0 & 0 & 0 & 0 & 0 & 0 & 0 & \alpha^3 & 0 & 0 & 0 & \cdots \\
0 & 0 & 0 & 0 & 0 & 0 & 0 & 0 & 0 & 0 & 0 & \alpha^4 & 0 & 0 & \cdots \\
0 & 0 & 0 & 0 & 0 & 0 & 0 & 0 & 0 & 0 & 0 & 0 & \alpha^5 & 0 & \cdots \\
0 & 0 & 0 & 0 & 0 & 0 & 0 & 0 & 0 & 0 & 0 & 0 & 0 & \alpha^6 & \cdots \\
0 & 0 & 0 & 0 & 0 & 0 & 0 & 1 & 0 & 0 & 0 & 0 & 0 & 0 & \cdots \\
0 & 0 & 0 & 0 & 0 & 0 & 0 & 0 & \alpha & 0 & 0 & 0 & 0 & 0 & \cdots \\
0 & 0 & 0 & 0 & 0 & 0 & 0 & 0 & 0 & \alpha^2 & 0 & 0 & 0 & 0 & \cdots \\
0 & 0 & \alpha^2 & 0 & 0 & 0 & 0 & 0 & 0 & 0 & 0 & 0 & 0 & 0 & \cdots \\
0 & 0 & 0 & \alpha^3 & 0 & 0 & 0 & 0 & 0 & 0 & 0 & 0 & 0 & 0 & \cdots \\
0 & 0 & 0 & 0 & \alpha^4 & 0 & 0 & 0 & 0 & 0 & 0 & 0 & 0 & 0 & \cdots \\
0 & 0 & 0 & 0 & 0 & \alpha^5 & 0 & 0 & 0 & 0 & 0 & 0 & 0 & 0 & \cdots \\
0 & 0 & 0 & 0 & 0 & 0 & \alpha^6 & 0 & 0 & 0 & 0 & 0 & 0 & 0 & \cdots \\
1 & 0 & 0 & 0 & 0 & 0 & 0 & 0 & 0 & 0 & 0 & 0 & 0 & 0 & \cdots \\
0 & \alpha & 0 & 0 & 0 & 0 & 0 & 0 & 0 & 0 & 0 & 0 & 0 & 0 & \cdots \\
\vdots & \vdots & \vdots & \vdots & \vdots & \vdots & \vdots & \vdots & \vdots & \vdots & \vdots & \vdots & \vdots & \vdots & \ddots
\end{bmatrix}.$$

6.5 Decoding Non-Binary LDPC Codes with the Sum–Product Algorithm

The sum–product algorithm for binary LDPC codes can be extended to decode non-binary LDPC codes, but with an increase in decoding complexity. Firstly, for a non-binary LDPC code defined over GF(q), each received symbol can be one of q different elements in GF(q). Secondly, the horizontal step becomes more complicated as there are now more possible non-binary sequences to satisfy the parity check equations. The matrices $\mathbf{Q}$ and $\mathbf{R}$ used in the horizontal and vertical steps of the sum–product algorithm are defined in (6.23) and (6.24) respectively:

$$
\mathbf{Q} =
\begin{bmatrix}
q_{11}(0) & q_{12}(0) & q_{13}(0) & \cdots & q_{1,n-1}(0) & q_{1n}(0) \\
q_{11}(1) & q_{12}(1) & q_{13}(1) & \cdots & q_{1,n-1}(1) & q_{1n}(1) \\
q_{11}(\alpha) & q_{12}(\alpha) & q_{13}(\alpha) & \cdots & q_{1,n-1}(\alpha) & q_{1n}(\alpha) \\
\vdots & \vdots & \vdots & \cdots & \vdots & \vdots \\
q_{11}(\alpha^{q-2}) & q_{12}(\alpha^{q-2}) & q_{13}(\alpha^{q-2}) & \cdots & q_{1,n-1}(\alpha^{q-2}) & q_{1n}(\alpha^{q-2}) \\
\vdots & \vdots & \vdots & \ddots & \vdots & \vdots \\
\vdots & \vdots & \vdots & \ddots & \vdots & \vdots \\
\vdots & \vdots & \vdots & \ddots & \vdots & \vdots \\
\vdots & \vdots & \vdots & \ddots & \vdots & \vdots \\
\vdots & \vdots & \vdots & \ddots & \vdots & \vdots \\
q_{m1}(0) & q_{m2}(0) & q_{m3}(0) & \cdots & q_{m,n-1}(0) & q_{mn}(0) \\
q_{m1}(1) & q_{m2}(1) & q_{m3}(1) & \cdots & q_{m,n-1}(1) & q_{mn}(1) \\
q_{m1}(\alpha) & q_{m2}(\alpha) & q_{m3}(\alpha) & \cdots & q_{m,n-1}(\alpha) & q_{mn}(\alpha) \\
\vdots & \vdots & \vdots & \cdots & \vdots & \vdots \\
q_{m1}(\alpha^{q-2}) & q_{m2}(\alpha^{q-2}) & q_{m3}(\alpha^{q-2}) & \cdots & q_{m,n-1}(\alpha^{q-2}) & q_{mn}(\alpha^{q-2})
\end{bmatrix}
$$

$$(6.23)$$

$$
\mathbf{R} = \begin{bmatrix}
r_{11}(0) & r_{12}(0) & r_{13}(0) & \cdots & r_{1,n-1}(0) & r_{1n}(0) \\
r_{11}(1) & r_{12}(1) & r_{13}(1) & \cdots & r_{1,n-1}(1) & r_{1n}(1) \\
r_{11}(\alpha) & r_{12}(\alpha) & r_{13}(\alpha) & \cdots & r_{1,n-1}(\alpha) & r_{1n}(\alpha) \\
\vdots & \vdots & \vdots & \cdots & \vdots & \vdots \\
r_{11}(\alpha^{q-2}) & r_{12}(\alpha^{q-2}) & r_{13}(\alpha^{q-2}) & \cdots & r_{1,n-1}(\alpha^{q-2}) & r_{1n}(\alpha^{q-2}) \\
\vdots & \vdots & \vdots & \ddots & \vdots & \vdots \\
\vdots & \vdots & \vdots & \ddots & \vdots & \vdots \\
\vdots & \vdots & \vdots & \ddots & \vdots & \vdots \\
\vdots & \vdots & \vdots & \ddots & \vdots & \vdots \\
\vdots & \vdots & \vdots & \ddots & \vdots & \vdots \\
r_{m1}(0) & r_{m2}(0) & r_{m3}(0) & \cdots & r_{m,n-1}(0) & r_{mn}(0) \\
r_{m1}(1) & r_{m2}(1) & r_{m3}(1) & \cdots & r_{m,n-1}(1) & r_{mn}(1) \\
r_{m1}(\alpha) & r_{m2}(\alpha) & r_{m3}(\alpha) & \cdots & r_{m,n-1}(\alpha) & r_{mn}(\alpha) \\
\vdots & \vdots & \vdots & \cdots & \vdots & \vdots \\
r_{m1}(\alpha^{q-2}) & r_{m2}(\alpha^{q-2}) & r_{m3}(\alpha^{q-2}) & \cdots & r_{m,n-1}(\alpha^{q-2}) & r_{mn}(\alpha^{q-2})
\end{bmatrix}
$$

$$(6.24)$$

It can be observed that each nonzero element defined over GF(q) in the parity check matrix **H** has q probabilities associated with it, instead of two probabilities as in the binary case.

6.5.1 Received Symbol Likelihoods

In (6.8) it was shown how to determine the likelihoods of a demodulated symbol for the AWGN channel. Assuming a non-binary LDPC code defined over GF(q) and that M-PSK modulation is chosen, then provided that $q = M$ the likelihoods of the demodulated symbols are also the likelihoods of the received symbols. However, for the case where $q > M$ and M divides q we must concatenate the likelihoods of the demodulated symbols. As an example, take an LDPC code defined over GF(16) and QPSK modulation. We know that a finite field element in GF(16) is made up of four bits and a QPSK modulated symbol is made up of two bits. Therefore, two consecutive QPSK modulated symbols contain a finite field element. The element α in GF(16) is represented as 0010 in binary. Therefore, the likelihood of a received symbol being α is the product of the likelihood that one demodulated symbol is 00 and the likelihood that the neighbouring demodulated symbol is 10, that is, from (6.8), $P(r_i = \alpha) = f^\alpha = g^{00}g^{10}$.

6.5.2 Permutation of Likelihoods

In general, the parity check equations are of the form:

$$z_i = \sum_{j=1}^{n} h_{ij}c_j, \tag{6.25}$$

where h_{ij} and $c_j \in GF(q)$. The parity check equation is satisfied when:

$$h_{11}c_1 + h_{12}c_2 + \cdots + h_{1n}c_n = 0.$$

For the horizontal step we calculate the probabilities $r_{ij}(x)$ from (6.10) by first substituting all possible non-binary elements into the coded symbols that satisfy the parity check equation when $c_j = x$, that is:

$$h_{i1}c_1 + h_{i2}c_2 + \cdots + h_{in}c_n = h_{ij}c_j,$$

and then determining the probability of each sequence. This is more complicated than the binary case since we must now consider the non-binary parity check matrix elements. Each coded symbol has q likelihoods associated with it. When the coded symbol is multiplied by a non-binary parity check element we can compensate for this by cyclically shifting downwards this column vector of likelihoods, with the exception of the first likelihood, corresponding to the probability of the coded symbol being zero. The number of cyclic shifts is equal to the power of the primitive element that is multiplied with the coded symbol. This is illustrated below:

$$c_j \rightarrow \begin{bmatrix} q_{ij}(0) \\ q_{ij}(1) \\ \vdots \\ q_{ij}(\alpha^{q-2}) \end{bmatrix} \quad \alpha c_j \rightarrow \begin{bmatrix} q_{ij}(0) \\ q_{ij}(\alpha^{q-2}) \\ q_{ij}(1) \\ \vdots \end{bmatrix} \quad \alpha^2 c_j \rightarrow \begin{bmatrix} q_{ij}(0) \\ \vdots \\ q_{ij}(\alpha^{q-2}) \\ q_{ij}(1) \end{bmatrix} \quad \text{etc.}$$

This cyclic shift of the likelihoods is known as a *permutation* [8] and transforms the parity check equation from (6.25) to:

$$c_1 + c_2 + \cdots + c_n = c_j,$$

which is more similar to the binary parity check equations. The inverse of a permutation is a *depermutation*, where the likelihoods are cyclically shifted upwards, again with the exception of the first likelihood.

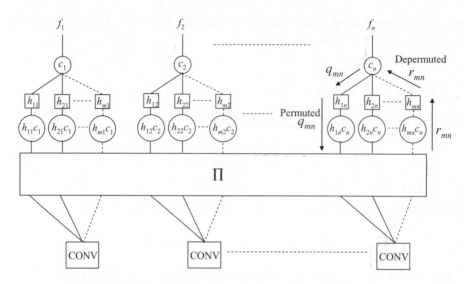

Figure 6.4   Generalized factor graph of a non-binary LDPC code [8].

6.5.3 Factor Graph of a Non-Binary LDPC Code

The factor graph of a non-binary LDPC code graphically shows the operation of the sum–product decoding algorithm (Figure 6.4).

The factor graph for non-binary LPDC codes is similar to that for binary LPDC codes, but we must now take into account the non-binary elements in the parity check matrix, which are denoted as h_{ij}, $i = 1, 2, \ldots, m$ and $j = 1, 2, \ldots, n$. In Figure 6.4 the number of parity check matrix elements connected to a coded symbol c_j is the column weight of the code and the number of connections to each parity check z_i is the row weight of the code. The likelihoods of each coded symbol f_j are column vectors containing the q likelihoods of the coded symbol being an element in GF(q). The block labelled Π connects the non-binary elements in each row to the parity checks. In Figure 6.5 a generalized factor graph is shown, with FFT blocks replacing the convolutional blocks, which reduces complexity.

A factor graph of the parity check matrix in (6.26) is show in Figure 6.6.

$$\mathbf{H} = \begin{bmatrix} \alpha & 0 & 1 & \alpha & 0 & 1 \\ \alpha^2 & \alpha & 0 & 1 & 1 & 0 \\ 0 & \alpha & \alpha^2 & 0 & \alpha^2 & 1 \end{bmatrix}. \tag{6.26}$$

6.5.4 The Fast Fourier Transform for the Decoding of Non-Binary LDPC Codes

In Section 6.3.4, the FFT was used to reduce the complexity of the horizontal step in the sum–product algorithm. For binary decoding of LDPC codes, the matrix $\mathbf{Q}$ contains

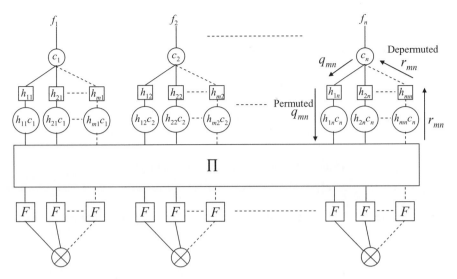

Figure 6.5 Generalized factor graph of a non-binary LDPC code showing the replacement of the convolution operations with FFTs.

1×2 column vectors containing two probabilities, $q_{mn}(0)$ and $q_{mn}(1)$. The Fourier transform of this column vector is then obtained by multiplying by the Hadamard matrix defined in (6.21).

However, for non-binary LDPC codes defined over GF(q), $\mathbf{Q}$ contains $1 \times q$ column vectors and the Fourier transform is obtained by multiplying by the tensor product of

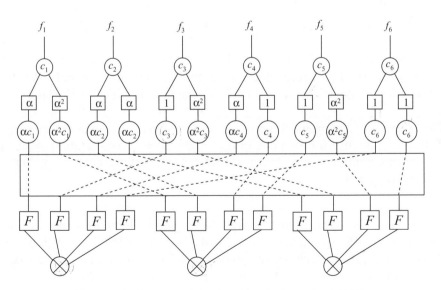

Figure 6.6 Factor graph for the parity check matrix of (6.26).

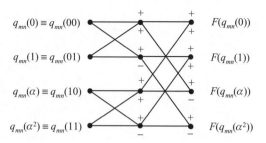

Figure 6.7   Radix-2 butterfly diagram for GF(4).

Hadamard matrices. An example of the tensor product of two Hadamard matrices is given below:

$$
\mathbf{W_4} = \mathbf{W_2} \otimes \mathbf{W_2} = \begin{bmatrix} \mathbf{W_2} & \mathbf{W_2} \\ \mathbf{W_2} & -\mathbf{W_2} \end{bmatrix} = \frac{1}{2} \begin{bmatrix} 1 & 1 & 1 & 1 \\ 1 & -1 & 1 & -1 \\ 1 & 1 & -1 & -1 \\ 1 & -1 & -1 & 1 \end{bmatrix}.
$$

Therefore, for an LDPC code defined over GF(4), the Fourier transform of the 1×4 column vector of probabilities $q_{mn}(0)$, $q_{mn}(1)$, $q_{mn}(\alpha)$ and $q_{mn}(\alpha^2)$ is:

$$
\begin{bmatrix} F\left(q_{mn}(0)\right) \\ F\left(q_{mn}(1)\right) \\ F\left(q_{mn}(\alpha)\right) \\ F\left(q_{mn}(\alpha^2)\right) \end{bmatrix} = \frac{1}{2} \begin{bmatrix} 1 & 1 & 1 & 1 \\ 1 & -1 & 1 & -1 \\ 1 & 1 & -1 & -1 \\ 1 & -1 & -1 & 1 \end{bmatrix} \begin{bmatrix} q_{mn}(0) \\ q_{mn}(1) \\ q_{mn}(\alpha) \\ q_{mn}(\alpha^2) \end{bmatrix}.
$$

As before, the inverse FFT is achieved by multiplying by the Hadamard matrix, that is $\mathbf{W_4}\mathbf{W_4} = \mathbf{I_4}$. The FFT operation can be expressed in terms of its radix-2 butterfly diagram. An example of the radix-2 butterfly diagram for GF(4) is shown in Figure 6.7.

The ordering of the probabilities $q_{mn}(x)$ is very important when determining the FFT. It is essential that for each pair of $q_{mn}(x)$, the binary representation of the finite field elements x differ by *one bit*. In Figure 6.8, the radix-2 butterfly diagram for GF(16) is also given.

Example 6.5: Decoding a non-binary LDPC code with the FFT sum–product algorithm: In this example we decode the (6, 3) LDPC code defined over GF(4) with a parity check matrix given by (6.23). We assume that the transmitted code word was $\mathbf{c} = [\,2 \quad 1 \quad 0 \quad 3 \quad 0 \quad 2\,]$ and the received word is $\mathbf{r} = [\,2 \quad 1 \quad 0 \quad \mathbf{2} \quad \mathbf{2} \quad 2\,]$, with two errors highlighted. Furthermore, it is assumed that

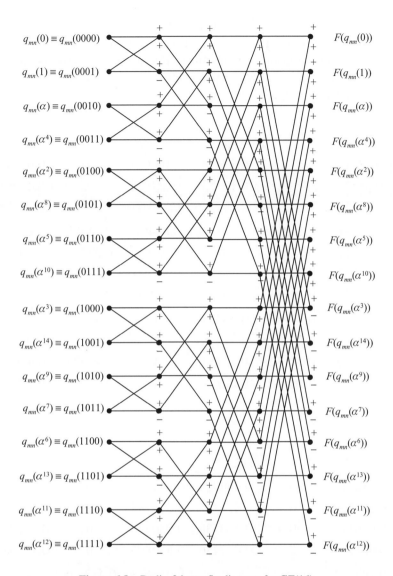

Figure 6.8 Radix-2 butterfly diagram for GF(16).

the received symbols have the following likelihoods:

$$\mathbf{f} = \begin{bmatrix} 0.182512 & 0.149675 & 0.444988 & 0.044678 & 0.412355 & 0.320530 \\ 0.046118 & 0.723055 & 0.324187 & 0.030350 & 0.073539 & 0.079952 \\ 0.615774 & 0.021827 & 0.133538 & 0.550805 & 0.436298 & 0.479831 \\ 0.155596 & 0.105443 & 0.097286 & 0.374167 & 0.077809 & 0.119687 \end{bmatrix},$$

(6.27)

where each column in **f** contains the likelihoods of a received symbol being 0, 1, α or α^2. The matrix **Q** is then initialized as shown in (6.21). However, as described earlier, the elements in **Q** must be permuted due to the received coded symbols being multiplied by a non-binary element in the parity check matrix. The permuted version of **Q** is given in (6.28):

$$
\mathbf{Q} = \begin{bmatrix}
0.182512 & 0 & 0.444988 & 0.044678 & 0 & 0.320530 \\
0.046118 & 0 & 0.324187 & 0.030350 & 0 & 0.079952 \\
0.615774 & 0 & 0.133538 & 0.550805 & 0 & 0.479831 \\
0.155596 & 0 & 0.097286 & 0.374167 & 0 & 0.119687 \\
0.182512 & 0.149675 & 0 & 0.044678 & 0.412355 & 0 \\
0.046118 & 0.723055 & 0 & 0.030350 & 0.073539 & 0 \\
0.615774 & 0.021827 & 0 & 0.550805 & 0.436298 & 0 \\
0.155596 & 0.105443 & 0 & 0.374167 & 0.077809 & 0 \\
0 & 0.149675 & 0.444988 & 0 & 0.412355 & 0.320530 \\
0 & 0.723055 & 0.324187 & 0 & 0.073539 & 0.079952 \\
0 & 0.021827 & 0.133538 & 0 & 0.436298 & 0.479831 \\
0 & 0.105443 & 0.097286 & 0 & 0.077809 & 0.119687
\end{bmatrix}.
$$

$$(6.28)$$

Permuted **Q**

$$
= \begin{bmatrix}
0.182512 & 0 & 0.444988 & 0.044678 & 0 & 0.320530 \\
0.155596 & 0 & 0.324187 & 0.374167 & 0 & 0.079952 \\
0.046118 & 0 & 0.133538 & 0.030350 & 0 & 0.479831 \\
0.615774 & 0 & 0.097286 & 0.550805 & 0 & 0.119687 \\
0.182512 & 0.149675 & 0 & 0.044678 & 0.412355 & 0 \\
0.615774 & 0.105443 & 0 & 0.030350 & 0.073539 & 0 \\
0.155596 & 0.723055 & 0 & 0.550805 & 0.436298 & 0 \\
0.046118 & 0.021827 & 0 & 0.374167 & 0.077809 & 0 \\
0 & 0.149675 & 0.444988 & 0 & 0.412355 & 0.320530 \\
0 & 0.105443 & 0.133538 & 0 & 0.436298 & 0.079952 \\
0 & 0.723055 & 0.097286 & 0 & 0.077809 & 0.479831 \\
0 & 0.021827 & 0.324187 & 0 & 0.073539 & 0.119687
\end{bmatrix}.
$$

$$(6.29)$$

We now perform the FFT operation on each of the 1×4 column vectors. For example, the Fourier transform of the column vector in the top-left-hand corner of the permuted $\mathbf{Q}$ matrix is determined by:

$$
F\begin{pmatrix} q_{11}(0) \\ q_{11}(1) \\ q_{11}(\alpha) \\ q_{11}(\alpha^2) \end{pmatrix} = \frac{1}{2} \begin{bmatrix} 1 & 1 & 1 & 1 \\ 1 & -1 & 1 & -1 \\ 1 & 1 & -1 & -1 \\ 1 & -1 & -1 & 1 \end{bmatrix} \begin{bmatrix} 0.182512 \\ 0.155596 \\ 0.046118 \\ 0.615774 \end{bmatrix}
$$

$$
= \frac{1}{2} \begin{bmatrix} 0.182512 + 0.155596 + 0.046118 + 0.615774 \\ 0.182512 - 0.155596 + 0.046118 - 0.615774 \\ 0.182512 + 0.155596 - 0.046118 - 0.615774 \\ 0.182512 - 0.155596 - 0.046118 + 0.615774 \end{bmatrix}
$$

$$
F\begin{pmatrix} q_{11}(0) \\ q_{11}(1) \\ q_{11}(\alpha) \\ q_{11}(\alpha^2) \end{pmatrix} = \frac{1}{2} \begin{bmatrix} 1 \\ -0.542740 \\ -0.323783 \\ 0.596571 \end{bmatrix}.
$$

The FFT of each column vector in the permuted $\mathbf{Q}$ matrix is given in (6.30):

$F(\mathbf{Q}) =$

1	0	1	1	0	1
−0.542740	0	0.157053	−0.849944	0	0.600723
−0.323783	0	0.538351	−0.162310	0	−0.199036
0.596571	0	0.084549	0.190965	0	−0.119566
1	1	0	1	1	0
−0.323783	0.745461	0	0.190965	0.697305	0
0.596571	−0.489765	0	−0.849944	−0.028213	0
−0.542740	−0.656996	0	−0.162310	−0.019673	0
0	1	1	0	1	1
0	0.745461	0.084549	0	−0.019673	0.600723
0	−0.489765	0.157053	0	0.697305	−0.199036
0	−0.656996	0.538351	0	−0.0282143	−0.119566

$$(6.30)$$

To complete the horizontal step we apply (6.20) to determine the probabilities $r_{mn}(x)$. To determine $r_{11}(x)$ we take the product of the Fourier transformed column vectors in $F(\mathbf{Q})$ corresponding to the third, fourth and sixth received symbols and then apply the inverse FFT to obtain a 1×4 column vector containing $r_{11}(x)$:

$$r_{11}(x) = F^{-1}\left(\prod_{n' \in N_1/1} F\left(q_{1n'}(x)\right) \right)$$

$$= F^{-1}\left(\frac{1}{2} \begin{bmatrix} 1 \\ 0.157053 \\ 0.538351 \\ 0.084549 \end{bmatrix} \times \begin{bmatrix} 1 \\ -0.849944 \\ -0.162310 \\ 0.190965 \end{bmatrix} \times \begin{bmatrix} 1 \\ 0.600723 \\ -0.199036 \\ -0.119566 \end{bmatrix} \right)$$

$$= F^{-1}\left(\frac{1}{2} \begin{bmatrix} 1 \\ -0.080188 \\ 0.0173917 \\ -0.003222 \end{bmatrix} \right)$$

$$= \frac{1}{4} \begin{bmatrix} 1 & 1 & 1 & 1 \\ 1 & -1 & 1 & -1 \\ 1 & 1 & -1 & -1 \\ 1 & -1 & -1 & 1 \end{bmatrix} \begin{bmatrix} 1 \\ -0.080188 \\ 0.0173917 \\ -0.003222 \end{bmatrix}$$

$$= \frac{1}{4} \begin{bmatrix} 0.933982 \\ 1.100802 \\ 0.905642 \\ 1.059574 \end{bmatrix} = \begin{bmatrix} 0.233818 \\ 0.274878 \\ 0.226088 \\ 0.265216 \end{bmatrix}$$

The remaining probabilities $r_{mn}(x)$ are given in (6.31):

$$\mathbf{R} = \begin{bmatrix}
0.233818 & 0 & 0.313258 & 0.244365 & 0 & 0.277593 \\
0.274878 & 0 & 0.181512 & 0.272982 & 0 & 0.236553 \\
0.226088 & 0 & 0.325298 & 0.230033 & 0 & 0.258631 \\
0.265216 & 0 & 0.179931 & 0.252620 & 0 & 0.227223 \\
\hline
0.271356 & 0.242364 & 0 & 0.208230 & 0.286092 & 0 \\
0.222772 & 0.264788 & 0 & 0.295891 & 0.338076 & 0 \\
0.278277 & 0.236078 & 0 & 0.207616 & 0.190862 & 0 \\
0.227595 & 0.256769 & 0 & 0.288262 & 0.184970 & 0 \\
\hline
0 & 0.244755 & 0.264237 & 0 & 0.273865 & 0.238776 \\
0 & 0.244347 & 0.269750 & 0 & 0.233789 & 0.234406 \\
0 & 0.254745 & 0.231358 & 0 & 0.245066 & 0.260604 \\
0 & 0.256153 & 0.234655 & 0 & 0.247279 & 0.266214
\end{bmatrix}$$

$$(6.31)$$

The matrix $\mathbf{R}$ is now complete, but it must be depermuted before it can be used in the vertical step. The depermuted $\mathbf{R}$ is given in (6.32):

Depermuted $\mathbf{R}$

$$
= \begin{bmatrix}
0.233818 & 0 & 0.313258 & 0.244365 & 0 & 0.277593 \\
0.226088 & 0 & 0.181512 & 0.230033 & 0 & 0.236553 \\
0.265216 & 0 & 0.325298 & 0.252620 & 0 & 0.258631 \\
0.274878 & 0 & 0.179931 & 0.272982 & 0 & 0.227223 \\
0.271356 & 0.242364 & 0 & 0.208230 & 0.286092 & 0 \\
0.227595 & 0.236078 & 0 & 0.295891 & 0.338076 & 0 \\
0.222772 & 0.256769 & 0 & 0.207616 & 0.190862 & 0 \\
0.278277 & 0.264788 & 0 & 0.288262 & 0.184970 & 0 \\
0 & 0.244755 & 0.264237 & 0 & 0.273865 & 0.238776 \\
0 & 0.254745 & 0.234655 & 0 & 0.247279 & 0.234406 \\
0 & 0.256153 & 0.269750 & 0 & 0.233789 & 0.260604 \\
0 & 0.244347 & 0.231358 & 0 & 0.245066 & 0.266214
\end{bmatrix}.
$$

$$(6.32)$$

The vertical step is carried out in the same way as for the binary case. To determine $q_{11}(x)$ we apply (6.13):

$$q_{11}(0) = \beta_{11} f_1^0 \prod_{m' \in M_1/1} r_{m'1}(0) = \beta_{11} \times 0.182512 \times 0.271356 = 0.049526\beta_{11}$$

$$q_{11}(1) = \beta_{11} f_1^1 \prod_{m' \in M_1/1} r_{m'1}(1) = \beta_{11} \times 0.046118 \times 0.227595 = 0.010496\beta_{11}$$

$$q_{11}(\alpha) = \beta_{11} f_1^\alpha \prod_{m' \in M_1/1} r_{m'1}(\alpha) = \beta_{11} \times 0.615774 \times 0.222772 = 0.137177\beta_{11}$$

$$q_{11}(\alpha^2) = \beta_{11} f_1^{\alpha^2} \prod_{m' \in M_1/1} r_{m'1}(\alpha^2) = \beta_{11} \times 0.155596 \times 0.278277 = 0.043299\beta_{11}$$

$$\beta_{11} = \frac{1}{0.049256 + 0.010496 + 0.137177 + 0.043299} = 4.162712$$

$$q_{11}(0) = 0.049526\beta_{11} = 0.205930$$
$$q_{11}(1) = 0.010496\beta_{11} = 0.043644$$
$$q_{11}(\alpha) = 0.137177\beta_{11} = 0.570388$$
$$q_{11}(\alpha^2) = 0.043299\beta_{11} = 0.180039$$

The complete matrix $\mathbf{Q}$ is given in (6.33):

$$
\mathbf{Q} =
\begin{bmatrix}
0.205930 & 0 & 0.46625 & 0.038683 & 0 & 0.303488 \\
0.043644 & 0 & 0.301653 & 0.037341 & 0 & 0.074315 \\
0.570388 & 0 & 0.142839 & 0.475497 & 0 & 0.495852 \\
0.180039 & 0 & 0.089252 & 0.448479 & 0 & 0.126345 \\
0.164650 & 0.145266 & 0 & 0.042123 & 0.447806 & 0 \\
0.040229 & 0.730398 & 0 & 0.026937 & 0.072108 & 0 \\
0.630104 & 0.022170 & 0 & 0.536854 & 0.404473 & 0 \\
0.165017 & 0.102165 & 0 & 0.394068 & 0.075612 & 0 \\
0 & 0.150837 & 0.537825 & 0 & 0.490530 & 0.343296 \\
0 & 0.709767 & 0.227035 & 0 & 0.103376 & 0.072970 \\
0 & 0.023304 & 0.167601 & 0 & 0.346251 & 0.478806 \\
0 & 0.116092 & 0.067538 & 0 & 0.059844 & 0.104928
\end{bmatrix}.
$$

$$(6.33)$$

Finally, the pseudo posterior probabilities are determined using (6.18) and are given in (6.34):

$$
\mathbf{Q}' =
\begin{bmatrix}
0.205930 & 0.145266 & 0.466255 & 0.038683 & 0.447806 & 0.303488 \\
0.043644 & 0.730398 & 0.301653 & 0.037341 & 0.072108 & 0.074315 \\
0.570388 & 0.022170 & 0.142839 & 0.475497 & 0.404473 & 0.495852 \\
0.180039 & 0.102165 & 0.089252 & 0.448479 & 0.075612 & 0.126345
\end{bmatrix}.
$$

$$(6.34)$$

From (6.19), taking the highest likelihood in each column of (6.34) gives a decoded code word of $\hat{\mathbf{c}} = [\alpha \quad 1 \quad 0 \quad \alpha \quad 0 \quad \alpha]$. The first iteration of the sum–product algorithm has corrected the fifth received symbol but there is still the error in the fourth received symbol. It turns out that only one more iteration is required to correct this error too, and the relevant matrices from the second iteration are given. The updated matrix $\mathbf{Q}$ in (6.33) is permuted and the FFT is applied to each column vector, as shown in (6.35). The matrix $\mathbf{R}$ is depermuted and shown in (6.36), and the updated matrix $\mathbf{Q}$ and pseudo posterior probabilities are given in

(6.37) and (6.38) respectively:

$F(\mathbf{Q}) =$

$$
\begin{bmatrix}
1 & 0 & 1 & 1 & 0 & 1 \\
-0.500852 & 0 & 0.218190 & -0.847952 & 0 & 0.598679 \\
-0.228062 & 0 & 0.535817 & -0.025675 & 0 & -0.244394 \\
0.552635 & 0 & 0.111015 & 0.028361 & 0 & -0.140334 \\
1 & 1 & 0 & 1 & 1 & 0 \\
-0.340666 & 0.751328 & 0 & 0.157955 & 0.704559 & 0 \\
0.589508 & -0.505138 & 0 & -0.861880 & -0.039829 & 0 \\
-0.590243 & -0.665128 & 0 & -0.127581 & -0.046837 & 0 \\
0 & 1 & 1 & 0 & 1 & 1 \\
0 & 0.721208 & 0.210727 & 0 & 0.100747 & 0.644204 \\
0 & -0.466142 & 0.410854 & 0 & 0.673561 & -0.167468 \\
0 & -0.651719 & 0.529721 & 0 & 0.187811 & -0.103553
\end{bmatrix}
$$

(6.35)

Depermuted **R**

$$
=
\begin{bmatrix}
0.223039 & 0 & 0.312657 & 0.238958 & 0 & 0.274386 \\
0.221579 & 0 & 0.186628 & 0.228330 & 0 & 0.227183 \\
0.276740 & 0 & 0.314472 & 0.256737 & 0 & 0.271947 \\
0.278642 & 0 & 0.186243 & 0.275975 & 0 & 0.226484 \\
0.276232 & 0.236345 & 0 & 0.206548 & 0.291534 & 0 \\
0.225755 & 0.244699 & 0 & 0.287521 & 0.336792 & 0 \\
0.232438 & 0.265419 & 0 & 0.203285 & 0.188251 & 0 \\
0.265575 & 0.253537 & 0 & 0.302645 & 0.183422 & 0 \\
0 & 0.239258 & 0.278016 & 0 & 0.291432 & 0.205369 \\
0 & 0.267581 & 0.228322 & 0 & 0.226443 & 0.230132 \\
0 & 0.255591 & 0.248275 & 0 & 0.224605 & 0.302287 \\
0 & 0.237570 & 0.245388 & 0 & 0.257521 & 0.262212
\end{bmatrix}
$$

(6.36)

$$\mathbf{Q} = \begin{bmatrix} 0.205545 & 0 & 0.485609 & 0.037950 & 0 & 0.252543 \\ 0.042447 & 0 & 0.290544 & 0.035886 & 0 & 0.070589 \\ 0.583537 & 0 & 0.130139 & 0.460471 & 0 & 0.556467 \\ 0.168472 & 0 & 0.093708 & 0.465692 & 0 & 0.120401 \\ 0.153792 & 0.137779 & 0 & 0.040705 & 0.471531 & 0 \\ 0.038606 & 0.744379 & 0 & 0.026422 & 0.065340 & 0 \\ 0.643804 & 0.021464 & 0 & 0.539167 & 0.384507 & 0 \\ 0.163797 & 0.096378 & 0 & 0.393706 & 0.078622 & 0 \\ 0 & 0.144486 & 0.535638 & 0 & 0.498018 & 0.333508 \\ 0 & 0.722661 & 0.232931 & 0 & 0.102603 & 0.068878 \\ 0 & 0.023662 & 0.161675 & 0 & 0.340255 & 0.494822 \\ 0 & 0.109191 & 0.069757 & 0 & 0.059124 & 0.102792 \end{bmatrix}.$$

$$(6.37)$$

$$\mathbf{Q}' = \begin{bmatrix} 0.205545 & 0.137779 & 0.485609 & 0.037950 & 0.471531 & 0.252543 \\ 0.042447 & 0.744379 & 0.290544 & 0.035886 & 0.065340 & 0.070589 \\ 0.583537 & 0.021464 & 0.130139 & 0.460471 & 0.384507 & 0.556467 \\ 0.168472 & 0.096378 & 0.093708 & 0.465692 & 0.078622 & 0.120401 \end{bmatrix}.$$

$$(6.38)$$

From the matrix $\mathbf{Q}'$, the decoded code word is now $\hat{\mathbf{c}} = [\alpha \ \ 1 \ \ \ 0 \ \ \alpha^2 \ \ 0 \ \ \alpha]$, which matches the original transmitted code word.

6.6 Conclusions

A very important class of block code known as the low-density parity check (LDPC) code has been explained, with discussions on some structured construction methods for binary and non-binary LDPC codes. Additionally, a decoding algorithm called the sum–product algorithm has been studied in detail for use with binary and non-binary LDPC codes. A method to reduce the complexity of the sum–product algorithm has been given, using fast Fourier transforms based on the Hadamard matrix, which replaces the original horizontal step of the decoding algorithm.

The study of LDPC codes is very important as they are fast becoming one of the more popular coding schemes for a number of future applications in wireless communications and eventually magnetic storage. The binary LDPC codes are well known and perform as well as turbo codes, and in some cases can outperform turbo codes for large block lengths. Non-binary LDPC codes are less well known, but Mackay showed that these codes outperform binary LDPC codes with further increases in performance as the size of the finite field increases [10]. This is of course at the expense of higher complexity, but non-binary LDPC codes also have better convergence properties, requiring less iterations to decode a received word. With the inclusion of the FFT proposed by Barnault *et al.* [8], the complexity is reduced and non-binary LDPC codes can now be used practically in many future applications.

References

[1] Gallager, R.G. (1962) Low-density parity-check codes. *IRE Transactions on Information Theory*, **IT-8**, 21–8.

[2] Mackay, D.J. and Neal, R.M. (1995) Good codes based on very sparse matrices. *Cryptography and Coding*, 5th IMA Conference (ed. C. Boyd), Vol. **1025**, pp. 100–11 (Lecture Notes in Computer Science, Springer).

[3] Chung, S.Y., Forney, G.D. Jr., Richardson, T.J. and Urbanke, R. (2001) On the design of low-density parity-check codes within 0.0045 dB of the Shannon limit. *IEEE Communications Letters*, **5** (2), 58–60.

[4] Ammar, B., Honary, B., Kou, Y. *et al.* (2004) Construction of low-density parity-check codes based on balanced incomplete block designs. *IEEE Transactions on Information Theory*, **50** (6), 1257–69.

[5] Hu, X.-Y., Eleftheriou, E. and Arnold, D.-M. (2001) Progressive edge-growth tanner graphs. Proceedings of IEEE GlobeCom, San Antonio, Texas, pp. 995–1001.

[6] Kou, Y., Lin, S. and Fossorier, M.P.C. (2001) Low-density parity-check codes based on finite geometries: a rediscovery and new results. *IEEE Transactions on Information Theory*, **47** (7), 2711–36.

[7] Moon, T.K. (2005) *Error Correction Coding. Mathematical Methods and Algorithms*, Wiley Interscience, ISBN 0-471-64800-0.

[8] Barnault, L. and Declerq, D. (2003) Fast decoding algorithm for LDPC over GF(2^q). IEEE Information Theory Workshop, Paris, France, pp. 70–3.

[9] Sylvester, J.J. (1867) Thoughts on orthogonal matrices, simultaneous sign-successions, and tessellated pavements in two or more colours, with applications to newton's rule, ornamental tile-work, and the theory of numbers. *Phil. Mag.*, **34**, 461–75.

[10] Davey, M.C. and Mackay, D.J. (1998) Low density parity check codes over GF(q). IEEE Information Theory Workshop, Killarney, Ireland, pp. 70–1.

[11] Lin, S., Song, S., Zhou, B. *et al.* (2007) Algebraic constructions of non-binary quasi-cyclic LDPC codes: array masking and dispersion. 9th International Symposium on Communication Theory and Applications (ISCTA), Ambleside, Lake District, UK.

7

Non-Binary Convolutional Codes

7.1 Introduction

So far this book has dealt with many different types of binary and non-binary block code, from simple parity check codes to the powerful low-density parity check codes, but another important class of error-correcting code is the *convolutional code*, developed by Elias in 1955 [1]. Convolutional codes differ from block codes in that a block code takes a fixed message length and encodes it, whereas a convolutional code can encode a continuous stream of data. Convolutional codes also have much simpler trellis diagrams than block codes, and soft-decision decoding can easily be realized using the Viterbi algorithm, explained later. However, at higher code rates the performance of convolutional codes is poorer than that of block codes.

In this chapter, the principles of convolutional codes are introduced. The convolutional encoder is explained, followed by a description of its state table, tree diagram, signal flow graph and trellis diagram. From the signal flow graph we can determine the transfer function of the code, which gives us the Hamming distance of each code word, beginning with the minimum Hamming distance of the code and the number of nearest neighbours. This distance spectrum can be used to evaluate the performance of the convolutional code.

The idea of trellis coded modulation(TCM) is then presented, whereby a convolutional code is combined with the modulation process. By carefully partitioning the constellation diagram, a bandwidth-efficient TCM code can be constructed with a significant coding gain over an uncoded system.

Binary TCM coding is then extended to TCM codes defined over rings of integers, known as ring-TCM codes. Ring-TCM encoding is explained and a method to search for new ring-TCM codes using a genetic algorithm is presented. Finally, ring-TCM codes are combined with spatial-temporal diversity to create space-time ring-TCM (ST-RTCM) codes. The design criteria for good ST-RTCM codes are

Non-Binary Error Control Coding for Wireless Communication and Data Storage   Rolando Antonio Carrasco and Martin Johnston
© 2008 John Wiley & Sons, Ltd

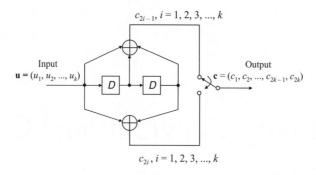

Figure 7.1   Nonsystematic convolutional encoder, R 1/2 with constraint length $K = 3$.

given and the performances of several ST-RTCM codes are presented on multiple-input–multiple-output (MIMO) fading channels, including urban environments such as indoor, pedestrian and vehicular.

An example of a convolutional code is shown in Figure 7.1, where D denotes a memory element. The encoder takes an input $\mathbf{u}$ of length k bits and for each bit produces two coded bits. These coded bits are collected using the switch to produce a code word $\mathbf{c}$ of length $n = 2k$ bits. In this case the input message cannot be easily recovered from the code word since this is an example of a nonsystematic convolutional code.

The two memory elements of Figure 7.1 can only have four possible pairs of binary values, 00, 01, 10 and 11. These values are called states and for a given input we can construct a state table showing what the next state will be from the current state, along with the corresponding output. The state table for the convolutional encoder of Figure 7.1 is shown in Table 7.1.

We can write the encoding process of the convolutional code in terms of poly-nomials, with $U(D)$, $G(D)$ and $C(D)$ as the input, generator and output polynomials respectively. For example, the message $\mathbf{u} = 110\,100$ is equivalent to $U(D) = 1 + D + D^3$. The convolutional encoder of Figure 7.1 has two generator polynomials, $G_1(D) = 1 + D + D^2$ and $G_2(D) = 1 + D^2$. Therefore, the code word polynomials are

Table 7.1   State table for the convolutional code of Figure 7.1.

Input	Current state	Next state	Output
0	00	00	00
1	00	10	11
0	01	00	11
1	01	10	00
0	10	01	10
1	10	11	01
0	11	01	01
1	11	11	10

$C_1(D) = U(D)G_1(D)$ and $C_2(D) = U(D)G_2(D)$. For the message polynomial defined previously, the output will be:

$$
\begin{aligned}
C_1(D) &= U(D)G_1(D) \\
&= (1 + D + D^3)(1 + D + D^2) = 1 + D + D^2 + D + D^2 + D^3 \\
&\quad + D^3 + D^4 + D^5 \\
&= 1 + D^4 + D^5 \equiv 100\,011 \\
C_2(D) &= U(D)G_2(D) \\
&= (1 + D + D^3)(1 + D^2) = 1 + D^2 + D + D^3 + D^3 + D^5 \\
&= 1 + D + D^2 + D^5 \equiv 111\,001
\end{aligned}
$$

Combining the two coded output gives $\mathbf{c} = 11\ 01\ 01\ 00\ 10\ 11$.

Graphical displays of convolutional codes have proven invaluable over the years for their understanding and analysis. A practically useful graphical representation is the code tree or tree diagram of the code. A tree diagram is created by assuming zero initial conditions for the encoder and considering all possible encoder input sequences. The tree diagram for the convolutional code of Figure 7.1 is given in Figure 7.2.

Since the output of the convolutional encoder is determined by the input and the state of the encoder, a more compact representation is the state diagram and signal flow graph [2]. The state diagram is simply a graph of the possible states of the encoder and the possible transmissions from one state to another. The signal flow graph for the encoder of Figure 7.1 is shown in Figure 7.3.

The signal flow graph will be used to explain the method for obtaining the distance properties of a convolutional code. We label the branches of the state diagram as D^0,

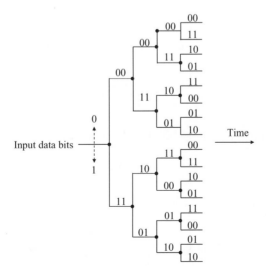

Figure 7.2   The tree diagram.

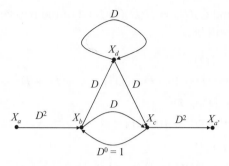

Figure 7.3   The signal flow graph.

D^1 or D^2, where the exponent of D denotes the Hamming distance between the coded bits of each branch in the trellis and the coded bits of the all-zeros branch. If we use the notation for S_1, S_2, S_3 and S_4 for state representation and replace $X_a = S_1$, $X_b = S_2$, $X_c = S_3$ and $X_d = S_2$, we can write four state equations:

$$X_b = D^2 X_a + X_c. \tag{7.1}$$

$$X_c = D X_b + D X_d. \tag{7.2}$$

$$X_d = D X_b + D X_d. \tag{7.3}$$

$$X_{a'} = D^2 X_c. \tag{7.4}$$

The transfer function for the code is defined as:

$$T(D) = \frac{X_{a'}}{X_a} = \frac{D^2 X_c}{X_a}. \tag{7.5}$$

Now we must express the numerator and denominator of $T(D)$ in terms of the same variable so that they will cancel out. First we can rearrange (7.3) to give:

$$X_d = \frac{D X_b}{1 - D}.$$

We also observe that (7.2) and (7.3) are identical, so:

$$X_c = X_d$$

and therefore $X_c = \frac{D X_b}{1-D}$.

We now rearrange (7.1) so that:

$$X_a = \frac{X_b - X_c}{D^2} = \frac{X_b - \dfrac{D X_b}{1 - D}}{D^2} = \frac{\left(\dfrac{X_b(1 - D)}{1 - D} - \dfrac{D X_b}{1 - D} \right)}{D^2} = \frac{(1 - 2D)X_b}{D^2 (1 - D)}.$$

We can also rewrite (7.4) as:

$$X_{a'} = D^2 X_c = \frac{D^3 X_b}{1 - D}.$$

Finally, substituting X_a and $X_{a'}$ into (7.5) gives:

$$T(D) = \frac{\left(\dfrac{D^3 X_b}{1 - D}\right)}{\left(\dfrac{(1 - 2D) X_b}{D^2 (1 - D)}\right)} = \frac{D^3 X_b}{\left(\dfrac{(1 - 2D) X_b}{D^2}\right)} = \frac{D^5 X_b}{(1 - 2D) X_b} = \frac{D^5}{(1 - 2D)}.$$

We have expressed X_a and $X_{a'}$ in terms of X_b so that they cancel out. Evaluating the transfer function gives:

$$
\begin{array}{r}
D^5 + 2D^6 + 4D^7 + 8D^8 + \cdots \\
1 - 2D \overline{\smash{)}\ \ D^5 \hspace{6cm}} \\
\underline{D^5 - 2D^6 \hspace{5cm}} \\
2D^6 \hspace{4.5cm} \\
\underline{2D^6 - 4D^7 \hspace{3.5cm}} \\
4D^7 \hspace{2.8cm} \\
\underline{4D^7 - 8D^8 \hspace{1.8cm}} \\
8D^8. \hspace{0.8cm}
\end{array}
$$

The transfer function tells us that there is one code word with a minimum Hamming distance of 5, two code words with a Hamming distance of 6, four code words with a Hamming distance of 7, and so on.

For a given input sequence, we can find the output by tracing through the trellis, but without the exponential growth in branches as in the tree. Just as the encoding operation for convolutional codes is quite different from the encoding operation for linear block codes, the decoding operation for linear block codes and the decoding process for convolutional codes proceed quite differently. Since we can represent a transmitted code word for a convolutional code as a path through the trellis, the decoding operation consists of finding that path through a trellis which is 'almost like' the received binary sequence. As before with the linear block codes, we are interested in hard-decision decoding and a decoder that immunizes the possibility of error, and therefore, for a given received binary vector, the decoder finds that path through the trellis which has minimum Hamming distance from the received sequence. Given a long received sequence of binary digits and a trellis similar to that in Figure 7.4, it would seem a quite a formidable task to search all possible paths for the best path. However, there exists an illustrative procedure called the Viterbi algorithm [3, 4] which simplifies the process. This algorithm is a special case of what is called forward dynamic programming.

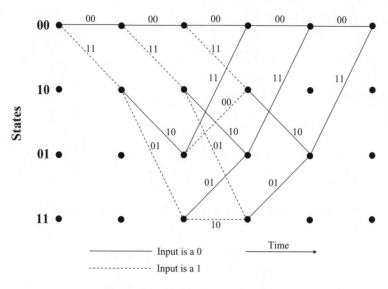

Figure 7.4 Trellis diagram for code.

A convolutional code can be made systematic [2] without affecting its minimum distance properties by feeding back one of the outputs to the input. Such a code is called a recursive systematic convolutional (RSC) code, and is the basic building block for turbo codes, described in Chapter 8. An example of an RSC encoder, derived from the nonsystematic encoder of Figure 7.1, is shown in Figure 7.5.

The decoder for the convolutional code finds the most probable sequence of data bits $\hat{u}$ given the received sequence y:

$$\hat{u} = \arg \left\{ \max_u \ P\left(u \mid y\right) \right\},$$

where y is the set of code symbols c observed through noise. The above equation can be solved using the Viterbi algorithm [3, 4], explained next, in which case the solution is termed the 'maximum likelihood' or ML solution.

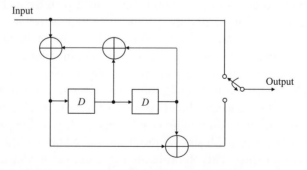

Figure 7.5 Recursive systematic convolutional (RSC) encoder.

7.1.1 The Viterbi Algorithm

The Viterbi algorithm was proposed in 1967 [3] for decoding convolutional codes. The application of the Viterbi algorithm to decoding convolutional codes is based on the trellis diagram representing the convolutional encoder. As applied to our particular problem, this principle states that the best (smallest Hamming distance) path through the trellis that includes a particular node necessarily includes the best path from the begging of the trellis to this node. What this means to us is that for each node in the trellis we need to retain only the single best path to a node, this limiting the number of retained paths at any time instant to the number of nodes in the trellis at that time.

The basic steps in the Viterbi algorithm are as follows [3, 4]:

1. Begin at time $t = 0$, at state 00.
2. Determine the distances between the first received symbols associated with each trellis path at the current time.
3. Select the path with the lowest distance at each state node and store both the path and the distance.
4. If there is more than one path with the lowest distance, select one arbitrarily.
5. Increment time t and return to step 2. Add the lowest distance to the previous lowest distance and store with the sequence of paths. This running score is called the survivor score, and the corresponding sequence of paths, the survivor path.
6. Continue this process until all the symbols in the code word have been compared.
7. The resulting survivor path with the lowest survivor score is the most likely transmitted code word.

7.1.2 Trellis Coded Modulation

In Chapter 3, the idea of coded modulation [5, 6] was introduced and applied to block codes to form block coded modulation (BCM) codes. However, it is much more common to combine convolutional codes with modulation, resulting in *trellis coded modulation (TCM)* codes [6, 7]. The basic idea of TCM codes is illustrated in Figure 7.6,

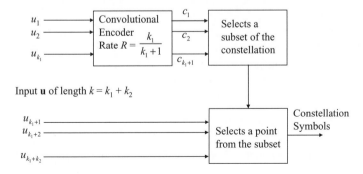

Figure 7.6   General idea of TCM.

which takes a binary message $\mathbf{u} = (u_1, u_2, \ldots, u_{k_1}, u_{k_1+1}, \ldots, u_{k_1+k_2})$ of length $k = k_1 + k_2$ bits.

The BCM codes from Chapter 3 consisted of a multi-level block encoder, with the component code with the largest minimum Hamming distance selecting the subset of the constellation where the Euclidean distance between neighbouring points was minimal. In the same way, a TCM code consists of a convolutional encoder, which also selects a subset where the distance between points is minimal. The uncoded bits then select points within the chosen subset in exactly the same way as the BCM code.

To find the free distance d_{free} of a TCM code we can draw its trellis diagram and find the path with minimum Euclidean distance. Once determined, the asymptotic coding gain (ACG) γ can be calculated by:

$$\gamma = \frac{d_{\text{free}}^2/\varepsilon}{d_{\text{min}}^2/\varepsilon'}, \tag{7.6}$$

where d_{min}^2 is the squared Euclidian distance between points in the uncoded constellation, ε is the energy of the coded constellation and ε' is the energy of the uncoded constellation. To maximize the free distance of the convolutional encoder, set partitioning [6] must be applied to the chosen constellation, as explained in Chapter 3. Figure 7.7 shows the set partitioning of the 8-PSK constellation.

An example of a TCM encoder for 8-PSK modulation is shown in Figure 7.8. In this case, coded bit c_1 selects a QPSK subset of the 8-PSK constellation, c_2 selects a BPSK subset of the QPSK subset and c_3 selects a point within the BPSK subset.

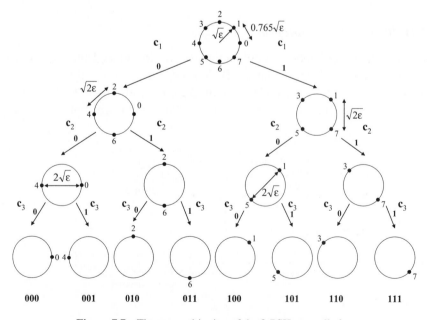

Figure 7.7   The set partitioning of the 8-PSK constellation.

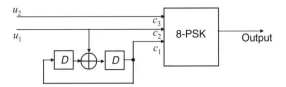

Figure 7.8   A four-state TCM code for 8-PSK modulation.

The state table for the TCM code of Figure 7.8 is given in Table 7.2. From this we can construct the trellis diagram of the TCM code, as shown in Figure 7.9.

From the trellis diagram we can see that there are two paths diverging from state 00 and converging at state 00 with minimum squared Euclidean distance $d_{\text{free}}^2 = 4\varepsilon'$. Since we are determining the ACG of a TCM code using 8-PSK modulation over uncoded QPSK, the energies of both constellations are equal, that is $\varepsilon = \varepsilon'$. The minimum squared Euclidean distance between the points of QPSK is $d_{\text{min}}^2 = 2\varepsilon$ and so substituting into (7.6) gives:

$$\gamma = \frac{d_{\text{free}}^2/\varepsilon}{d_{\text{min}}^2/\varepsilon'} = \frac{4\varepsilon'/\varepsilon}{2\varepsilon/\varepsilon'} = \frac{4}{2} = 2 \quad \text{or} \quad 3 \text{ dB.}$$

From it can be seen that the large free distance of the TCM code more than compensates for the smaller Euclidean distances between constellation points as a result of expanding the size of the constellation.

Table 7.2   State table for the TCM code of Figure 7.8.

Inputs		Initial States		Next States		Outputs		
u_1	u_2	S_1	S_2	S_1'	S_2'	c_1	c_2	c_3
0	0	0	0	0	0	0	0	0
1	0	0	0	0	1	0	1	0
0	1	0	0	0	0	0	0	1
1	1	0	0	0	1	0	1	1
0	0	0	1	1	0	1	0	0
1	0	0	1	1	1	1	1	0
0	1	0	1	1	0	1	0	1
1	1	0	1	1	1	1	1	1
0	0	1	0	0	1	0	0	0
1	0	1	0	0	0	0	1	0
0	1	1	0	0	1	0	0	1
1	1	1	0	0	0	0	1	1
0	0	1	1	1	1	1	0	0
1	0	1	1	1	0	1	1	0
0	1	1	1	1	1	1	0	1
1	1	1	1	1	0	1	1	1

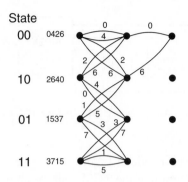

Figure 7.9   Trellis diagram of the TCM code of Figure 7.8.

7.1.3 TCM Based on Rings of Integers

Convolutional codes based on rings of integers modulo-M were first presented by
Massey and Mittelholzer [8, 9]. They were followed by Baldini and Farrell [10,
11], who presented TCM codes based on rings of integers. Baldini and Farrell have
developed a number of modulo-M ring-TCM codes for M-PSK constellations and have
concluded that, due to the similarities between M-PSK signal sets and the algebraic
structure of rings of integers modulo-M, modulo-M ring-TCM codes are the natural
linear codes for M-PSK modulation.

Assuming that m information bits are transmitted per baud, the general structure
of a ring-TCM encoder suitable for M-PSK modulation, with $M = 2^{m+1}$, is shown
in Figure 7.10. This ring-TCM encoder works as follows: first, $m + 1$ information
bits, b_i, are mapped into a modulo-M symbol, a_j, according to a mapping function f
(for instance, f can be a Gray mapping function). Next, m modulo-M a_j symbols are
introduced into a linear multi-level convolutional encoder (MCE), which generates
$m + 1$ modulo-M coded symbols, x_k. Finally, each one of these coded symbols x_k
is associated with a signal of the M-PSK signal set and is sent to the channel. As a
total of $m + 1$ modulo-M coded symbols x_k are transmitted per single trellis branch,
ring-TCM codes can be considered as $2(m + 1)$-dimensional TCM codes.

Propagation delays caused by a fading channel can result in phase shifting of the
transmitted signal. At the demodulator these phase shifts result in a rotation of the
received signals compared with the transmitted symbols. If, after these phase shifts,
the received word is another valid code word, that code is known as *phase invariant*.

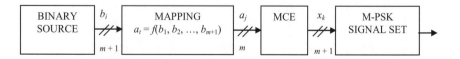

Figure 7.10   General structure of a ring-TCM encoder suitable for M-PSK modulation.

A multi-level convolutional code is $360/M$ phase invariant if and only if the all-one code word can be found in the code [11]. A code with the all-one code word is said to have a *transparent* encoder [11]. For example, an MCE defined over $\mathbb{Z}_4$ is $360/4 = 90°$ phase invariant if the all-one code word is present in the code.s

In general, transparent modulo-M ring-TCM codes can be readily designed, with neither additional difficulty nor significant decrease in performance with respect to nontransparent ring-TCM codes. Furthermore, in addition to the transparency property, ring-TCM codes present, in general, better coding gains than their nontransparent Ungerboeck counterparts. However, the major drawback of these codes is that they generally require a more computationally intensive decoding process. Therefore, although it has already been established that ring-TCM schemes can constitute a powerful alternative to conventional TCM schemes, their actual performance/complexity tradeoff should be investigated.

There is also the necessity to extend the use of ring-TCM codes to other modulation schemes than M-PSK constellations. In particular, the structure of a ring-TCM encoder based on rings of integers modulo-4, which is suitable for any rectangular M-QAM constellation, has been proposed [11]. This novel coded modulation scheme can result in ring-TCM codes with excellent coding gains which are, in addition, transparent to phase rotations of $360/4 = 90°$.

7.1.4 Ring-TCM Codes for M-PSK

This section describes a multi-level convolutional encoder (MCE) defined over the ring of integers modulo-M, $\mathbb{Z}_M$, which is especially suitable for combination with signals of an M-PSK constellation. It is important to notice that we would have preferred to define the MCE over a field of integers modulo-M rather than over a ring of integers modulo-M. However, due to the fact that the field requires M to be a prime integer, it is not suitable for most practical M-PSK constellations, which include a composite number of signal points $M = 2k$, with k some positive integer.

A rate (m/p) MCE defined over the ring of integers modulo-M, $\mathbb{Z}_M$, is a time-invariant linear finite-state sequential circuit that, having m information input symbols $(a_1, a_2, \ldots, a_m)$ at a time defined over $\mathbb{Z}_M$, generates p encoded output symbols $(x_1, x_2, \ldots, x_p)$ at a time defined over $\mathbb{Z}_M$, where the coefficients of the MCE also belong to $\mathbb{Z}_M$, and all arithmetic operations satisfy the properties of the ring of integers modulo-M. In order to achieve the same transmission rate as uncoded modulation, the value of p is chosen so that $p = m + 1$. The modulo-M ring-TCM MCE [11] suitable for M-PSK modulation is shown in Figure 7.11.

The information sequence of symbols, a_i, is shifted into the encoder, beginning at time zero, and continuing indefinitely. The stream of incoming information symbols, a_i (with $0 \leq i < \infty$), is broken into segments called *information frames*, each one containing m symbols of $\mathbb{Z}_M$. At discrete time n, an information frame can be represented

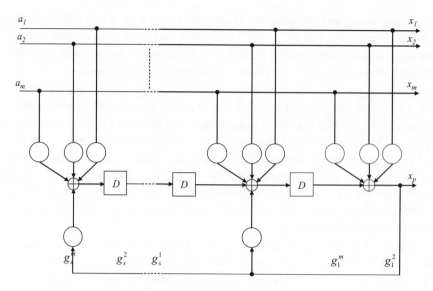

Figure 7.11   Ring-TCM MCE.

as the vector:

$$\bar{\mathbf{a}}_n = [a_1, a_2, \ldots, a_m]_n. \tag{7.7}$$

In order to keep the same transmission rate as uncoded $(M/2)$-PSK (that is, m information bits per channel symbol) by using the expanded rectangular M-PSK signal set, the rate of the MCE must be (m/p), where $2p = m$. Therefore, each information frame $\bar{\mathbf{a}}_n$ must generate an output coded frame $\bar{\mathbf{x}}_n$ containing p symbols of $\mathbb{Z}_M$:

$$\bar{\mathbf{x}}_n = [x_1, x_2, \ldots, x_p]_n. \tag{7.8}$$

The MCE is characterized by the *generator matrix* $G(D)$, which maps the input information frames $\bar{\mathbf{a}}_n$ into the output coded frames $\bar{\mathbf{x}}_n$ and which can be written in the functional form [12, 13]:

$$\bar{\mathbf{x}}_n(D) = \bar{\mathbf{a}}_n(D) \cdot G(D), \tag{7.9}$$

where D is the delay operator and:

$$G(D) = [I_m | P(D)] = \left[\begin{array}{c|c} & g^{(1)}(D)/f(D) \\ & g^{(2)}(D)/f(D) \\ I_m & \vdots \\ & g^{(m)}(D)/f(D) \end{array} \right], \tag{7.10}$$

where the entries of the matrix $G(D)$:

$$
\begin{aligned}
g^{(i)}(D) &= g_s^{(i)} D^s + \cdots + g_2^{(i)} D^2 + g_1^{(i)} D + g_0^{(i)} \quad \text{(Feed forward polynomial)} \\
f(D) &= f_s D^s + \cdots + f_2 D^2 + f_1 D + 1 \qquad \text{(Feedback polynomial)}
\end{aligned} \tag{7.11}
$$

are polynomials with coefficients g_i and f_i belonging to $\mathbb{Z}_M$, and I_m is the $(m \times m)$ identity matrix, as the MCE is *systematic*.

The *parity check* matrix $H(D)$ is a $(1 \times p)$ matrix that satisfies:

$$
G(D) \cdot H(D)^T = 0, \tag{7.12}
$$

where $H(D)$ is readily obtained from $G(D)$ following:

$$
H(D) = [-P(D) : I_1]. \tag{7.13}
$$

Every state q of the MCE is unique and can be defined as:

$$
q = \sum_{i=1}^{s} M^{i-1} \cdot v_i^{(n)}, \tag{7.14}
$$

where $v_i^{(n)}$, which belongs to $\mathbb{Z}_M$, represents the value of the ith *memory* element at discrete time n, and is given by:

$$
\begin{aligned}
v_i^{(n)} &= (\text{modulo} - \text{M}) \left(f_i \cdot x_p + \sum_{j=1}^{m} g_i^{(j)} \cdot a_j + v_{i+1}^{(n-1)} \right) \quad \text{for} \quad 1 \le i < s \\
v_i^{(n)} &= (\text{modulo} - \text{M}) \left(f_i \cdot x_p + \sum_{j=1}^{m} g_i^{(j)} \cdot a_j \right) \qquad\qquad \text{for} \quad s = 1
\end{aligned} \tag{7.15}
$$

with s denoting the *constraint length* of the MCE (number of memory elements). It can be determined by examination of $G(D)$ that:

$$
s = \max_i [\text{degree } g^{(i)}(D)]. \tag{7.16}
$$

The number of states of the MCE, n_{st}, represented in the trellis diagram as the number of nodes in each column, is given by [14]:

$$
n_{st} \le M^s. \tag{7.17}
$$

The reason for the inequality in (7.17) is that every $v_i^{(n)}$ in (7.15) can take on either M or $M/2$ or $M/4, \ldots$, or $M/2^m$ different values of $\mathbb{Z}_M$ at a time, depending on the coefficients g_i and f_i (with $1 \le i < s$). So, $n_{st} = M^s$ only in the case that all $v_i^{(n)}$ take on the M possible values of $\mathbb{Z}_M$; otherwise, $n_{st} < M^s$. For instance, as a clear example, if all coefficients g_i and f_i are even, then all $v_i^{(n)}$ will also be even, and $n_{st} < M^s$.

Each information frame $\bar{\mathbf{a}}_n$ causes the MCE to change state. This is represented by a branch or transition to the next node in the trellis. There are M^m branches entering each state and M^m branches leaving each state. Each branch is labelled with a coded frame $\bar{\mathbf{x}}_n$. Therefore, any encoded frame sequence can be found by walking through the appropriate path in the trellis. It becomes very common that trellis diagrams of MCEs have a sizeable number of parallel transitions between states (especially encoders with a low number of states), due to the sizeable number of total transitions of their trellis codes. In fact, the minimum number of parallel branches, n_p, occurs for a fully-connected trellis diagram; that is, every state of the trellis is connected to all others and itself, and is given by [14]:

$$n_p \geq M^m / M^S = M^{(m-S)}, \tag{7.18}$$

where $(m \geq s)$. If $s > m$, the trellis diagram can never be fully connected.

7.1.5 Ring-TCM Codes Using Quadrature Amplitude Modulation

In this section, a novel 4D TCM technique defined over $\mathbb{Z}_4$, suitable for rectangular M-QAM signal sets, is described [12]. There are many real channels that require the use of transparent, or at least RI TCM, codes. We have presented the design of transparent ring-TCM codes suitable for M-PSK constellations, which have a minimum phase ambiguity of $360/M°$ and require the proper combination of a $(360/M)°$ RI MCE with a $(360/M)°$ code-to-signal mapping. Now the design of RI and transparent ring-TCM codes for rectangular quadrature amplitude modulation (QAM) is explained. These have a phase ambiguity of $90°$. The functional block diagram of the 4D ring-TCM encoder [12] is shown in Figure 7.12. It is similar to the ring-TCM transmitter for M-PSK signal sets in Figure 7.11, but there are two important differences [12]:

1. 4D ring-TCM codes suitable for rectangular M-QAM constellation (where $M = 4^i$, with i a positive integer greater than 0) are always defined over the ring of integers modulo-4, $\mathbb{Z}_4$. That is, both bit-to-symbol mapping (i.e. Gray mapping) and MCE are defined over $\mathbb{Z}_4$. This particular ring definition is required in order to resolve the phase ambiguities of rectangular M-QAM signal sets.

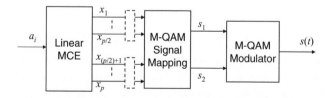

Figure 7.12   Block diagram of a 4D ring-TCM transmitter suitable for rectangular M-QAM.

2. The code-to-signal mapping must properly assign the output of the MCE to the signals of the 2D M-QAM constellation. In this particular case, the p encoded output symbols of the MCE must now be split into two sets containing $(p/2)$ symbols each, $x^{(1)}$ and $x^{(2)}$, where:

$$
\begin{aligned}
x^{(1)} &= \left(x_1, x_2, \ldots, x_{p/2}\right) \\
x^{(2)} &= \left(x_{(p/2)+1}, x_{(p/2)+2}, \ldots, x_p\right)
\end{aligned}
\tag{7.19}
$$

Each set of symbols is then mapped onto one signal of the 2D M-QAM signal set, $s_k = f(x^{(k)})$, according to some set partitioning rules that guarantee a certain minimum distance between coded sequences. The two coded 2D M-QAM signals, s_1 and s_2, are then modulated and transmitted on the channel using time division, forming the 4D coded signal, $s(t)$.

Let us define the isomorphism $\psi(\mathbb{Z}_4) = P$ as the mapping of the elements of the ring of integers modulo-4, $\mathbb{Z}_4 = \{0, 1, 2, 3\}$, onto the phase rotation they represent, $P = \{0°, 90°, 180°, 270°\}$. Then, following (7.19), groups of $(p/2)$ output symbols of the MCE $(x_1, x_2, \ldots, x_{(p/2)})$ must be mapped onto one signal of the expanded rectangular M-QAM signal set. Also, all the symbols must be used in the mapping process with equal frequency.

A 90° RI code-to-signal mapping requires that any pair of signals in the rectangular M-QAM constellation which have the same radius but are $\alpha \cdot 90°$ apart, where $\alpha \in \mathbb{Z}_4$, be assigned the groups of symbols $(x_1, x_2, \ldots, x_{(p/2)})$ and $\alpha.(x_1, x_2, \ldots, x_{(p/2)})$.

Some possible transparent mappings for 16-QAM used by Tarokh [15], and 16-QAM and 64-QAM from Carrasco and Farrell are shown in Figure 7.13a, b and c, respectively [13].

7.1.6 Searching for Good Ring-TCM Codes

The choice of feed-forward and feedback coefficients of the MCE can significantly change the parameters of the code, such as the free distance, d_{free}, and so it is important to have methods for finding good TCM codes. The most obvious method would be to apply an exhaustive search algorithm where all values of the coefficients are tried, but this would become too time-consuming for larger codes defined over larger rings.

There are two criteria for good TCM codes over a Gaussian channel: maximizing the free distance and minimizing the number of paths in the code trellis with Euclidean distance equal to the free distance, denoted as N_{free}, subject to the feed-forward and feedback coefficients [16]:

$$
\max \ d_{\text{free}}\left\{g_s^{(i)}, \ldots, g_0^{(i)}, f_s, \ldots, f_1\right\}
\tag{7.20}
$$

$$
\min \ N_{\text{free}}\left\{g_s^{(i)}, \ldots, g_0^{(i)}, f_s, \ldots, f_1\right\}.
\tag{7.21}
$$

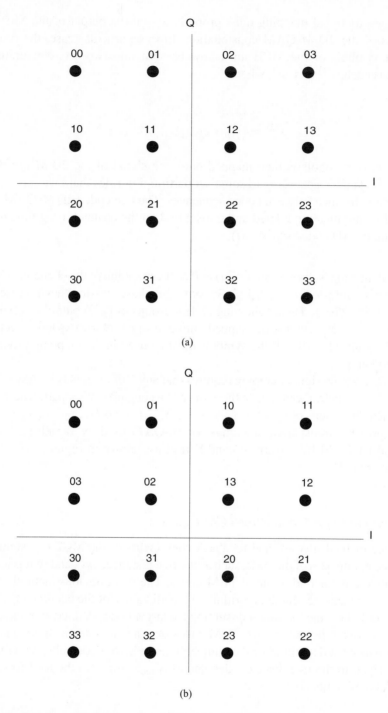

Figure 7.13 QAM constellation mapping (a) Tarokh's 16-QAM, (b) Carrasco and Farrell 16-QAM, (c) Carrasco and Farrell 64-QAM.

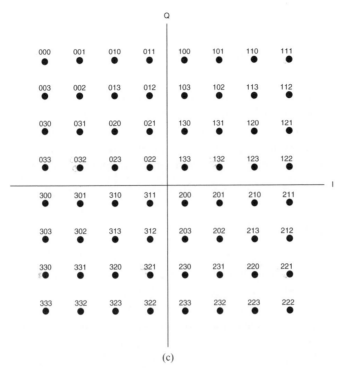

$$Q$$

000	001	010	011	100	101	110	111
003	002	013	012	103	102	113	112
030	031	020	021	130	131	120	121
033	032	023	022	133	132	123	122
300	301	310	311	200	201	210	211
303	302	313	312	203	202	213	212
330	331	320	321	230	231	220	221
333	332	323	322	233	232	223	222

(c)

Figure 7.13 *(Continued)*

In the following section, a much faster search algorithm for finding good ring-TCM codes for M-PSK and M-QAM is presented.

7.1.7 Genetic Algorithm

For the solution of optimization problems, a genetic algorithm [17] has been investigated for several applications in signal processing, such as speech compression, and has been shown to be effective at exploring a large and complex search space in an adaptive way. Next, a formal definition of genetic algorithms is given, and the operators used in this work are described.

A genetic algorithm can be defined as the nine-tuple:

$$\text{GA} = \left(p^0, S, \lambda, L, f, s, c, m, T\right), \text{ where}$$

$$p^t = (a_1^t, \ldots, a_\lambda^t) = \text{population in the } t\text{th generation}$$
$$S = \text{search space in which the chromosome is encoded}$$
$$a_\lambda^0 \in S = \text{elements of the search space}$$
$$\lambda = \text{population size}$$
$$L = \text{length of chromosome}$$
$$f : S \to R = \text{fitness function}$$

$$s : S^\lambda \to S = \text{selection operator}$$
$$c : S \times S \to S \times S = \text{crossover operator}$$
$$m : S \to S = \text{mutation operator}$$
$$T : S^\lambda \to \{0, 1\} = \text{termination criterion.}$$

In searching for TCM codes, the genetic algorithm basically selects and classifies the generator code word with a high degree of adaptation as parents generate a new generation by the combination of their components, and by the elimination of the weakest generators from the population [18]. In genetic algorithms there are many operators but the selection, crossover and mutation operators are only used for code searching.

7.1.7.1 Selection Operator

Assuming that the initial population $p^0 = (a_1^0, \ldots, a_\lambda^0)$ can be obtained in a heuristic or random way from the search space, S, in which the chromosome is encoded, the next generation is obtained from members of the previous generation using a stochastic process, which guarantees that the number of times that one structure is selected is dependent on its performance compared with the rest of the population.

The parent-selection operation s, $s : S^\lambda \to S$ produces an intermediate population $p^t = (a_1^t, \ldots, a_\lambda^t)$ from the population $p^t = (a_1^t, \ldots, a_\lambda^t)$ in the tth generation, $s(p^t) = p^t$. Any $a_i^t = a_q^t$ in the p^t is selected by a given random number α_i satisfying the following condition:

$$0 \leq \alpha_i \leq \sum_{j=1}^{\lambda} f\left(a_j^t\right). \tag{7.22}$$

The index q is obtained from:

$$q = \max \left\{ k \bigg/ \forall k \in \{1, \ldots, \lambda\}, s.t. \alpha_i \leq \sum_{k=1}^{\lambda} f\left(a_k^t\right) \right\}, \tag{7.23}$$

where $\sum_{k=1}^{\lambda} f\left(a_k^t\right)$ is the summation of all the fitness from population p^t for all members of the population picking up the first index k that reaches $\alpha_i \leq \sum_{k=1}^{\lambda} f(a_k^t)$.

7.1.7.2 Crossover Operator

For any selected chromosomes in a population p^t, an associated real value $0 \leq \rho \leq 1$ is generated randomly. If ρ is greater than the defined crossover threshold ρ_c, where $0 \leq \rho_c \leq 1$, the crossover operator $c : S \times S \to S \times S$ is applied to this pair of chromosomes. The strategy used in this work is the one-point crossover, c_{op}, which produces an intermediate population p^t from the population p^t and is defined below:

$$\begin{pmatrix} a_i^{\prime t} \\ a_{(i+1)}^{\prime t} \end{pmatrix} = C_{op} \left(\begin{pmatrix} a_i^t \\ a_{(i+1)}^t \end{pmatrix} \right) \forall a_i^t \in p^t / i \in \{1, \ldots, \lambda\}$$

$$= C_{op} \left(\begin{pmatrix} [a_{i,\rho 1}, a_{i,\rho 2}, \ldots, a_{i,\rho j}, a_{i,(\rho j+1)}, \ldots, a_{i,\rho L}]^T \\ [a_{(i+1),\rho 1}, a_{(i+1),\rho 2}, \ldots, a_{(i+1),\rho j}, a_{(i+1),(\rho j+1)}, \ldots, a_{(i+1),\rho L}]^T \end{pmatrix} \right).$$

$$= C_{op} \left(\begin{matrix} [a_{i,\rho 1}, a_{i,\rho 2}, \ldots, a_{i,\rho j}, a_{(i+1),(\rho j+1)}, \ldots, a_{(i+1),\rho L}]^T \\ [a_{(i+1),\rho 1}, a_{(i+1),\rho 2}, \ldots, a_{(i+1),\rho j}, a_{i,(\rho j+1)}, \ldots, a_{i,\rho L}]^T \end{matrix} \right)$$

$$(7.24)$$

The pair of chromosomes is separated into two subchromosomes at ρ_j, hence a new pair is composed by swapping the second subchromosome, where each one is crossed from ρ_{j+1} to ρ_L.

7.1.7.3 Mutation Operator

For any chromosome in a population p^t, an associated real value $0 \le \rho \le 1$ is generated randomly. If ρ is less than the defined mutation threshold ρ_m, where $0 \le \rho_m \le 1$, the mutation operator is applied to the chromosome. The mutation operator simply alters one bit in a chromosome from 0 to 1 (or 1 to 0). The mutation operator $m : S \to S$ produces an intermediate population p^t from the population p^t, as below:

$$a_i^{\prime t} = m(a_i^t) \forall_i \in \{1, \ldots, \lambda\}$$

$$a_{i,\rho k}^{\prime t} = \begin{cases} a_{i,\rho k} & \text{for } k \in \{1, 2, \ldots, p-1, p+1, \ldots, L\} \\ \bar{a}_{i,\rho k} & \text{for } k = p \end{cases} \tag{7.25}$$

The reproduction and crossover operators give to the genetic algorithms the bulk of their processing power. The mutation operator is needed because, even though reproduction and crossover are the main operators, occasionally they may become overzealous and lose some potentially useful genetic material.

Usually there are four parameters to control the evolution of the genetic algorithm. They are: the population size, λ; the crossover threshold, ρ_c; the mutation threshold, ρ_m; and the number of generations. If three of these are kept fixed, an optimum value for the free parameter can be found in order to produce the optimum code rate, d_{free} and N_{free}.

The disadvantage of the exhaustive search algorithm is the time taken to generate codes, especially when a high number of states and a high-order constellation are used. It is important to note that the exhaustive search for the rectangular M-QAM constellation is more computationally intensive than that for M-PSK constellations, because the rectangular M-QAM constellation requires that, to determine d_{free} and N_{free}, all paths in the trellis be examined and compared to each other. The fitness function used in this work is an *a priori* function based on the computation of the distances.

Once the code is obtained, a further check is required to investigate the performance in terms of symbol error rate as a function of signal-to-noise ratio. The application of the genetic algorithm to 16-QAM is shown with new codes for this particular modulation scheme to demonstrate its feasibility, when the number of states is increased to produce better constraint lengths. The application of the genetic algorithm to the code-to-signal mapping for 16-QAM has produced a range of 180° and 90° rotational invariant ring-TCM codes (RI ring-TCM), which are presented in Table 7.3 and compared with the exhaustive algorithm [18].

In Table 7.4, d_{free}^2 is the minimum squared Euclidean distance between coded sequences of the ring-TCM code suitable for 16-QAM; g_∞ is the asymptotic coding gain of the ring-TCM code suitable for 16-QAM over uncoded 8-AMPM modulation; * means that the exhaustive search was stopped; and ♦ means a new generated code using the genetic algorithm.

Table 7.3 RI ring-TCM codes for 4-PSK on AWGN channel.

n_{st}	Ring-TCM Code	Rot	d_{free}^2	N_{free}
2	320	90°	8	5
4	230	90°	8	5
4	032	90°	8	5
4	212	90°	8	1
8	030/02	90°	8	5
8	012/02	90°	8	1
16	032/31	90°	12	6
16	221/31	90°	12	2
16	232/33	90°	16	14
32	0302/130	90°	12	2
32	2032/110	90°	16	8
64	0232/303	90°	16	5
64	3103/213	90°	16	4
64	1013/332	90°	16	2
64	2122/321	90°	16	1
128	21 112/0222	90°	16	4
128	13 010/0112	90°	16	2
128	11 230/0312	90°	20	2
128	21 132/1210	90°	20	6
256	32 130/3113	90°	20	10
256	30 013/1122	90°	20	8
256	33 012/2101	90°	20	2
256	21 103/1221	90°	24	29
256	23 121/0332	90°	24	10

Table 7.4 RI ring-TCM codes for 16-QAM on AWGN channel.

n_{st}	Ring-TCM Code	Rot	d_{free}^2	N_{free}	g_∞ dB
2	(030 020/0)	90°	4.0	19.187	3.01
4	(220 030/2)	90°	4.0	5.937	3.01
♦4	(020 300/0)	90°	4.0	6.5	3.01
*8	(020 212 010/32)	90°	4.0	1.031	3.01
*8	(022 231 0130/03)	90°	4.0	0.75	3.01
*8	(010 232 020/21)	180°	5.0	5.234	3.98
♦8	(000 311 010/03)	180°	4.0	4.5	3.98
*16	(010 232 010/22)	90°	6.0	9.796	4.77
♦16	(233 311 213/20)	90°	10.0	0.5	4.77

Algorithm 7.1: Algorithm for select operator

select()
{

Randomly choose α_i, in the range $0 \leq \alpha_i \leq \sum_{j=1}^{\lambda} f\left(a_j^t\right)$

for $\{k/\forall\, k \in \{1, \ldots, \lambda\}\}$

    {

        if $\left(q = \max\{k.s.t.\alpha_i \leq \sum_{k=1}^{\lambda} f\left(a_k^t\right)\}\right)$

        {

            return(q)

        }

    }

}

Algorithm 7.2: Algorithm for crossover operator

crossing()
{

Randomly choose the coefficient from $g_s^{(i)}, \ldots, g_0^{(i)}, f_s, \ldots, f_1 \in S$

if($\rho \geq \rho_c$)

    {

        increment *crossover pointer*;

        compute *crossover pointer*;

    }

else

    crossover pointer; = 0; /* no cross, just reproduction*/

    gene_counter = 0;

    while(*gene_counter* < *crossover pointer*)

    {

        Copy to child1 the genetic information of father;

        Copy to child2 the genetic information of mother;
        increment the *gene_counter*;
    }
    while(*gene_counter* < L)
    {
        Copy to child1 the genetic information of mother;
        Copy to child2 the genetic information of father;
        increment the *gene_counter*;
    }
    share parentage information();
}

Algorithm 7.3: Algorithm for crossover operator

```
mutate()
{
    k = ring_elements() ;
    for(k/∀k ∈ {1, ..., λ})
    {
        if(ρ ≥ ρₘ)
        {
            increment mutation counter;
            alter one bit k = p in the chromosome;
        }
    }
}
```

Example 7.1: A 4-state ring-TCM code defined over 4: A good 4-state ring-TCM code over $\mathbb{Z}_4$ is the 23/1 ring-TCM code shown in Figure 7.14.

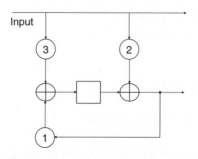

Figure 7.14   The 23/1 ring-TCM code.

Table 7.5   State table for 23/1 ring-TCM code over $\mathbb{Z}_4$.

Input	Initial state	Next state	Output
0	0	0	00
1	0	1	12
2	0	2	20
3	0	3	32
0	1	1	01
1	1	2	13
2	1	3	21
3	1	0	33
0	2	2	02
1	2	3	10
2	2	0	22
3	2	1	30
0	3	3	03
1	3	0	11
2	3	1	23
3	3	2	31

Its state table is given in Table 7.5 and the corresponding trellis diagram is shown in Figure 7.15. From the state table we are able to obtain the signal flow graph for the 23/1 ring-TCM code, as shown in Figure 7.16. Comparing it to the signal flow graph of Figure 7.3, we can see that the number of states is the same but there are more connections. As before, state 0 is denoted X_a, state 1 is denoted X_b, state 2 is denoted X_c and state 3 is denoted X_d. From the signal flow graph, it is possible to determine the transfer function of the 23/1 ring-TCM code and hence generate the weight distribution of the code.

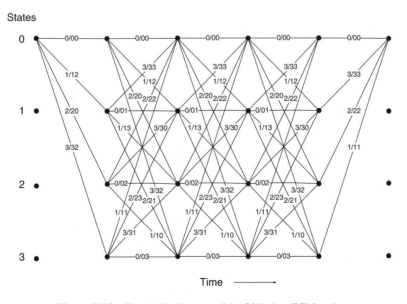

Figure 7.15   The trellis diagram of the 23/1 ring-TCM code.

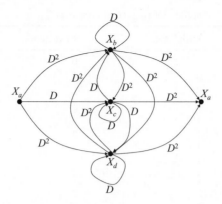

Figure 7.16   The signal flow graph of the 23/1 ring-TCM code.

From Figure 7.16, we can define the following set of equations:

$$X_b = D^2 X_a + D X_b + D X_c + D^2 X_d. \tag{7.26}$$
$$X_c = D X_a + D^2 X_b + D X_c + D^2 X_d. \tag{7.27}$$
$$X_d = D^2 X_a + D^2 X_b + D X_c + D^2 X_d. \tag{7.28}$$
$$X_{a'} = D^2 X_b + D^2 X_c + D^2 X_d. \tag{7.29}$$

An important step is to first notice that $X_b = X_d$. Subtracting (7.28) from (7.29):

$$X_b - X_d = (D - D^2) X_b + (D^2 - D) X_d$$
$$\left(1 - D + D^2\right) X_b = \left(1 - D + D^2\right) X_d$$
$$\therefore X_b = X_d.$$

The transfer function $T(D)$ of the signal flow diagram is:

$$T(D) = \frac{X_{a'}}{X_a} = \frac{D^2 \left(X_b + X_c + X_d\right)}{X_a} = \frac{D^2 \left(2X_b + X_c\right)}{X_a}. \tag{7.30}$$

$$T(D) = \frac{D^2 \left(\left(2D^2 + D\right) X_a + 2\left(2D^2 + D\right) X_b + 3 D X_c\right)}{X_a}$$

$$= 2D^4 + D^3 + 2D^2 \left(2D^2 + D\right) \frac{X_b}{X_a} + 3D^3 \frac{X_c}{X_a}. \tag{7.31}$$

We can show that:

$$X_b = \frac{D^2 X_a + D X_c}{1 - D - D^2}. \tag{7.32}$$

Also:

$$X_c = \frac{D X_a + 2 D^2 X_b}{1 - D}. \tag{7.33}$$

Substituting X_c into X_b gives:

$$X_b = \frac{D^2 X_a}{1 - D - D^2} + \frac{D}{1 - D - D^2}\left(\frac{DX_a + 2D^2 X_b}{1 - D}\right)$$

$$\therefore \frac{X_b}{X_a} = \frac{2D^2 - D^3}{1 - 2D - D^3}. \tag{7.34}$$

Similarly:

$$\frac{X_c}{X_a} = \frac{D}{1 - D} + \frac{2D^2 X_b}{(1 - D)X_a} = \frac{D - D^2 + D^3}{1 - 2D - D^3}. \tag{7.35}$$

Now substituting (7.34) and (7.35) into the transfer function (7.31) gives:

$$\begin{aligned} T(D) &= 2D^4 + D^3 + 2D^2\left(2D^2 + D\right)\frac{X_b}{X_a} + 3D^3\frac{X_c}{X_a} \\ &= \frac{D^3 + 3D^4 - 3D^5 + 8D^6 - 6D^7}{1 - 2D - D^3}. \end{aligned} \tag{7.36}$$

Evaluating (7.36) gives:

$$\begin{array}{r} D^3 + 5D^4 + 7D^5 + 23D^6 + \cdots \\ 1 - 2D - D^3 \overline{\smash{\big)}\, D^3 + 3D^4 - 3D^5 + 8D^6 - 6D^7} \\ \underline{D^3 - 2D^4 \qquad\quad - D^6} \\ 5D^4 - 3D^5 + 9D^6 - 6D^7 \\ \underline{5D^4 - 10D^5 \qquad\quad - 5D^7} \\ 7D^5 + 9D^6 - D^7 \\ \underline{7D^5 - 14D^6 \qquad\quad - 7D^8} \\ 23D^6 - D^7 + 7D^8 \\ \vdots \end{array}$$

Therefore, the 23/1 ring-TCM code over $\mathbb{Z}_4$ has one path with a weight of 3, five paths with a weight of 4, seven paths with a weight of 5, and so on.

7.1.8 Performance of Ring-TCM Codes on Urban Fading Channels

Three simulation results are presented, showing the performances of three ring-TCM codes on an indoor, a pedestrian and a vehicular channel, as defined in Chapter 1. The ring-TCM codes tested were transparent, and decoded using the soft-decision Viterbi algorithm. The codes used were the 4-state 21/2, 8-state 213/30 and 16-state 212/31 ring-TCM codes defined over $\mathbb{Z}_4$. The user velocities were 0 mph for indoor, 4 mph for pedestrian and 70 mph for vehicular. The channel scenarios are easily modified

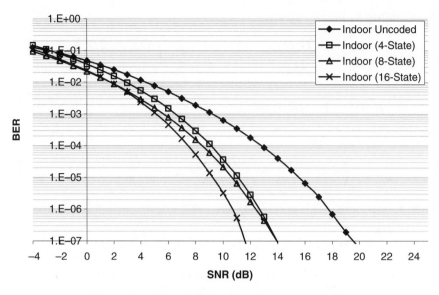

Figure 7.17   Ring-TCM codes on the indoor channel.

by altering the delay, power profile and Doppler spectra to create virtually any single-input–single-output (SISO) environment based on measured data. This gives a much more flexible channel model, which corresponds to actual measured data and produces a time-varying frequency-selective channel that is much more realistic and is essential for testing certain distortion mitigation techniques.

Figure 7.17 shows the performance of the ring-TCM codes on the indoor channel. This is a slow-fading channel, but the least harsh of the three urban channel models and good results are achieved, with the 16-state 212/31 ring-TCM code achieving a coding gain of 1.2 dB over the 21/2 and 213/30 ring-TCM codes.

Figure 7.18 shows the performance of the ring-TCM codes on the pedestrian channel. This channel is more harsh but the relative performance of the three ring-TCM codes has not changed. The 212/31 ring-TCM code achieves a coding gain of approximately 14 dB over uncoded performance.

Figure 7.19 shows the performance of the ring-TCM codes on the vehicular channel. This channel is a time-varying fading channel and is very harsh. The performance of all three ring-TCM codes is poor and there is no significant coding gain over an uncoded system.

7.2 Space-Time Coding Modulation

7.2.1 Introduction

Future-generation wireless communication systems require high-speed transmission rate for both indoor and outdoor applications. Space-time (ST) coding and

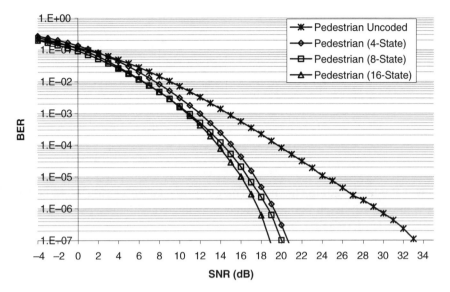

Figure 7.18 Ring-TCM codes on the pedestrian channel.

multiple-input–multiple-output (MIMO) channels have been envisaged as the solution for high-capacity levels [15]. A reliable high-speed communication is guaranteed due to diversity, provided by space-time codes. This section describes the realization of a space-time TCM coding scheme suitable for M-PSK [19] and M-QAM [20] over fading channels. In this section, we also provide ST-ring TCM codes suitable for

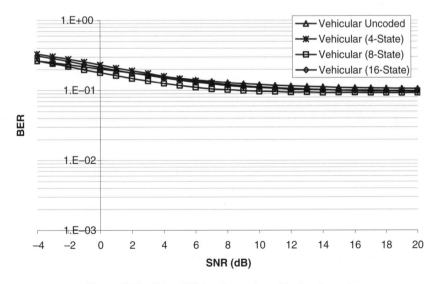

Figure 7.19 Ring-TCM codes on the vehicular channel.

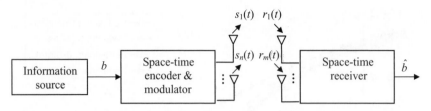

Figure 7.20 Space-time system model.

QAM over Rayleigh fading channels. The performance of ST-ring TCM codes with 16-QAM in slow Rayleigh fading channels is evaluated with designs that involve rate 4 and 6 bits/s/Hz of 16 and 4 states.

7.2.2 Space-Time Codes Model

Consider a mobile communication system where the transmitter consists of n antennas and the receiver has m antennas, as shown by Figure 7.20. The binary source data are encoded by a space-time encoder that generates n streams, and each stream is modulated and transmitted through a different antenna. Let $s_i(t)$ be the symbol transmitted by antenna i at time t for $1 \leq i \leq n$ and let T be the symbol period. At a given instant t, all $s_i(t)$ symbols are transmitted simultaneously in space-time coding.

The signal received at each receiving antenna is the superposition of the n transmitted signals corrupted by both Rayleigh and Rician fading and additive white Gaussian noise (AWGN). Assuming that the elements of the signal constellations have an average energy of E_s equal to 1, the signal $r_j(t)$ received by antenna j at time t can be described as [15]:

$$r_j(t) = \sum_{i=1}^{n} h_{ji}(t)s_i(t) + \eta_j(t) \quad \text{with} \quad 1 \leq j \leq m, 1 \leq i \leq n, 1 \leq t \leq l, \quad (7.37)$$

where the noise $\eta_j(t)$ at time t is modelled as independent samples of a zero-mean complex Gaussian random variable with variance $N_0/2$ per dimension. The coefficients $h_{ji}(t)$ are the channel gains from the transmit antenna, i, to the receive antenna, j. These channel gains are usually modelled as independent samples of complex Gaussian random variables with variance 0.5 per dimension.

(7.37) can be expanded to give a clear insight into the system model:

$$r_j(t) = h_{1j}(t)s_1(t) + h_{2j}(t)s_2(t) + h_{3j}(t)s_3(t) + \cdots + \eta_j(t). \quad (7.38)$$

Likewise, the received signal sequence may be written as an array $\boldsymbol{R}$ of dimension $m \times l$, as well as the random AWGN variables in a matrix $\boldsymbol{N}$ of:

$$\boldsymbol{R} = \begin{bmatrix} r_1(1) & r_1(2) & \cdots & r_1(l) \\ r_2(1) & r_2(2) & \cdots & r_2(l) \\ \vdots & & \ddots & \vdots \\ r_m(1) & r_m(2) & \cdots & r_m(l) \end{bmatrix},$$

$$\boldsymbol{N} = \begin{bmatrix} \eta_1(1) & \eta_1(2) & \cdots & \eta_1(l) \\ \eta_2(1) & \eta_2(2) & \cdots & \eta_2(l) \\ \vdots & & \ddots & \vdots \\ \eta_m(1) & \eta_m(2) & \cdots & \eta_m(l) \end{bmatrix}.$$

(7.39)

The system model expressed in (7.37) may also be expressed in matrix form:

$$\boldsymbol{R} = \boldsymbol{H}\boldsymbol{S} + \boldsymbol{N}, \tag{7.40}$$

where $\boldsymbol{H}$ is the $m \times n$ channel coefficient matrix.

The information symbol $b(t)$ at time t is encoded by the ST encoder as the sequence of code symbols $s_1(t), s_2(t), \ldots, s_n(t)$. In the receiver a maximum-likelihood decoder receives the signals $r_1(t), r_2(t), \ldots, r_m(t)$, each at a different receive antenna. For a given code of length l, the transmitted sequence was:

$$\boldsymbol{S} = \begin{bmatrix} s_1(1) & s_1(2) & \cdots & s_1(l) \\ s_2(1) & s_2(2) & \cdots & s_2(l) \\ \vdots & & \ddots & \vdots \\ s_n(1) & s_n(2) & \cdots & s_n(l) \end{bmatrix}. \tag{7.41}$$

The code symbols are generated by the ring-TCM encoder shown in Figure 7.14. The outputs $x_i (i = 1, 2, 3, \ldots, d + 1)$ are $\mathbb{Z}_4$ symbols and are mapped according to the constellation used and the number of transmitting antennas, n. To decode the ST-RTCM code using the Viterbi algorithm, the accumulated metrics are determined by:

$$M\left(r_j(t), s_i(t) | h_{ij}\right) = \sum_{t=1}^{l} \sum_{j=1}^{m} \left| r_j(t) - \sum_{i=1}^{n} h_{ij} s_i(t) \right|. \tag{7.42}$$

Example 7.2: Construction and Viterbi decoding of a 4-state space-time ring-TCM code: In this example, the 23/1 ring-TCM code from Figure 7.14 is combined with spatial diversity to form a space-time ring-TCM code defined over $\mathbb{Z}_4$ with QPSK modulation. The The two QPSK outputs are transmitted from two antennas simultaneously. The trellis diagram in Figure 7.21 shows how the Viterbi algorithm can be used for the soft-decision decoding of the 23/1 space-time ring-TCM code, with $n = 2$ and $m = 2$.

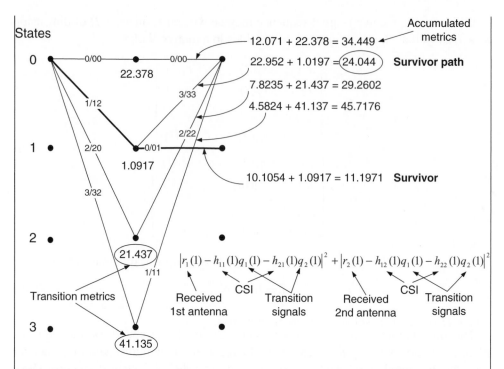

Figure 7.21   Soft-decision decoding example of a space-time ring-TCM code using the Viterbi algorithm.

Observe how the transition metric is calculated for every state based on (7.38). All the information required in the calculation is from the received signals, the CSI and the transmitted signals specified by the respective branch label. At $t = 2$, there are four competitive paths arriving at every state, whose accumulated metrics are computed according to (7.42). In state 0, for instance, the second competitor has the lowest accumulated metric, of 24.044; hence this will be the survivor path. This procedure is repeated for every state and carried on to the continuing symbol intervals. When the sequence finalizes there will be four survivors, one for each state. The survivor with the lowest resultant metric will be selected and the input sequences traced back so that the original data can be reconstructed.

7.2.3 Performance of ST-RTCM Codes Using QPSK for MIMO Urban Environments

Simulation results for ST-RTCM codes are now presented, evaluating their performance on the indoor, pedestrian and vehicular MIMO fading channels. The codes are the 4-state 21/3, the 16-state 212/31 and the 64-state 2103/132 ring-TCM codes. Figure 7.22 shows that, as for the case of the ring-TCM codes in Figure 7.19, the performance of the ST-RTCM codes on the vehicular MIMO channel is poor. The

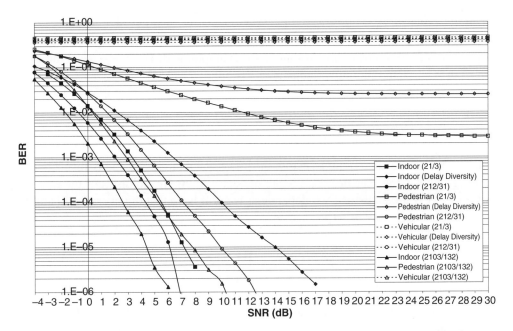

Figure 7.22   Performance of a 4-state, 16-state and 64-state ring-TCM code on urban MIMO fading channels.

4-state and 16-state ST-RTCM codes on the pedestrian MIMO channel also perform poorly with an error floor, but the 64-state ST-RTCM code performs well with no error floor at a BER = 10^{-6}. The best performance is on the indoor MIMO channel, as expected, with the 2103/132 ST-RTCM code performing very well with a BER = 10^{-6} at a SNR of 6 dB.

7.2.4 Ideal CSI and Slow Fading

It is assumed here that the elements h_{ij} of matrix H are time-independent and the receiver has full knowledge of them, so that there is no need to carry out any sort of estimation process within the receiver. The probability of having received the sequence R at the receiver when the code matrix S has been transmitted through a fading channel H follows a multidimensional Gaussian distribution and shall be described by the pdf:

$$p(R|S, H) = \prod_{t=1}^{l} \prod_{j=1}^{m} \frac{1}{\sqrt{2\pi N_0}} \exp \left[-\frac{|r_j(t) - \sum_{i=1}^{n} h_{ij} s_i(t)|^2}{2N_0} \right]. \tag{7.43}$$

The above is due to the fact that, at a given symbol interval, the m-dimensional noise array has a multivariate Gaussian complex distribution with zero mean and covariance matrix $\sum_\eta$ equal to $N_0 \cdot I_m$, where I_m is the m-order identity matrix. So, if we consider the random vector $\vec{\eta}(t) = \begin{bmatrix} \eta_1(t_0) & \eta_2(t_0) & \cdots & \eta_m(t_0) \end{bmatrix}$ as a column of the noise matrix N for a given time instant t_0, for example, with $1 \leq t_0 \leq l$, then the joint pdf of the Gaussian random variables $\eta_j(t_0)$ is defined as [15]:

$$p\left[\vec{\eta}(t_0)\right] = \frac{1}{(2\pi)^{m/2}\sqrt{\det \sum_\eta}} \exp\left[-\frac{1}{2}(\vec{r} - H \cdot \vec{s})' \sum_\eta^{-1} (\vec{r} - H \cdot \vec{s})\right], \quad (7.44)$$

where $\vec{r}$ and $\vec{s}$ are the column vectors at time t_0 of matrices R and S respectively, and $(\vec{x})'$ denotes the transpose conjugate of vector $\vec{x}$. As the columns of the random noise matrix N are statistically independent, we will have that the overall probability $p(R|S, H)$, which may be expressed as:

$$p(R|S, H) = \prod_{t_0=1}^{l} p\left[\vec{\eta}(t_0)\right]. \quad (7.45)$$

When combined with 7.44, and after some mathematical processing, this will yield expression given in (7.43).

7.2.5 Log-Likelihood Function

The next step is to define the likelihood function that shall be used in the maximum likelihood sequence decoding of ST codes. It is desirable that this function, f, has the following additive property: the likelihood of the complete sequence is the sum of the partial symbol likelihoods; that is to say [15]:

$$f = \sum_{t=1}^{l} f_n, \quad (7.46)$$

where f_n represents an individual instant likelihood. Having this in mind, the most suitable function is the natural logarithm of $p(R|S, H)$, namely the log-likelihood function, which is obtained by taking the logarithm of (7.43):

$$\ln[p(R|S, H)] = \sum_{t=1}^{l} \sum_{j=1}^{m} \left[\ln\left(\frac{1}{\sqrt{2\pi N_0}}\right) - \frac{\left| r_j(t) - \sum_{i=1}^{n} h_{ij} s_i(t) \right|^2}{2N_0} \right], \quad (7.47)$$

$$= -\frac{ml}{2} \ln(2\pi N_0) - \sum_{t=1}^{l} \sum_{j=1}^{m} \frac{\left| r_j(t) - \sum_{i=1}^{n} h_{ij} s_i(t) \right|^2}{2N_0}$$

with the first term being a constant, and the second dependent on the received signal sequence, R, the transmitted code matrix, S, and the channel matrix, H. This function is of vital important for decoding ST codes.

7.2.6 Maximum-Likelihood Sequence Detection

The maximum-likelihood (ML) sequence detection in ST codes is performed by maximizing the above log-likelihood function, that is the task is to find the sequence $\hat{s}_i(t)$ that maximizes $\ln[p(R|S, H)]$:

$$\hat{s}_i(t) = \arg \max_{s_i} \ln[p(R|S, H)]. \tag{7.48}$$

The first term of (7.47) is constant, so it is rendered irrelevant in the maximization process. Furthermore, the denominator of the second term is also unnecessary. The maximization of (7.48) turns into the following minimization problem:

$$\hat{s}_i(t) = \arg \min_{s_i} \sum_{t=1}^{l} \sum_{j=1}^{m} \left| r_j(t) - \sum_{i=1}^{n} h_{ij} s_i(t) \right|^2, \tag{7.49}$$

which can be easily achieved with the Viterbi algorithm, using the following metric to compute the ML path through the trellis:

$$m(r_j(t), s_i(t)|h_{ij}) = \sum_{t=1}^{l} \sum_{j=1}^{m} \left| r_j(t) - \sum_{i=1}^{n} h_{ij} s_i(t) \right|^2, \tag{7.50}$$

with $1 \leq i \leq n$, $1 \leq j \leq m$ and $1 \leq t \leq l$.

7.2.7 ST Codes Pairwise Error Probability

Consider that the code word s shown in (7.43) is initially transmitted but the decoder decodes erroneously in favour of the following decoded sequence e:

$$e = e_1(1)e_2(1) \ldots e_n(1)e_1(2)e_2(2) \ldots e_n(2) \ldots e_1(l)e_2(l) \ldots e_n(l). \tag{7.51}$$

The probability of making such an erroneous decision, namely the pairwise error probability (PEP), is given by [7]:

$$P(s \to e|h_{ij}) = P\{m(r_j(t), e_i(t)|h_{ij}) \leq m(r_j(t), s_i(t)|h_{ij})\}, \tag{7.52}$$

since the metric calculated for the erroneous sequence along the trellis must be less than the metric worked out for the correct sequence in order to make the incorrect decision.

By applying a Chernoff bound [21] to the above expression, it turns out that the PEP is bounded as follows [7]:

$$P(s \rightarrow e|h_{ij}) \leq E\{\exp(\mu[m(r_j(t), s_i(t)|h_{ij}) - m(r_j(t), e_i(t)|h_{ij})])\}, \quad (7.53)$$

where $E\{\cdot\}$ represents the expectation and μ is a nonnegative parameter that must be optimized in order to produce the tightest upper bound possible.

Replacing (7.50) into the above expression will yield:

$$P(s \rightarrow e|h_{ij}) \leq \prod_{t=1}^{l} \prod_{j=1}^{m} E\{\exp(\mu|r_j(t) - \sum_{i=1}^{n} h_{ij}s_i(t)|^2 - \mu|r_j(t) - \sum_{i=1}^{n} h_{ij}e_i(t)|^2)\}.$$
$$(7.54)$$

After lengthy mathematical simplifications, which are presented in full detail in [7] for the SISO channel (the only difference in this case is the different metric being used), it is possible to reduce inequality (7.54) to an upper bound dependent on the signal-to-noise ratio E_s/N_0 and the squared Euclidean distance between sequences s and e only [15]:

$$P(s \rightarrow e|h_{ij}) \leq \exp\left(-d^2(s, e)\frac{E_s}{4N_0}\right), \quad (7.55)$$

where $d^2(s, e)$ will from now on be called the space-time squared Euclidean distance (ST-SED) between code words s and e and is given by [15]:

$$d^2(s, e) = \sum_{j=1}^{m} \sum_{t=1}^{l} \left| \sum_{i=1}^{n} h_{ij}(s_i(t) - e_i(t)) \right|^2. \quad (7.56)$$

By defining $v_j = (h_{1j}, h_{2j}, \ldots, h_{nj})$ as the vector of associated channel gains of receive antenna j, (7.56) can be rewritten as follows:

$$d^2(s, e) = \sum_{j=1}^{m} \sum_{i=1}^{n} \sum_{\tilde{i}=1}^{n} h_{ij}\overline{h_{\tilde{i}j}} \sum_{t=1}^{l} (s_i(t) - e_i(t))\overline{(s_{\tilde{i}}(t) - e_{\tilde{i}}(t))}, \quad (7.57)$$

where $\overline{x}$ indicates the conjugate of complex number x. After some manipulation, the distance turns out to be:

$$d^2(s, e) = \sum_{j=1}^{m} v_j A(s, e)v'_j, \quad (7.58)$$

where $v\prime_j$ is the transpose conjugate of complex matrix v_j, the matrix $A(s, e)$ has dimension $n \times n$ and the element at row p and column q is given by:

$$A_{pq} = \sum_{t=1}^{l} (s_p(t) - e_p(t))\overline{(s_q(t) - e_q(t))}. \tag{7.59}$$

We have from (7.55):

$$P(s \rightarrow e|h_{ij}) \leq \prod_{j=1}^{m} \exp\left(-v_j A(s, e)v'_j \frac{E_s}{4N_0}\right). \tag{7.60}$$

Matrix A has some linear algebra properties [22] that are extremely helpful in this context and will allow a mathematical interpretation of this expression for the PEP. First, A is a Hermitian matrix because it is equal to its transpose conjugate matrix, that is to say:

$$A = A'. \tag{7.61}$$

Furthermore, there is a unitary matrix U whose rows are eigenvectors of A and constitute an orthonormal basis of C^n, the n-dimensional set of complex numbers. There also exists a nondiagonal matrix, D, whose elements $\lambda_i(i = 1, \ldots, n)$ are eigenvalues of A, counting multiples, and which verifies the following equality test:

$$UA(s, e)U' = D. \tag{7.62}$$

At last, the matrix A can be decomposed in the following form:

$$A(s, e) = B(s, e)B'(s, e), \tag{7.63}$$

where matrix $B(s,e)$ has dimension nxl, and is made by the symbol difference between the transmitted sequence s and the received code word e, as follows:

$$B(s, e) = \begin{bmatrix} e_1(1) - s_1(1) & e_1(2) - s_1(2) & \cdots & \cdots & e_1(l) - s_1(l) \\ e_2(1) - s_2(1) & e_2(2) - s_2(2) & \cdots & \cdots & e_2(l) - s_2(l) \\ e_3(1) - s_3(1) & e_3(2) - s_3(2) & \cdots & \cdots & e_3(l) - s_3(l) \\ \vdots & \vdots & \ddots & \ddots & \vdots \\ e_n(1) - s_n(1) & e_n(2) - s_n(2) & \cdots & \cdots & e_n(l) - s_n(l) \end{bmatrix} = E - S. \tag{7.64}$$

B is called the square root matrix of A. Another important linear algebra property is that any matrix with a square root is always nonnegative-definite, so in this context matrix A will verify the inequality:

$$x Ax' \geq 0, \quad \forall x \in C^n. \tag{7.65}$$

It also follows that the eigenvalues of a nonnegative definite Hermitian matrix are nonnegative, therefore resulting in A's eigenvalues λ_i being all nonnegative.

It is easy to see that the vectors vj as well as eigenvalues λ_i are time-dependant. Now let's define the vector $\beta_j = (\mu_{1j}, \mu_{2j}, \ldots, \mu_{nj}) = v_j U'$. We can see that μ_{ij} are also independent complex Gaussian random variables of variance 0.5 per dimension. Therefore, using (7.62):

$$v_j A(s, v)v'_j = v_j(U'DU)v'_j = \beta_j D\beta'_j$$

$$= \left(\mu_{ij}\, \mu_{2j} \cdots \mu_{nj}\right) \begin{pmatrix} \lambda_1 & 0 & \cdots & 0 \\ 0 & \lambda_2 & \cdots & 0 \\ \vdots & \vdots & \ddots & \vdots \\ 0 & 0 & \cdots & \lambda_n \end{pmatrix} \begin{pmatrix} \bar{\mu}_{1j} \\ \bar{\mu}_{2j} \\ \vdots \\ \bar{\mu}_{nj} \end{pmatrix} = \sum_{i=1}^{n} \lambda_i \left|\mu_{ij}\right|^2.$$

(7.66)

Thus the probability of error is:

$$P(s \rightarrow e|h_{ij}) \leq \prod_{j=1}^{m} \exp\left(-\frac{Es}{4N_0} \sum_{i=1}^{n} \lambda_i \left|\mu_{ij}\right|^2\right).$$

(7.67)

We now examine two different cases:

1. If the random variables μ_{ij} have nonzero mean, then their modulus $\left|\mu_{ij}\right|$ follows a Rician distribution and, by averaging inequality (7.67) with respect to these random variables μ_{ij}, we get the following upper bound on the probability of error:

$$P(s \rightarrow e|h_{ij}) \leq \prod_{j=1}^{m} \left(\prod_{i=1}^{n} \frac{1}{1 + \frac{E_s}{4N_0}\lambda_i} \exp\left(\frac{K_{ij}\frac{E_s}{4N_0}\lambda_i}{1 + \frac{E_s}{4N_0}\lambda_i} - \frac{Es}{4N_0} \sum_{i=1}^{n} \lambda_i \left|\mu_{ij}\right|^2 \right) \right),$$

(7.68)

where K_{ij} is the Rician factor of $\left|\mu_{ij}\right|$. When taking the mean $E(\cdot)$ of inequality (7.67), it is necessary to work out the 'moment-generating function' of the random variable $\left|\mu_{ij}\right|^2$ which has a noncentral chi-square distribution.

2. If the random variables h_{ij} have zero mean then the random variables μ_{ij} also have zero mean. This is the case of independent Rayleigh fading and $K_{ij} = 0$ for all i and j. Hence, by replacing in 7.68, we now get:

$$P(s \rightarrow e|h_{ij}) \leq \left(\prod_{i=1}^{n} \frac{1}{1 + \frac{E_s}{4N_0}\lambda_i} \right)^{m}.$$

(7.69)

Let ρ be the rank of the matrix A; it follows that exactly $n - \rho$ eigenvalues of A are zero. Considering that the nonzero eigenvalues of A are $\lambda_1, \lambda_2, \ldots, \lambda_\rho$ and assuming the signal-to-noise ratio E_s/N_0 is high:

$$P(s \rightarrow e|h_{ij}) \leq \left(\prod_{i=1} \lambda_i \right)^{-m} \left(\frac{E_s}{N_0} \right)^{-\rho m}. \tag{7.70}$$

Let us define the diversity advantage of an ST code as the power of the SNR in the denominator of the expression given for the PEP bound. So a diversity advantage of ρm is achieved in the above bound. Let us also define the coding gain of an ST code as the approximate measure of the gain over an uncoded system that bears the same diversity advantage. We can see that a coding gain of $(\lambda_1 \lambda_2 \ldots \lambda_\rho)^{1/\rho}$ is obtained by the multi-antenna system.

For the first case where Rician fading is present, the upper bound for the pairwise error probability is very similar to the above, except for a correction factor in the coding gain that depends on K_{ij}. The same diversity advantage is achieved. So, for high SNR values in Rician channels, 7.68 is now reduced to:

$$P(s \rightarrow e|h_{ij}) \leq \left(\prod_{i=1} \lambda_i \right)^{-m} \left(\frac{E_s}{N_0} \right)^{-\rho m} \left(\prod_{j=1}^{m} \prod_{i=1}^{\rho} e^{-K_{ij}} \right). \tag{7.71}$$

Both matrices A and B have equal rank ρ, so there is no need to calculate matrix A; it is enough to know the code word-difference matrix B in order to get ρ and hence the diversity advantage.

7.2.8 ST Code Design Criteria for Slow Fading Channels

The code design criteria for slow Rayleigh fading channels can be summarized as follows:

- Rank Criterion: In order to achieve the maximum diversity advantage nm, the matrix $B(s, e)$ must have full rank for any code words s and e. This will guarantee a tighter upper bound on the PEP of inequality (7.70).
- Product Criterion: If the system has a diversity advantage of ρm, the minimum of the ρth roots of the product of the product of the nonzero eigenvalues of $A(s, e)$ taken over all pairs of different code words s and e must be maximized. That is to say, the target is to maximize the coding gain G, defined as:

$$G = \min_{\substack{\forall s,e \\ s \neq e}} \left(\sqrt[\rho]{\prod_{i=1}^{\rho} \lambda_i} \right). \tag{7.72}$$

These criteria are valid only when full channel state information is available at the decoder. They will assure the tightest upper bound on the PEP possible. The product criterion is often called the determinant criterion.

In slow Rician fading channels the criteria are the same, with the only difference being that the coding advantage to be maximized according to the product criterion is (from (7.70)):

$$
G = \min_{\substack{\forall s,e \\ s \neq e}} \left(\sqrt[\rho]{\prod_{i=1}^{\rho} \lambda_i} \right) \left(\prod_{j=1}^{m} \prod_{j=1}^{\rho} e^{-K_{ij}} \right)^{1/\rho m}. \tag{7.73}
$$

These criteria are only intended to be a guideline for ST code design. They are not laws, are based only on upper bounds, and tighter bounds may be found later on. Besides, the criteria work on the basis of making the bound as tight as possible, which only happens when a given transmitted sequence s is confused with a decoded sequence e; it could be that this occurrence seldom happens in practice and it might be necessary to study the PEP bounds of sequence pairs (s, e) that are actually more likely to be produced.

7.2.9 ST Code Design Criteria for Fast Fading Channels

The code design criteria for time-varying Rayleigh fading channels are similar to those shown for slow fading and are described as follows:

- Rank Criterion: Identical to criterion for slow Rayleigh fading channels.
  Product Criterion: Slightly changes in the case of rapid fading. Here, in order to achieve the highest coding gain possible, the minimum of the products:

$$
\prod_{t \in \Theta(s,e)} \left(\sum_{i=1}^{n} |e_i(t) - s_i(t)|^2 \right), \quad 1 \leq t \leq l
$$

taken over all distinct code words s and e must be maximized, where $\Theta(s, e)$ is the set of time instances in which the code words s and e differ in at least one symbol. If the system has a diversity advantage of ρm, the coding gain G is:

$$
G = \min_{\substack{\forall s,e \\ s \neq e}} \prod_{t \in \Theta(s,e)} \left(\sum_{i=1}^{n} |e_i(t) - s_i(t)|^2 \right)^{1/\rho}. \tag{7.74}
$$

7.2.10 Space-Time Ring-TCM Codes for QAM in Fading Channels

Space-Time Coded Modulation (STCM) has been implemented with a wide variety of modulation schemes. Most technical journals/books available nowadays only describe the case where binary STCM uses phase shift keying (PSK) modulation, especially QPSK and 8-PSK. In this book we consider non-binary coding (ring-codes) combined with quadrature amplitude modulation. We describe techniques to search for good ST codes for 16-QAM for slow fading channels. The search was developed according to a set of extended designed rules proposed in [7]. We assume QAM modulation with sizes of 16 and 64 waveforms. Naguib et al. [23] employed the signal mapping shown in Figure 7.13a for 16-QAM.

However, there are other mappings that yield increased Euclidean distances for ST codes, as they provide higher values for the minimum determinant of the differential code matrix taken over all pairs of possible code words. This improved 16-QAM signal mapping is shown in Figure 7.13b and was originally proposed by Carrasco and Farrell [13]. Figure 7.13c presents their proposed mapping for 64-QAM. The improved mappings were designed by taking into account the isomorphism produced by Z-numbers when considering phase rotations.

In Figure 7.13 every signal point is composed of a pair of $\mathbb{Z}$-symbols, where the leftmost symbol is the same for the four constituent signals of a given quadrant. Within all quadrants the rightmost $\mathbb{Z}$-number follows the same isomorphism. The count goes up in one unit clockwise, starting at the upper-left corner point. The design criteria for ST codes have been extended to refine computer code search and make it more accurate. This complementary set of rules takes into account the calculation of other parameters that are relevant in ST code performance. These parameters are the average number of competing paths with minimum determinant $\Delta_{\min}$ associated with a given path in the trellis N_D, and the average determinant $\bar{\Delta}$ for all pairs of code words. The design rules now have two new extra criteria. The four rules are presented in hierarchical structure and are practical for use in code searches:

1. The minimum rank must be maximized. Ideally the code must have full rank n, which is the number of transmitters. ST codes that do not possess full rank are therefore discarded.

2. The codes selected are the ones that provide the highest minimum determinant $\Delta_{\min}$ of the matrix $A(s, e)$ taken over all pairs of distinct code words, where the matrix $A(s, e)$ has dimension $n \times n$ and contains the symbol differences between the transmitted sequence s and the received code e. The element at row p and column q is given by [21]:

$$A_{pq} = \sum_{i=1}^{l} (s_p(t) - e_p(t)) \overline{(s_q(t) - e_q(t))}, \tag{7.75}$$

where ($\bullet$) represents complex conjugation. A hierarchical code list can thus be made, ordering them according to $\Delta_{\min}$ values.

3. In the case that two or more codes have the same $\Delta_{\min}$, the best code is the one with the smallest N_Δ, so codes are prioritized based on N_Δ.

Table 7.6 Optimum ST-RTCM codes of modulo-16 for 16-QAM.

Code	States	Δ_{min}	N_Δ	$\overline{\Delta}$
8 9/8	16	4	1.625	61.3
0 9/0	16	4	1.625	27.2
0 9/8	16	4	1.625	27.2
8 9/0	16	4	1.625	27.2
8 3/8	16	1	0.094	62.4
7 12/3	16	1	0.316	51.7
Tarokh *et al.*	16	1	1.25	27.2

4. If any two good codes still have identical Δ_{min} and N_Δ, the one with the highest average $\overline{\Delta}$ is finally chosen.

An exhaustive *a priori* search was conducted among all 16-state ST-RTCM codes to find those that best satisfy these rules. Table 7.6 shows the best codes found with their respective Δ_{min}, N_Δ and $\overline{\Delta}$; they are ordered from best at the top of the table to worst at the bottom. All codes shown are of full rank, therefore yielding the maximum diversity advantage nm, equal to four. All codes are written in the form $g_1 g_2/f_1$, where g_1 and g_2 are the feed-forward coefficients and f_1 is the feedback coefficient.

Note that codes 0 9/0, 0 9/8 and 8 9/0 have the same parameter values and must have identical performance as well. This is an indication of the existence of multiple optimum codes. They are the best codes, together with 8 9/8, because their minimum determinant is four. No other codes can achieve this, all having a minimum determinant of one. In spite of it not being an optimum code from the determinant criterion point of view, the code 7 12/3 has been included for performance comparison with Naguib et al. [23] and optimum ST-RTCM codes. It has a smaller N_Δ than Naguib's code and a higher average determinant.

Figure 7.23 presents the simulation results for Naguib et al.'s code and the ST-RTCM 8 9/8 and 7 12/3 codes for values of m equal to 2 and 4. The uncoded single-channel 16-QAM curve is also indicated. For two receive antennas and a BER of 10^{-3}, the 8 9/8 code has a coding gain of 1.8 dB in relation to Naguib et al.'s, while keeping a gain of approximately 1.2 dB when compared with the 7 12/3 code. With four receive antennas, the 8 9/8 now has a coding gain of 4 dB in relation to Naguib's at 10^{-4} BER. This proves that the higher Δ_{min} is for a particular code, the higher the coding advantage will be when m increases. Another conclusion drawn from the graph is that 7 12/3 performance is better than Naguib et al.'s, accounting for a 3.3 dB margin at SNRs greater than 20 dB in a system with two transmitters and two receivers.

Both codes have full rank and minimum determinant, however 7 12/3 performs better due to its smaller N_D and bigger $\overline{\Delta}$, originating from the different code-to-signal mapping used in the modulation of ST-RTCM codes. The ST-RTCM 8 9/8 is the one with the best performance as it fully complies with the extended criteria: it has the maximum Δ_{min} of four and higher determinant average than the other three

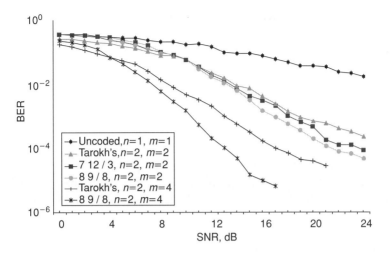

Figure 7.23 Error performance of ST-RTCM codes with 16-QAM over slow Rayleigh fading channels, 130 symbols per frame, 4 bits/s/Hz.

codes that also present the highest Δ_{min}. When more receiving antennas are employed, Figure 7.23 shows a lower probability of error for a given high SNR value due to the increased diversity. Therefore the respective curve sinks sooner, with a pronounced slope.

Figure 7.24 shows that the ST-RTCM code 213 132/3 with $n = m = 2$ performs on the Gaussian bound until an SNR of 18 dB. It outperforms code 25 12/47 by a

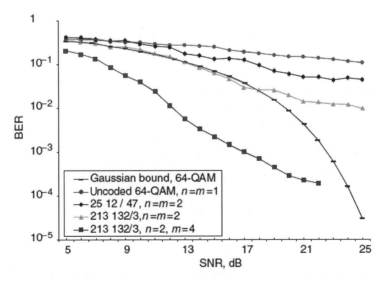

Figure 7.24 BER performance of ST-RTCM codes with QAM over slow Rayleigh fading channels, 65 symbols per frame, 6 bits/Hz, 25 12/47, 64-state, 64-QAM, 213 132/3, 4-state, 16-QAM.

Table 7.7   Comparison of ST codes with 16-QAM, 4 bit/s/Hz.

Code	SNR (dB)
Delay diversity, $m = 2$	26
ST-RTCM 7 12/3, $m = 2$	23.6
ST-RTCM 8 9/8, $m = 2$	21.4
Delay diversity, $m = 4$	17.2
ST-RTCM 8 9/8, $m = 4$	13.2

margin of 7 dB. This is due to the high coding gain provided by the former, being an optimized code; the latter is not an optimum code. The 4-state code is not only superior in performance but also has a lower complexity than the 64-state code. There must be 64-QAM ST-RTCM codes of better performance than 25 12/47, but the computer search is quite long due to the large number of different codes and is not of practical interest since these 64-state codes are much more complex than 213 132/3.

Finally, 213 132/3 performance is also presented in a four-receiver configuration, which is even better due to higher diversity. It is 11.5 dB better off than the same code used with $m = 2$.

Table 7.7 gives a brief summary of the simulation results obtained for ST-ring TCM codes with QAM modulation over slow Rayleigh fading channels. Table 7.7 compares different ST codes that have a bandwidth efficiency of 4 bit/s/Hz, providing SNR values that reach a BER of 10^{-4} in 16-QAM simulations.

7.3 Conclusions

In this chapter we have introduced the concepts of convolutional coding and trellis coded modulation. We have shown how TCM codes are not only bandwidth efficient but also achieve good performance. The idea of TCM has been extended for symbols defined over rings of integers, which outperform binary TCM codes with only a small increase in decoding complexity. The criteria for constructing good ring-TCM codes have been given and a genetic algorithm has been presented as a search method for finding good codes. We have shown how Ring-TCM codes are suitable for combining with spatial-temporal diversity, resulting in space-time ring-TCM codes. The performance of ST-RTCM codes were evaluated with simulation results over urban mobile radio channels, and these performed well in the slow fading indoor and pedestrian environments. Finally, the design criteria were presented for constructing good ST-RTCM codes and the performance of some new ST-RTCM codes were evaluated with simulation results for QAM on slow fading channels.

References

[1] Elias, P. (1955) *Coding for Noise Channels*, IRE Convention Report, Part 4, pp. 37–47.
[2] Moon, T.K. (2005) *Error Correction Coding. Mathematical Methods and Algorithms*, Wiley Inter-science, ISBN 0-471-64800-0.

[3] Viterbi, A.J. (1967) Error bounds for convolutional codes and asymptotically optimum decoding algorithm. *IEEE Transactions on Information Theory*, **13**, 260–9.

[4] Forney, G.D. (1973) The Viterbi algorithm. *Proceedings of IEEE*, **61**, 268–78.

[5] Massey, J.L. (1974) Coding and modulation in digital communications. Presented at International Zurich Seminar on Digital Communications, Zurich, Switzerland.

[6] Ungerboeck, G. (1982) Channel coding with multilevel/phase signals. *Information Theory, IEEE Transactions*, **28**, 55–67.

[7] Biglieri, E., Divisar, D., Melane, P.J. and Simon, M.K. (1991) *Introduction to Trellis Coded Modulation with Application*, MacMillan, New York.

[8] Massey, J.L. and Mittelholzer, T. (1984) Codes over rings – a practical necessity. Presented at AAECC7 International Conference, Universite' P. Sabatier, Toulouse, France.

[9] Massey, J. and Mittelholzer, T. (1989) Convolutional codes over rings. Proceedings of the 4th Joint Swedish Soviet Workshop Information Theory, pp. 14–8.

[10] Baldini, R.F. and Farrell, P.G. (1990) Coded modulation wit convolutional codes over rings. Presented at Second IEE Bangor Symposium on Communications, Bangor, Wales.

[11] Baldini, R.F. and Farrell, P. (1994) Coded modulation based on rings of integers modulo-q part 2: convolutional codes. *IEE Proceedings: Communications*, **141**, 137–42.

[12] Carrasco, R., Lopez, F. and Farrell, P. (1996) Ring-TCM for M-PSK modulation: AWGN channels and DSP implementation. *Communications, IEE Proceedings*, **143**, 273–80.

[13] Carrasco, R. and Farrell, P. (1996) Ring-TCM for fixed and fading channels: land-mobile satellite fading channels with QAM. *Communications, IEE Proceedings*, **143**, 281–8.

[14] Lopez, F.J., Carrasco, R.A. and Farrell, P.G. (1992) Ring-TCM Codes over QAM. *IEE Electronics Letter*, **28**, 2358–9.

[15] Tarokh, V., Seshadri, N. and Calderbank, A. (1998) Space-time codes for high data rate wireless communication: performance criterion and code construction. *Information Theory, IEEE Transactions on*, **44**, 744–65.

[16] Benedetto, S., Mondin, M. and Montorsi, G. (1994) Performance evaluation of trellis-coded modulation schemes. *Proceedings of the IEEE*, **82**, 833–55.

[17] Goldberg, D. (1989) *Genetic Algorithms in Search, Optimization and Machine Learning*, Addison-Wesley Longman Publishing Co., Inc., Boston, MA, USA.

[18] Soto, I. and Carrasco, A. (1997) Searching for TCM codes using genetic algorithms. *Communications, IEE Proceedings*, **144**, 6–10.

[19] Pereira, A. and Carrasco, R. (2001) Space-time ring TCM codes for QPSK on time-varying fast fading channels. *Electronics Letters*, **37**, 961–2.

[20] Carrasco, R. and Pereira, A. (2004) Space-time ring-TCM codes for QAM over fading channels. *IEE Proceedings Communications*, **151** (4), 316–21.

[21] Pereira, A. (2003) *Space-Time Ring TCM Codes on Fading Channels*. PhD Thesis, Staffordshire University, Staffordshire, UK.

[22] Horn, R. and Johnson, C. (1990) *Matrix Analysis*, Cambridge University Press, Cambridge, UK, ISBN 0-521-38632-2.

[23] Naguib, A., Tarokh, V., Seshadri, N. and Calderbank, A. (1998) A space-time coding modem for high-data-rate wireless communications. *Selected Areas in Communications, IEEE Journal*, **16**, 1459–78.

8

Non-Binary Turbo Codes

8.1 Introduction

One of the most important breakthroughs in coding theory was the development of *turbo codes* by Berrou, Glavieux and Thitimajshima [1] in 1993. Turbo codes are a parallel concatenation of recursive systematic convolutional codes separated by an interleaver. They provide a practical way of achieving near-Shannon limit performance by using an iterative decoder that contains two soft-input–soft-output component decoders in series, passing reliability information between them. Originally, the component decoders used the *BCJR* algorithm [2, 3] or *MAP* algorithm, but this was considered too complex for practical applications. A reduction in the complexity of the MAP algorithm with only a small degradation in performance was achieved by working in the logarithmic domain, resulting in the *log MAP* algorithm. Simpler trellis decoding algorithms also include the *max-log MAP* algorithm and the *Soft Output Viterbi Algorithm* (SOVA). Both algorithms perform the same as one another, but do not perform as well as the log MAP algorithm. Turbo codes are used for error-correction in 3G mobile communications and are now also used for deep-space communications.

In this chapter, the binary turbo encoding and decoding processes are introduced, with detailed descriptions of log-likelihood ratios, the MAP algorithm and the iterative log MAP algorithm. This is followed by an introduction to non-binary turbo coding, which is an area of research that has received very little attention, with only a few papers published. Finally, we complete this chapter with an explanation of the non-binary turbo encoder and decoder structure.

8.2 The Turbo Encoder

A turbo code is a parallel concatenation of two recursive systematic convolutional (RSC) codes separated by an interleaver, denoted by Π. A general turbo encoder is shown in Figure 8.1.

Non-Binary Error Control Coding for Wireless Communication and Data Storage   Rolando Antonio Carrasco and Martin Johnston
© 2008 John Wiley & Sons, Ltd

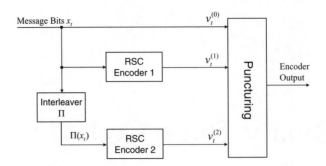

Figure 8.1 Turbo encoder.

The two encoders have code rates R_1 and R_2 respectively, and the overall code rate R of the turbo encoder is given by:

$$R = \frac{R_1 R_2}{R_1 + R_2 - R_1 R_2}.$$ (8.1)

Usually the code rates of both encoders are the same and a parallel concatenation of codes reduces the overall code rate. It is common to choose an RSC with code rate 1/2 as the component encoders result in an overall code rate of 1/3. We are not limited to convolutional codes but could also choose block codes as the component encoders, known as block turbo codes [2].

The upper encoder receives the data directly, while the lower encoder receives the data after it has been interleaved by a permutation function, Π. The interleaver is in general a pseudo random interleaver, that is it rearranges bits according to a prescribed but randomly-generated rule. The interleaver operates in a block-wise fashion, interleaving L bits at a time, and thus turbo codes can be viewed as block codes. Since both encoders are systematic, the systematic output of the top encoder is sent, while the systematic output of the lower encoder is not sent. However, the parity bits of both encoders are transmitted. The overall code rate can be made higher by puncturing, which operates only on the parity sequences – the systematic bits are not punctured. The code rate of a rate 1/3 turbo code is typically increased to 1/2 by only transmitting the odd indexed parity bits from the upper encoder and the even indexed parity bits from the lower encoder.

8.2.1 Notation

In this chapter, we use the following notation for the turbo encoding and decoding processes:

- $v_t^{(0)} \in \{0, 1\}$ is a turbo encoder output corresponding to an information bit at time t.
- $v_t^{(1)} \in \{0, 1\}$ is a parity bit from the first component RSC encoder.

- $v_t^{(2)} \in \{0, 1\}$ is a parity bit from the second component RSC encoder.
- $a_t^{(i)} \in \{-1, +1\}, i = 0, 1, 2$, is the mapping of $v_t^{(i)}$ to the BPSK constellation, where $a_t^{(i)} = 2v_t^{(i)} - 1$.
- $v^{(0,s',s)} \in \{0, 1\}$ is the information bit corresponding to a transition from an initial state s' to the next state s.
- $v^{(1,s',s)} \in \{0, 1\}$ is the parity bit from the first RSC encoder corresponding to a transition from state s' to state s.
- $v^{(2,s',s)} \in \{0, 1\}$ is the parity bit from the second RSC encoder corresponding to a transition from state s' to state s.
- $a^{(i,s',s)} \in \{-1, +1\}$ is equal to $2v^{(i,s',s)} - 1, i = 0, 1, 2$.
- $r_t^{(0)}$ is the received information bit, where $r_t^{(0)} = a_t^{(0)} + \eta_t$ and η_t is an additive white Gaussian noise sample at time t.
- $r_t^{(1)}$ is the received parity bit from the first RSC encoder, where $r_t^{(1)} = a_t^{(1)} + \eta_t$.
- $r_t^{(2)}$ is the received information bit from the second RSC encoder, where $r_t^{(2)} = a_t^{(2)} + \eta_t$.

Example 8.1: Binary turbo encoding: Figure 8.2 shows a turbo encoder with (1, 5/7) RSC component codes with a pseudo random interleaver of length $N = 5$ bits. Therefore, the turbo encoder receives a 5 bit message x_t, which is encoded by the first component encoder to obtain a 5 bit parity output $v_t^{(1)}$. The message is also interleaved $\Pi(x_t)$ and encoded by the second component encoder to obtain the 5 bit parity output $v_t^{(2)}$. The turbo encoder output is therefore 15 bits long.

Let the message be:

$$x_t = [x_1 x_2, x_3, x_4, x_5] = [1, 0, 1, 0, 1].$$

The state table for the first component RSC encoder is given in Table 8.1.

The input sequence runs through the first RSC encoder with transfer function $G(D)$, resulting in a parity sequence:

$$v_t^{(1)} = \left[v_1^{(1)}, v_2^{(1)}, v_3^{(1)}, v_4^{(1)}, v_5^{(1)} \right] = [1, 1, 0, 1, 1].$$

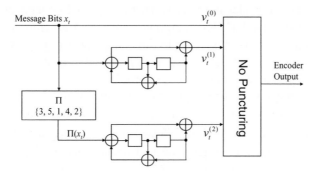

Figure 8.2 Turbo encoder with (1, 5/7) RSC component codes.

Table 8.1   State table for the first component RSC encoder

x_t	s'	s	$v_t^{(0,s',s)}$	$v_t^{(1,s',s)}$	$a_t^{(0,s',s)}$	$a_t^{(1,s',s)}$
0	00	00	0	0	-1	-1
1	00	10	1	1	$+1$	$+1$
0	01	10	0	1	-1	$+1$
1	01	00	1	0	$+1$	-1
0	10	11	0	1	-1	$+1$
1	10	01	1	0	$+1$	-1
0	11	01	0	0	-1	-1
1	11	11	1	1	$+1$	$+1$

The sequence x is also passed through an interleaver or permuter of length $N = 5$, denoted by Π, which produces the permuted output sequence $x' = \Pi(x) = [1, 1, 1, 0, 0]$. The sequence x' is passed through another convolutional encoder with transfer function $G(D)$, which produces the output sequence:

$$v_t^{(2)} = \left[v_1^{(2)}, v_2^{(2)}, v_3^{(2)}, v_4^{(2)}, v_5^{(2)}\right] = [1, 0, 1, 1, 1].$$

The three output sequences are multiplexed together to form the output sequence:

$$\mathbf{v}_t = \left[\left(v_1^{(0)}, v_1^{(1)}, v_1^{(2)}\right), \left(v_2^{(0)}, v_2^{(1)}, v_2^{(2)}\right), \ldots, \left(v_5^{(1)}x, v_5^{(1)}, v_5^{(2)}\right)\right]$$
$$= [(1, 1, 1), (0, 1, 0), (1, 0, 1), (0, 1, 1), (1, 1, 1)]$$

Assuming BPSK modulation with $E_s = 1$, the mapped output $\mathbf{a}_t$ will be:

$$\mathbf{a}_t = [(+1, +1, +1), (-1, +1, -1), (+1, -1, +1), (-1, +1, +1), (+1, +1, +1)].$$

8.3 The Turbo Decoder

The near-Shannon limit performance of turbo codes is accomplished by their iterative decoding algorithm. A general block diagram of the turbo decoder is shown in Figure 8.3. The idea of the turbo decoding process is to extract extrinsic information from the output of one decoder and pass it on to the second decoder in order to improve the reliability of the second decoder's output. Extrinsic information is then extracted from the second decoder and passed on to the first decoder to improve the reliability of first decoder's output. This process is then repeated until no further gains in the performance of the turbo decoder can be achieved.

The turbo decoder takes as its input the reliability values of the systematic information, $r_t^{(0)}$, the parity bits from encoder 1, $r_t^{(1)}$, and the interleaved parity bits from encoder 2, $r_t^{(2)}$. Turbo decoding is a two-stage process consisting of two soft-input–soft-output decoders in series. The first decoder takes $r_t^{(0)}$, $r_t^{(1)}$ and *prior* information from decoder 2, which initially is zero. The output of decoder 1 is the reliability of the systematic information. The prior information from decoder 2 and

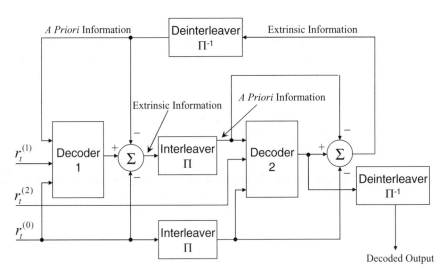

Figure 8.3 Turbo decoder.

the reliability values of the systematic information are subtracted from the decoder output, leaving *extrinsic* information, which is interleaved and becomes the prior information for decoder 2.

Decoder 2 now takes the interleaved systematic information, the reliability values of the interleaved parity bits $r_t^{(2)}$ and the prior information from decoder 1. The output is the reliability of the interleaved systematic information. The prior information and the reliability values of the interleaved systematic information are subtracted from the decoder output, resulting in extrinsic information, which is now deinterleaved to become once again the prior information for decoder 1. This is called a decoder iteration, where after each information the prior information to both decoders becomes more reliable, thus improving the reliabilities of the decoder outputs. After each iteration, the improvement in performance becomes less until it converges and no further improvements can be achieved. Alternatively, instead of feeding back the extrinsic information to decoder 1, the output of decoder 2 can be deinterleaved to give the reliability of $r_t^{(0)}$. A hard decision can then be made to recover the systematic information.

8.3.1 Log-Likelihood Ratios (LLRs)

One measure of the reliability of a variable is to calculate its log-likelihood ratio (LLR). For some variable u, its LLR $L(u)$ is defined as:

$$L(u) = \ln\left(\frac{P(u = +1)}{P(u = -1)}\right), \tag{8.2}$$

where $P(u = +1) = 1 - P(u = -1)$. We can see that if $P(u = +1) = P(u = -1) = 0.5$ then $L(u) = 0$, that is the reliability of u is zero. As stated previously, the turbo decoder takes as its inputs the reliabilities of the systematic information and parity bits from encoders 1 and 2. If we define the received systematic information bit at time t as $r_t^{(0)}$ then:

$$L\left(r_t^{(0)} \mid a_t^{(0)}\right) = \ln\left(\frac{P\left(r_t^{(0)} \mid a_t^{(0)} = +1\right)}{P\left(r_t^{(0)} \mid a_t^{(0)} = -1\right)}\right). \tag{8.3}$$

For the AWGN channel and BPSK modulation with constellation points at $\pm\sqrt{E_s}$:

$$P\left(r_t^{(0)} \mid a_t^{(0)} = +1\right) = \frac{\frac{1}{\sigma\sqrt{2\pi}} e^{-\left(r_t^{(0)} - \sqrt{E_s}\right)^2 / 2\sigma^2}}{\frac{1}{\sigma\sqrt{2\pi}} e^{-\left(r_t^{(0)} + \sqrt{E_s}\right)^2 / 2\sigma^2} + \frac{1}{\sigma\sqrt{2\pi}} e^{-\left(r_t^{(0)} - \sqrt{E_s}\right)^2 / 2\sigma^2}}$$

$$P\left(r_t^{(0)} \mid a_t^{(0)} = -1\right) = \frac{\frac{1}{\sigma\sqrt{2\pi}} e^{-\left(r_t^{(0)} + \sqrt{E_s}\right)^2 / 2\sigma^2}}{\frac{1}{\sigma\sqrt{2\pi}} e^{-\left(r_t^{(0)} + \sqrt{E_s}\right)^2 / 2\sigma^2} + \frac{1}{\sigma\sqrt{2\pi}} e^{-\left(r_t^{(0)} - \sqrt{E_s}\right)^2 / 2\sigma^2}}.$$

Therefore:

$$\ln\left(\frac{P\left(r_t^{(0)} \mid a_t^{(0)} = +1\right)}{P\left(r_t^{(0)} \mid a_t^{(0)} = -1\right)}\right) = \ln\left(\frac{e^{-\left(r_t^{(0)} - \sqrt{E_s}\right)^2 / 2\sigma^2}}{e^{-\left(r_t^{(0)} + \sqrt{E_s}\right)^2 / 2\sigma^2}}\right) = \ln\left(e^{-\left(r_t^{(0)} - \sqrt{E_s}\right)^2 / 2\sigma^2}\right)$$

$$- \ln\left(e^{-\left(r_t^{(0)} + \sqrt{E_s}\right)^2 / 2\sigma^2}\right)$$

$$= \frac{-\left(r_t^{(0)} - \sqrt{E_s}\right)^2}{2\sigma^2} - \frac{-\left(r_t^{(0)} + \sqrt{E_s}\right)^2}{2\sigma^2}$$

$$= \frac{1}{2\sigma^2}\left(-\left(r_t^0\right)^2 + 2\sqrt{E_s}r_t^{(0)} - 1 + \left(r_t^{(0)}\right)^2 + 2\sqrt{E_s}r_t^{(0)} + 1\right)$$

$$= \frac{2\sqrt{E_s}r_t^{(0)}}{\sigma^2}.$$

The term $\frac{2\sqrt{E_s}}{\sigma^2}$ is called the *channel reliability* and denoted by L_c. Therefore:

$$L\left(r_t^{(0)} \mid a_t^{(0)}\right) = L_c r_t^{(0)} = \frac{2\sqrt{E_s}r_t^{(0)}}{\sigma^2}. \tag{8.4}$$

The output of decoder 1 is defined as:

$$L\left(a_t^{(0)} \mid \mathbf{r}\right) = \ln\left(\frac{P\left(a_t^{(0)} = +1 \mid \mathbf{r}\right)}{P\left(a_t^{(0)} = -1 \mid \mathbf{r}\right)}\right), \tag{8.5}$$

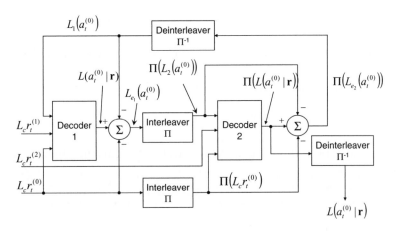

Figure 8.4   A more detailed diagram of the turbo decoder.

where **r** is the received vector. Finally, the prior information into decoder 1 is the deinterleaved extrinsic information from decoder 2.

$$L_1\big(a_t^{(0)}\big) = L_{e_2}\big(\Pi\big(a_t^{(0)}\big)\big) = \ln\left(\frac{P\big(a_t^{(0)} = +1\big)}{P\big(a_t^{(0)} = -1\big)}\right). \tag{8.6}$$

Therefore, the extrinsic information from decoder 1 is:

$$L_{e_1}\big(a_t^{(0)}\big) = L_1\big(a_t^{(0)}\,|\,r\big) - L_1\big(a_t^{(0)}\big) - L_c r_t^{(0)}. \tag{8.7}$$

Similarly, the extrinsic information from decoder 2 is:

$$\Pi\big(L_{e_2}\big(a_t^{(0)}\big)\big) = \Pi\big(L_2\big(a_t^{(0)}\,|\,\mathbf{r}\big)\big) - \Pi\big(L_2\big(a_t^{(0)}\big)\big) - \Pi\big(L_c r_t^{(0)}\big). \tag{8.8}$$

A more detailed block diagram of the turbo decoder showing the different LLRs is given in Figure 8.4.

When no more iterations are required, a hard decision is made on $L(a_t^{(0)}\,|\,\mathbf{r})$ from the deinterleaved output of decoder 2, depending on the *sign* of the LLR.

$$\begin{aligned}\text{If}\quad &\text{sign}\big(L\big(a_t^{(0)}\,|\,r\big)\big) < 0, \quad a_t^{(0)} = -1 \quad \therefore v_t^{(0)} = 0\\\text{If}\quad &\text{sign}\big(L\big(a_t^{(0)}\,|\,r\big)\big) \geq 0, \quad a_t^{(0)} = +1 \quad \therefore v_t^{(0)} = 1\end{aligned}. \tag{8.9}$$

8.3.2 *Maximum* A Posteriori *(MAP) Decoding*

The BCJR or MAP algorithm was originally presented coincidentally in [2] and [5], and later in [3]. It is a modified version of this that is used by each of the component decoders of the MAP turbo decoder. This algorithm is used to perform the symbol-by-symbol MAP decoding referred to earlier. The turbo decoder's decision is whether $\hat{v}_t^{(0)} = 1$ or 0.

It is presumed that $\hat{v}_t^{(0)} = 1$ if $P(a_t^{(0)} = +1 \mid \mathbf{r}) > P(a_t^{(0)} = -1 \mid \mathbf{r})$ and that $\hat{v}_t^{(0)} = 0$ if $P(a_t^{(0)} = -1 \mid \mathbf{r}) > P(a_t^{(0)} = +1 \mid \mathbf{r})$.

Put more simply, it can be said that the decision $\hat{v}_t^{(0)}$ is given by:

$$\hat{v}_t^{(0)} = \text{sign}\left[L\left(a_t^{(0)} \mid \mathbf{r}\right)\right]. \tag{8.10}$$

The state transitions within the encoder's trellis are governed by the transitional probabilities:

$$P(s, s') = P(s_{t+1} = s \mid s_t = s'), \quad 0 \le s, s' \le M_s - 1. \tag{8.11}$$

The posterior probabilities can be defined as:

$$P\left(a_t^{(0)} = -1 \mid \mathbf{r}\right) = \sum_{(s',s)\in S^-} P(s_t = s', s_{t+1} = s \mid \mathbf{r})$$

$$P\left(a_t^{(0)} = +1 \mid \mathbf{r}\right) = \sum_{(s',s)\in S^+} P(s_t = s', s_{t+1} = s \mid \mathbf{r}),$$

which can be modified using Bayes' theorem to:

$$P\left(a_t^{(0)} = -1 \mid \mathbf{r}\right) = \sum_{s^-} \frac{P(s_t = s', s_{t+1} = s, \mathbf{r})}{P(\mathbf{r})}$$

$$P\left(a_t^{(0)} = +1 \mid \mathbf{r}\right) = \sum_{s^+} \frac{P(s_t = s', s_{t+1} = s, \mathbf{r})}{P(\mathbf{r})}, \tag{8.12}$$

where $s_t \in S$ is the state of the encoder at time t, S^+ is the set of ordered pairs (s', s) corresponding to all state transitions $(s_t = s') \to (s_{t+1} = s)$ brought about by the data input $v_t^{(0)} = 1$, and S^- is defined in the same way for $v_t^{(0)} = 0$. The received noisy code word is expressed as $\mathbf{r} = (r_1, r_2, \ldots, r_N)$. Therefore, (8.5) can be rewritten as:

$$L\left(a_t^{(0)} \mid \mathbf{r}\right) = \ln\left(\frac{\sum_{(s',s)\in S^+} (P(s_t = s', s_{t+1} = s, \mathbf{r}))}{\sum_{(s',s)\in S^{--}} (P(s_t = s', s_{t+1} = s, \mathbf{r}))}\right). \tag{8.13}$$

Consequently, the joint probability:

$$P(s, s', \mathbf{r}) = P(s_{t+1} = s, s_t = s', \mathbf{r}), \quad 0 \le s, s' \le M_s - 1. \tag{8.14}$$

(8.13) can then be rewritten in the following form:

$$P(s_t = s', s_{t+1} = s, \mathbf{r}) = P(s_t = s', s_{t+1} = s, \mathbf{r}_{<t}, \mathbf{r}_t, \mathbf{r}_{>t}). \tag{8.15}$$

Applying Bayes' theorem, this becomes:

$$
\begin{aligned}
P(s_t = s', s_{t+1} = s, \mathbf{r}_{<t}, \mathbf{r}_t, \mathbf{r}_{>t}) &= P(s_t = s', s_{t+1} = s, \mathbf{r}_{<t}, \mathbf{r}_t) \\
&\quad \times P(\mathbf{r}_{>t} \mid s_t = s', s_{t+1} = s, \mathbf{r}_{<t}, \mathbf{r}_t) \\
&= P(s_t = s', s_{t+1} = s, \mathbf{r}_{<t}, \mathbf{r}_t) P(\mathbf{r}_{>t} \mid s_t = s'). \\
&= P(s_{t+1} = s, \mathbf{r}_t \mid s_t = s') P(s_t = s', \mathbf{r}_{<t}) \\
&\quad \times P(\mathbf{r}_{>t} \mid s_t = s', s_{t+1} = s, \mathbf{r}_{<t}, \mathbf{r}_t)
\end{aligned}
$$

$$(8.16)$$

(8.16) is now expressed in terms of three probabilities:

$$
\begin{aligned}
\alpha_t(s') &= P(s_t = s', \mathbf{r}_{<t}), & (8.17) \\
\beta_{t+1}(s) &= P(\mathbf{r}_{>t} \mid s_t = s'), & (8.18) \\
\gamma_t(s, s') &= P(s_{t+1} = s, \mathbf{r}_t \mid s_t = s'). & (8.19)
\end{aligned}
$$

This gives:

$$
P(s_t = s', s_{t+1} = s, r_{<t}, r_t, r_{>t}) = \alpha_t(s') \beta_{t+1}(s) \gamma_t(s, s').
$$

Therefore (8.12) can be written as:

$$
L\left(a_t^{(0)} \mid \mathbf{r}\right) = \ln \left(\frac{\displaystyle\sum_{(s's) \in S^+} \alpha_t(s') \cdot \beta_{t+1}(s) \cdot \gamma_t(s', s)}{\displaystyle\sum_{(s', s) \in S^{--}} \alpha_t(s') \cdot \beta_{t+1}(s) \cdot \gamma_t(s', s)} \right). \tag{8.20}
$$

The values of $\alpha_t(s')$, $\beta_{t+1}(s)$ and $\gamma_t(s', s)$ can be determined by forming recursive relationships. To determine $\alpha_{t+1}(s')$:

$$
\begin{aligned}
\alpha_{t+1}(s) = P(s_{t+1} = s', \mathbf{r}_{<t+1}) &= P(s_{t+1} = s', \mathbf{r}_{<t}, \mathbf{r}_t) \\
&= \sum_{s'=0}^{M_s-1} P(s_{t+1} = s, \mathbf{r}_t, s_t = s', \mathbf{r}_{<t}) \\
&= \sum_{s'=0}^{M_s-1} P(s_t = s', \mathbf{r}_{<t}) P(s_{t+1} = s, \mathbf{r}_t \mid s_t = s', \mathbf{r}_{<t}) \\
&= \sum_{s'=0}^{M_s-1} P(s_t = s', \mathbf{r}_{<t}) P(s_{t+1} = s, \mathbf{r}_t \mid s_t = s') \\
&= \sum_{s'=0}^{M_s-1} \alpha_t(s') \gamma_t(s', s)
\end{aligned}
$$

$$(8.21)$$

To determine $\beta_t(s)$:

$$
\begin{aligned}
\beta_t(s') &= P(r_{>t-1} \mid s_t = s') = P(r_{>t}, r_t \mid s_t = s') \\
&= \sum_{s=0}^{M_s-1} P(\mathbf{r}_{>t}, \mathbf{r}_t, s_{t+1} = s \mid s_t = s') \\
&= \sum_{s=0}^{M_s-1} P(\mathbf{r}_t, s_{t+1} = s \mid s_t = s') P(r_{>t} \mid \mathbf{r}_t, s_{t+1} = s, s_t = s') \\
&= \sum_{s=0}^{M_s-1} P(\mathbf{r}_t, s_{t+1} = s \mid s_t = s') P(\mathbf{r}_{>t} \mid s_{t+1} = s) \\
&= \sum_{s=0}^{M_s-1} \gamma_t(s', s) \beta_{t+1}(s)
\end{aligned}
\tag{8.22}
$$

Finally, to determine $\gamma_t(s', s)$:

$$
\begin{aligned}
\gamma_t(s, s') &= P(s_{t+1} = s, \mathbf{r}_t \mid s_t = s') = P(\mathbf{r}_t \mid s_t = s', s_{t+1} = s) \\
&\quad \cdot P(s_t = s', s_{t+1} = s),
\end{aligned}
\tag{8.23}
$$

$$
P(\mathbf{r}_t \mid s_t = s, s_{t+1} = s') = \frac{1}{\sqrt{2\pi\sigma^2}} e^{\left(-\frac{1}{2\sigma^2}\left|r_t - \sqrt{E_s}a^{(s',s)}\right|^2\right)},
\tag{8.24}
$$

$$
\gamma_t(s', s) = P(s_t = s', s_{t+1} = s) \frac{1}{\sqrt{2\pi\sigma^2}} e^{\left(-\frac{1}{2\sigma^2}\left|r_t - \sqrt{E_s}a^{(s',s)}\right|^2\right)}.
\tag{8.25}
$$

We can simplify (8.23) by assuming that if each information bit is equally likely to be a 0 or a 1 then all states are also equally likely and $P(s_t = s', s_{t+1} = s)$ is a constant. Therefore:

$$
\gamma(s', s) = \frac{1}{\sqrt{2\pi\sigma^2}} e^{\left(-\frac{1}{2\sigma^2}\left|r_t - \sqrt{E_s}a^{(s',s)}\right|^2\right)}.
\tag{8.26}
$$

8.3.3 Max-Log MAP

The amount of memory required, and the number of operations involving exponential values and multiplicative procedures, means that the complexity of the MAP algorithm becomes prohibitive when implementing a turbo decoder. The system can be simplified by employing the logarithms of the probabilities defined in (8.16), (8.17) and (8.18) and thus transforming any multiplicative operations to summations. This gives the following:

$$
\bar{\alpha}_t(s') = \ln \alpha_t(s'),
\tag{8.27}
$$

$$
\bar{\beta}_t(s) = \ln \beta_t(s),
\tag{8.28}
$$

$$
\bar{\gamma}_t(s', s) = \ln \gamma_t(s', s).
\tag{8.29}
$$

This means that, with reference to (8.20), $\overline{\alpha_{t+1}}(s')$ becomes:

$$\overline{\alpha_{t+1}}(s') = \ln\left(\sum_{s'=0}^{M_s-1} \alpha_t(s')\gamma_t(s', s)\right) = \ln\left(\sum_{s'=0}^{M_s-1} e^{(\bar{\alpha}_t(s')+\bar{\gamma}_t(s',s))}\right). \tag{8.30}$$

To evaluate (8.28) we can use the following approximation:

$$\ln(e^{\delta_1} + e^{\delta_2} + \cdots + e^{\delta_n}) \approx \max_{i=1,2,\ldots,n} \delta_i. \tag{8.31}$$

Therefore:

$$\overline{\alpha_{t+1}}(s') = \ln\left(\sum_{s'=0}^{M_s-1} e^{(\bar{\alpha}_t(s')+\bar{\gamma}_t(s',s))}\right) \approx \max_{s'}\left\{\bar{\alpha}_t(s') + \bar{\gamma}_t(s', s)\right\}. \tag{8.32}$$

Similarly, for $\overline{\beta}_t(s)$:

$$\overline{\beta}_t(s) = \ln\left(\sum_{s'=0}^{M_s-1} \gamma_t(s', s)\beta_{t+1}(s)\right) = \ln\left(\sum_{s'=0}^{M_s-1} e^{(\bar{\gamma}_t(s',s)+\bar{\beta}_{t+1}(s))}\right)$$

$$\approx \max_{s}\left\{\bar{\gamma}_t(s', s) + \bar{\beta}_{t+1}(s)\right\}. \tag{8.33}$$

Finally, for $\overline{\gamma}_t(s', s)$:

$$\bar{\gamma}(s', s) = \ln\left(\frac{1}{\sqrt{2\pi\sigma^2}}e^{\left(-\frac{1}{2\sigma^2}\left|r_t-\sqrt{E_s}a^{(s',s)}\right|^2\right)}\right)$$

$$= \ln\left(\frac{1}{\sqrt{2\pi\sigma^2}}\right) + \ln e^{\left(-\frac{1}{2\sigma^2}\left|r_t-\sqrt{E_s}a^{(s',s)}\right|^2\right)}$$

$$= \ln\left(\frac{1}{\sqrt{2\pi\sigma^2}}\right) - \frac{1}{2\sigma^2}\left|r_t - \sqrt{E_s}a^{(s',s)}\right|^2.$$

By neglecting constants we obtain:

$$\bar{\gamma}_t(s', s) = -\left|r_t - \sqrt{E_s}a^{(s',s)}\right|^2. \tag{8.34}$$

Inserting these values into (8.20) yields:

$$L\left(a_t^{(0)} \mid \mathbf{r}\right) = \ln\frac{\sum\limits_{(s',s)\in S^+} e^{\bar{\alpha}_t(s')+\bar{\gamma}_t(s',s)+\bar{\beta}_{t+1}(s)}}{\sum\limits_{(s',s)\in S^-} e^{\bar{\alpha}_t(s')+\bar{\gamma}_t(s',s)+\bar{\beta}_{t+1}(s)}}. \tag{8.35}$$

(8.34) can be estimated as:

$$L\left(a_t^{(0)} \mid \mathbf{r}\right) \approx \max_{(s',s)\in S^+}\left[\bar{\gamma}_t(s', s) + \bar{\alpha}_t(s') + \bar{\beta}_{t+1}(s)\right]$$

$$- \max_{(s',s)\in S^-}\left[\bar{\gamma}_t(s', s) + \bar{\alpha}_t(s') + \bar{\beta}_{t+1}(s)\right]. \tag{8.36}$$

8.3.4 Log MAP Algorithm

The max-log MAP algorithm approximates the log-likelihood ratio $L(a_t^{(0)} \mid \mathbf{r})$, and is therefore suboptimal. The log MAP algorithm instead uses the Jacobian logarithm [6] to improve the log-likelihood ratio and therefore improve the decoding abilities of the algorithm with only a small increase in complexity:

$$\ln(e^{\delta_1} + e^{\delta_2}) = \max(\delta_1, \delta_2) + \ln(1 - e^{-|\delta_2 - \delta_1|}). \tag{8.37}$$

8.3.5 Iterative Log MAP Decoding

If we assume that the encoder is systematic with a rate $R = 0.5$ then the received vector $\mathbf{r}$ at time t can be expressed as $\mathbf{r}_t = (r_t^{(0)}, r_t^{(1)})$, where $r_t^{(0)}$ is the information bit and $r_t^{(1)}$ is the parity bit.

(8.22) can then be written as [7]:

$$\begin{aligned}
\gamma_t(s, s') &= P(\mathbf{r}_t \mid s_t = s', s_{t+1} = s) P(s_{t+1} = s, s_t = s') \\
&= P\big(r_t^{(0)}, r_t^{(1)} \mid s_t = s', s_{t+1} = s\big) P(s_{t+1} = s, s_t = s'). \\
&= P\big(r_t^{(0)}, r_t^{(1)} \mid s_t = s', s_{t+1} = s\big) P\big(a_t^{(0)} = a^{(0,s',s)}\big)
\end{aligned} \tag{8.38}$$

Since s_{t+1} is determined by $a^{(0,s',s)}$, $P(s_{t+1} = s \mid s_t = s') = P(a_t^{(0)} = a^{(0,s',s)})$. Therefore:

$$P\big(r_t^{(0)}, r_t^{(1)} \mid s_t = s', s_{t+1} = s\big) = P\big(r_t^{(0)}, r_t^{(1)} \mid s_t = s', a_t^{(0)} = a^{(0,s',s)}\big). \tag{8.39}$$

Since the encoder is systematic, $r_t^{(0)}$ is independent of the state transitions and only dependent on the input:

$$\begin{aligned}
P\big(r_t^{(0)}, r_t^{(1)} \mid s_t = s', a_t^{(0)} = a^{(0,s',s)}\big) &= P\big(r_t^{(0)} \mid a_t^{(0)}\big) P\big(r_t^{(1)} \mid s_t = s', s_{t+1} = s\big) \\
&= P\big(r_t^{(0)} \mid a_t^{(0)}\big) P\big(r_t^{(1)} \mid a_t^{(1)} = a^{(1,s',s)}\big)
\end{aligned}. \tag{8.40}$$

Substituting (8.38) into (8.36) gives:

$$\begin{aligned}
\gamma_t(s', s) &= P\big(r_t^{(0)}, r_t^{(1)} \mid s_t = s', s_{t+1} = s\big) P\big(a_t^{(0)} = a^{(0,s',s)}\big) \\
&= P\big(r_t^{(0)} \mid v_t^{(0)}\big) P\big(r_t^{(1)} \mid a_t^{(1)} = a^{(1,s',s)}\big) P\big(a_t^{(0)} = a^{(0,s',s)}\big),
\end{aligned} \tag{8.41}$$

$$\begin{aligned}
\bar{\gamma}_t(s', s) &= \ln\big[P\big(r_t^{(0)} \mid a_t^{(0)}\big) P\big(r_t^{(1)} \mid a_t^{(1)} = a^{(1,s',s)}\big) P\big(a_t^{(0)} = a^{(0,s',s)}\big)\big] \\
&= \ln P\big(r_t^{(0)} \mid a_t^{(0)} = a^{(0,s',s)}\big) + \ln P\big(r_t^{(1)} \mid a_t^{(1)} = a^{(1,s',s)}\big) \\
&\quad + \ln P\big(a_t^{(0)} = a^{(0,s',s)}\big)
\end{aligned}. \tag{8.42}$$

From (8.5):

$$L_1\big(a_t^{(0)}\big) = \ln\left(\frac{P\big(a_t^{(0)} = +1\big)}{P\big(a_t^{(0)} = -1\big)} \right) = \ln\left(\frac{P\big(a_t^{(0)} = +1\big)}{1 - P\big(a_t^{(0)} = -1\big)} \right),$$

$$e^{L_1\left(a_t^{(0)}\right)} = \frac{P\left(a_t^{(0)} = +1\right)}{1 - P\left(a_t^{(0)} = +1\right)}, \tag{8.43}$$

$$P\left(a_t^{(0)} = +1\right) = \frac{e^{L_1\left(a_t^{(0)}\right)}}{1 + e^{L_1\left(a_t^{(0)}\right)}} = \frac{1}{1 + e^{-L_1\left(a_t^{(0)}\right)}}. \tag{8.44}$$

Similarly:

$$L_1\left(v_t^{(0)}\right) = \ln\left(\frac{P\left(a_t^{(0)} = +1\right)}{P\left(a_t^{(0)} = -1\right)}\right) = \ln\left(\frac{1 - P\left(a_t^{(0)} = -1\right)}{P\left(a_t^{(0)} = -1\right)}\right), \tag{8.45}$$

$$e^{L_1\left(a_t^{(0)}\right)} = \frac{1 - P\left(a_t^{(0)} = -1\right)}{P\left(a_t^{(0)} = -1\right)}$$

$$P\left(a_t^{(0)} = -1\right) = \frac{1}{1 + e^{L_1\left(a_t^{(0)}\right)}}. \tag{8.46}$$

In general:

$$P\left(a_t^{(0)} = a^{(0,s',s)}\right) = \frac{e^{\frac{-L_1\left(a_t^{(0)}\right)}{2}}}{1 + e^{\frac{-L_1\left(a_t^{(0)}\right)}{2}}} e^{\frac{a^{(0,s',s)}L_1\left(a_t^{(0)}\right)}{2}}. \tag{8.47}$$

The term $\dfrac{e^{\frac{-L_1(a_t^{(0)})}{2}}}{1+e^{\frac{-L_1(a_t^{(0)})}{2}}}$ is independent of $a^{(0,s',s)}$ and can be taken out so we obtain:

$$\ln\left(P\left(a_t^{(0)} = a^{0,s',s}\right)\right) = \frac{a^{(0,s',s)}L_1\left(a_t^{(0)}\right)}{2}. \tag{8.48}$$

We can derive similar expressions to determine $\ln P(r_t^{(0)} \mid a_t^{(0)} = a^{(0,s',s)})$ and $\ln P(r_t^{(1)} \mid a_t^{(1)} = a^{(1,s',s)})$ using (8.3):

$$\ln\left(\frac{P\left(r_t^{(0)} \mid a_t^{(0)} = 1\right)}{P\left(r_t^{(0)} \mid a_t^{(0)} = 0\right)}\right) = \ln\left(\frac{P\left(r_t^{(0)} \mid a_t^{(0)} = 1\right)}{1 - P\left(r_t^{(0)} \mid a_t^{(0)} = 1\right)}\right) = L_c r_t^{(0)}.$$

$$P\left(r_t^{(0)} \mid a_t^{(0)} = 1\right) = \frac{e^{L_c r_t^{(0)}}}{1 + e^{L_c r_t^{(0)}}} = \frac{1}{1 + e^{-L_c r_t^{(0)}}}$$

Also, $P(r_t^{(0)} \mid a_t^{(0)} = 0) = \frac{1}{1+e^{L_c r_t^{(0)}}}$. Therefore:

$$P\left(r_t^{(0)} \mid a_t^{(0)} = a^{(0,s',s)}\right) = \frac{e^{\frac{-L_c r_t^{(0)}}{2}}}{1 + e^{\frac{-L_c r_t^{(0)}}{2}}} e^{\frac{a^{(0,s',s)}L_c r_t^{(0)}}{2}}.$$

Again, ignoring the constant term we obtain:

$$\ln\left(P\left(r_t^{(0)} \mid a_t^{(0)} = a^{0,s',s}\right)\right) = \frac{a^{(0,s',s)} L_c r_t^{(0)}}{2}. \tag{8.49}$$

In the same way:

$$\ln\left(P\left(r_t^{(1)} \mid a_t^{(1)} = a^{1,s',s}\right)\right) = \frac{a^{(1,s',s)} L_c r_t^{(1)}}{2}. \tag{8.50}$$

Substituting (8.46), (8.47) and (8.48) into (8.40) gives [7]:

$$\bar{\gamma}_t(s',s) = \frac{a^{(0,s',s)} L_1\left(a_t^{(0)}\right)}{2} + \frac{a^{(0,s',s)} L_c r_t^{(0)}}{2} + \frac{a^{(1,s',s)} L_c r_t^{(1)}}{2}. \tag{8.51}$$

8.4 Non-Binary Turbo Codes

Extending binary turbo codes to non-binary turbo codes [8, 9, 10] can be considered less complicated than the extension of binary LDPC codes to non-binary LDPC codes. In particular, the principle of the non-binary turbo decoding algorithm remains the same. One of the main differences is the trellis diagram associated with a non-binary convolutional code, which has more branches leaving and entering nodes in the trellis, resulting in more paths and higher decoding complexity. Secondly, an increase in the size of the alphabet means that the reliabilities of these extra symbols must also be considered. The non-binary turbo encoder has the same structure as the binary turbo encoder in Figure 8.1, with the component encoders being replaced by RSC codes defined over a ring of integers $\mathbb{Z}_q$, where q is the cardinality of the ring. The non-binary turbo encoder is given in Figure 8.5.

The message symbols x_t and the turbo encoder output symbols $v_t^{(0)}$, $v_t^{(1)}$, $v_t^{(2)}$ are now defined over $\mathbb{Z}_q$.

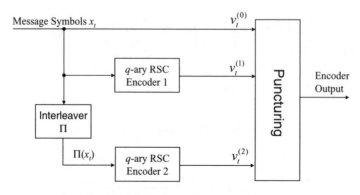

Figure 8.5   The non-binary turbo encoder.

Example 8.2: Non-binary turbo encoding: A non-binary turbo encoder defined over $\mathbb{Z}_4$ with 23/1 RSC component encoders is shown in Figure 8.6.

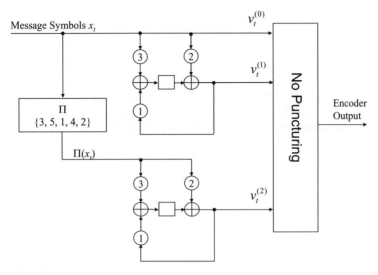

Figure 8.6   Non-binary turbo encoder with 23/1 RSC component codes.

Let the message be $x_t = [x_1\, x_2, x_3, x_4, x_5] = [1, 2, 3, 3, 3]$. The state table for the first component 23/1 RSC encoder defined over $\mathbb{Z}_4$ is given in Table 8.2.

Table 8.2   State table for the 23/1 RSC code.

x_t	s'	s	$v^{(0,s',s)}$	$v^{(1,s',s)}$
0	0	0	0	0
1	0	1	1	2
2	0	2	2	0
3	0	3	3	2
0	1	1	0	1
1	1	2	1	3
2	1	3	2	1
3	1	0	3	3
0	2	2	0	2
1	2	3	1	0
2	2	0	2	2
3	2	1	3	0
0	3	0	0	3
1	3	0	1	1
2	3	1	2	3
3	3	2	3	1

The input sequence runs through the first RSC encoder, resulting in a parity sequence:

$$v_t^{(1)} = \left[v_1^{(1)}, v_2^{(1)}, v_3^{(1)}, v_4^{(1)}, v_5^{(1)}\right] = [2, 1, 1, 0, 3].$$

The sequence x_t is also passed through the interleaver of length $N = 5$, which produces the permuted output sequence $x_t' = \Pi(x_t) = [3, 3, 1, 3, 2]$. The sequence x_t' is passed through the second convolutional encoder, which produces the output sequence:

$$v_t^{(2)} = \left[v_1^{(2)}, v_2^{(2)}, v_3^{(2)}, v_4^{(2)}, v_5^{(2)}\right] = [2, 1, 0, 1, 2].$$

The three output sequences are multiplexed together to form the output sequence:

$$\begin{aligned} v_t &= \left[\left(v_1^{(0)}, v_1^{(1)}, v_1^{(2)}\right), \left(v_2^{(0)}, v_2^{(1)}, v_2^{(2)}\right), \dots, \left(v_5^{(0)}, v_5^{(1)}, v_5^{(2)}\right)\right] \\ &= [(1, 2, 2), (2, 1, 1), (3, 1, 0), (3, 0, 1), (3, 3, 2)] \end{aligned}.$$

8.4.1 Multi-Dimensional Log-Likelihood Ratios

The binary log-likelihood ratio is a measure of reliability and is defined as the natural logarithm of the ratio of the likelihood of an event being 1 to the likelihood of the same event being 0. However, if we expand to a ring of integers $\mathbb{Z}_q$ we must consider the reliabilities of the other symbols too. The LLRs for an event u being an element in $\mathbb{Z}_4$ are [10]:

$$\begin{aligned} L^{(1)}(u) &= \ln\left(\frac{P(u = 1)}{P(u = 0)}\right) \\ L^{(2)}(u) &= \ln\left(\frac{P(u = 2)}{P(u = 0)}\right) \\ L^{(3)}(u) &= \ln\left(\frac{P(u = 3)}{P(u = 0)}\right). \end{aligned} \tag{8.52}$$

The output of decoder 1 is therefore defined by the multi-dimensional LLR:

$$\begin{aligned} \left\{ L^{(1)}\left(v_t^{(0)} \mid \mathbf{r}\right) \right. &= \ln\left(\frac{P\left(v_t^{(0)} = 1 \mid \mathbf{r}\right)}{P\left(v_t^{(0)} = 0 \mid \mathbf{r}\right)}\right), \quad L^{(2)}\left(v_t^{(0)} \mid \mathbf{r}\right) = \ln\left(\frac{P\left(v_t^{(0)} = 2 \mid \mathbf{r}\right)}{P\left(v_t^{(0)} = 0 \mid \mathbf{r}\right)}\right), \\ L^{(3)}\left(v_t^{(0)} \mid r\right) &= \left. \ln\left(\frac{P\left(v_t^{(0)} = 3 \mid \mathbf{r}\right)}{P\left(v_t^{(0)} = 0 \mid \mathbf{r}\right)}\right) \right\}, \end{aligned} \tag{8.53}$$

where $\mathbf{r}$ is the received vector. In the binary case, the sign of the LLR was used to make a hard decision on the decoded symbol. A negative LLR corresponded to a 0 and a positive LLR corresponded to a 1. This is also true for multi-dimensional LLRs.

If $\ln(\frac{P(u=2)}{P(u=0)})$ is negative, the hard decision will be $u = 0$; if it is positive the hard decision will be $u = 2$.

In the non-binary turbo decoding procedure, the prior information into decoder 1 is the deinterleaved extrinsic information from decoder 2.

$$\{L_1^{(1)}(v_t^{(0)}), L_1^{(2)}(v_t^{(0)}), L_1^{(3)}(v_t^{(0)})\}$$
$$= \{L_{e_2}^{(1)}(\Pi^{-1}(v_t^{(0)})), L_{e_2}^{(2)}(\Pi^{-1}(v_t^{(0)})), L_{e_2}^{(3)}(\Pi^{-1}(v_t^{(0)}))\}. \tag{8.54}$$

Therefore, the extrinsic information from decoder 1 is:

$$L_{e_1}^{(1)}(v_t^{(0)}) = L_1^{(1)}(v_t^{(0)} \mid \mathbf{r}) - L_1^{(1)}(v_t^{(0)}) - L^{(1)}(r_t^{(0)} \mid v_t^{(0)})$$
$$L_{e_1}^{(2)}(v_t^{(0)}) = L_1^{(2)}(v_t^{(0)} \mid \mathbf{r}) - L_1^{(2)}(v_t^{(0)}) - L^{(2)}(r_t^{(0)} \mid v_t^{(0)}) \tag{8.55}$$
$$L_{e_1}^{(3)}(v_t^{(0)}) = L_1^{(3)}(v_t^{(0)} \mid \mathbf{r}) - L_1^{(3)}(v_t^{(0)}) - L^{(3)}(r_t^{(0)} \mid v_t^{(0)}).$$

Similarly, the extrinsic information from decoder 2 is:

$$L_{e_2}^{(1)}(v_t^{(0)}) = L_2^{(1)}(v_t^{(0)} \mid \mathbf{r}) - L_2^{(1)}(v_t^{(0)}) - L_2^{(1)}(\Pi(r_t^{(0)} \mid v_t^{(0)}))$$
$$L_{e_2}^{(2)}(v_t^{(0)}) = L_2^{(2)}(v_t^{(0)} \mid \mathbf{r}) - L_2^{(2)}(v_t^{(0)}) - L_2^{(2)}(\Pi(r_t^{(0)} \mid v_t^{(0)})) \tag{8.56}$$
$$L_{e_2}^{(3)}(v_t^{(0)}) = L_2^{(3)}(v_t^{(0)} \mid \mathbf{r}) - L_2^{(3)}(v_t^{(0)}) - L_2^{(3)}(\Pi(r_t^{(0)} \mid v_t^{(0)})).$$

8.4.2 Non-Binary Iterative Turbo Decoding

Now that all the necessary multi-dimensional LLRs are defined we can present the non-binary turbo decoder. The complete non-binary turbo decoder over $\mathbb{Z}_4$ is shown in Figure 8.7.

A hard decision is made on the multi-dimensional LLRs from the deinterleaved output of decoder 2.

$$\begin{array}{ll} \text{If} & \text{sign}(L^{(i)}(v_t^{(0)} \mid r)) < 0, \quad v_t^{(0)} = 0 \\ \text{If} & \text{sign}(L^{(i)}(v_t^{(0)} \mid r)) \geq 0, \quad v_t^{(0)} = i, \quad \text{where} \quad i = 1, 2, \ldots, q-1 \end{array}. \tag{8.57}$$

Hence, there are $q - 1$ candidate values for the decoded symbol. The most likely element is determined by comparing each LLR value $L^{(i)}(v_t \mid \mathbf{r})$ and choosing the LLR with the largest magnitude (the highest reliability).

One concern with regards to using $q - 1$ LLR values of the form $\ln(\frac{P(u=i)}{P(u=0)})$, $i = 1, 2, \ldots, q-1$ is that we only have a measure of reliability of the nonzero elements with respect to zero, and not to each other. For the case of element in $\mathbb{Z}_4$ we would have the extra LLRs:

$$\ln\left(\frac{P(u=2)}{P(u=1)}\right) \quad \ln\left(\frac{P(u=3)}{P(u=1)}\right) \quad \ln\left(\frac{P(u=3)}{P(u=2)}\right).$$

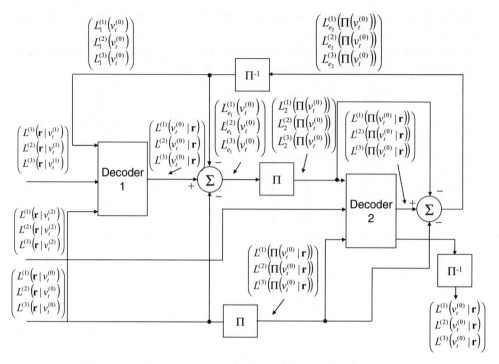

Figure 8.7   The non-binary turbo decoder defined over $\mathbb{Z}_4$.

Including these extra LLRs in the decoding procedure will provide more information about the reliability of each symbol, but the number of extra LLRs will increase exponentially with increasing alphabet size.

8.5 Conclusions

This chapter introduced the concept of non-binary turbo encoding and decoding defined over rings. It can be seen that the non-binary turbo encoder structure is the same as the binary turbo encoder structure, with non-binary RSC codes replacing the binary RSC codes. Non-binary turbo decoding was explained by first introducing multi-dimensional log-likelihood ratios. Essentially, the principle of non-binary turbo decoding is the same, with the exception that we must now process an array of LLR values for each nonzero element in the ring, instead of just one.

Non-binary turbo decoding is an area of coding theory that has not received much attention in the literature, most likely due to the extra complexity in implementation. However, with non-binary LDPC codes recently becoming more popular, we would expect non-binary turbo codes to perform just as well and this would be an interesting area of research for the future.

References

[1] Berrou, C., Glaviuex, A. and Thitimijshima, P. (1993) Near shannon limit error-correcting coding and decoding: turbo codes. Proceedings of the IEEE International Conference on Communications, Geneva, Switzerland, pp. 1064–70.

[2] Bahl, L.R., Cocke, J., Jelinek, F. and Raviv, J. (1972) Optimal decoding of linear codes for minimising symbol error rate. Abstracts of Papers, International Symposium on Information Theory, p. 50.

[3] Bahl, L.R., Cocke, J., Jelinek, F. and Raviv, J. (1974) Optimal decoding of linear codes for minimising symbol error rate. *IEEE Transactions on Information Theory*, **IT-20** (2), 284–7.

[4] Pyndiah, R., Glavieux, A., Picart, A. and Jacq, S. (1994) Near optimum decoding of products codes. Proceedings of the IEEE GlobeCom'94 Conference San Francisco, CA, Vol. **1/3**, pp. 339–43.

[5] McAdam, P.L., Welch, L. and Weber, C. (1972) MAP bit decoding of convolutional codes. Abstracts of Papers, International Symposium on Information Theory, p. 9.

[6] Erfanian, J.A., Pasupathy, S. and Gulot, G. (1994) Reduced complexity symbol detectors with parallel structures for ISI channels. *IEEE Transactions on Communications*, **42** (234, Part 3), 1661–71.

[7] Moon, T.K. (2005) *Error Correction Coding. Mathematical Methods and Algorithms*, Wiley Interscience, ISBN 0-471-64800-0.

[8] Berrou, C., Jezequel, M., Douillard, C. and Kerouedan, S. (2001) The Advantages of Non-Binary Turbo Codes, IEEE Information Theory Workshop, Australia.

[9] Berrou, C. and Jezequel, M. (1999) Non-binary convolutional codes for turbo coding. *IET Electronics Letters*, **35** (1), 39–40.

[10] Berkmann, J. (1998) On turbo decoding of non-binary turbo codes. *IEEE Communications Letters*, **2** (4), 94–6.

Index